KEY PAPERS IN PHYSICS

NMR in Biomedicine: The Physical Basis

Edited by

Eiichi Fukushima
Lovelace
Medical Foundation

NUMBER TWO • 1989

Consulting Editor: L. Henlein

Printed in the United States of America.

Library of Congress Cataloging-in-Publication Data

NMR in biomedicine.

(Key papers in physics; no. 2)
1. Nuclear magnetic resonance. 2. Nuclear magnetic resonance spectroscopy. 3. Magnetic resonance imaging. I. Fukushima, Eiichi. II. Series.
QP519.9.N83N68 1989 88-35138
616.07'57
ISBN 0-88318-603-9
ISBN 0-88318-609-8 (pbk.)

NMR in Biomedicine: The Physical Basis

Contents

RF Field Effects on Biological Samples

NMR in Biomedicine: The Physical Basis

Preface

EDITED BY EIICHI FUKUSHIMA

There is a special need for a reprint collection in nuclear magnetic resonance, covering the physical basis of NMR in medicine: a rapidly evolving field which straddles a wide range of disciplines such as physics, chemistry, biology, engineering, and medicine.

The developments and applications in NMR are occurring rapidly, and many of the scientists who are now entering this field from fields other than the traditional disciplines associated with it, are possibly unfamiliar with the historical and technical papers in the development of NMR. In addition, the physical science literature is often not easily accessible to workers in biomedical NMR. It is my hope that this reprint volume will alleviate some of these difficulties. This collection is also sufficiently general for it to be a useful reference for physical scientists.

To choose papers for a collection such as this is difficult. There is no way that anyone can choose all the appropriate papers—there is no such well-defined set—and even if there were, there are page limitations. I did not choose the reprints for priority, but for usefulness in obtaining background information necessary to perform, understand, or present NMR experiments. Although not a complete set, I hope that this collection will be helpful in either giving primary information or in pointing the reader to other references.

I have omitted some major areas of NMR (e.g., imaging, special pulses, and pulse sequences) as well as the most recent papers. Many papers in these categories are or will become suitable for other volumes of collections like this one.

Besides the authors and publishers who gave me permission to reprint these articles, I am indebted to Alan Rath who, as a graduate student, found a need for such a volume and suggested it to me. I also thank many colleagues for their suggestions regarding the content of this collection.

Eiichi Fukushima
November 1988

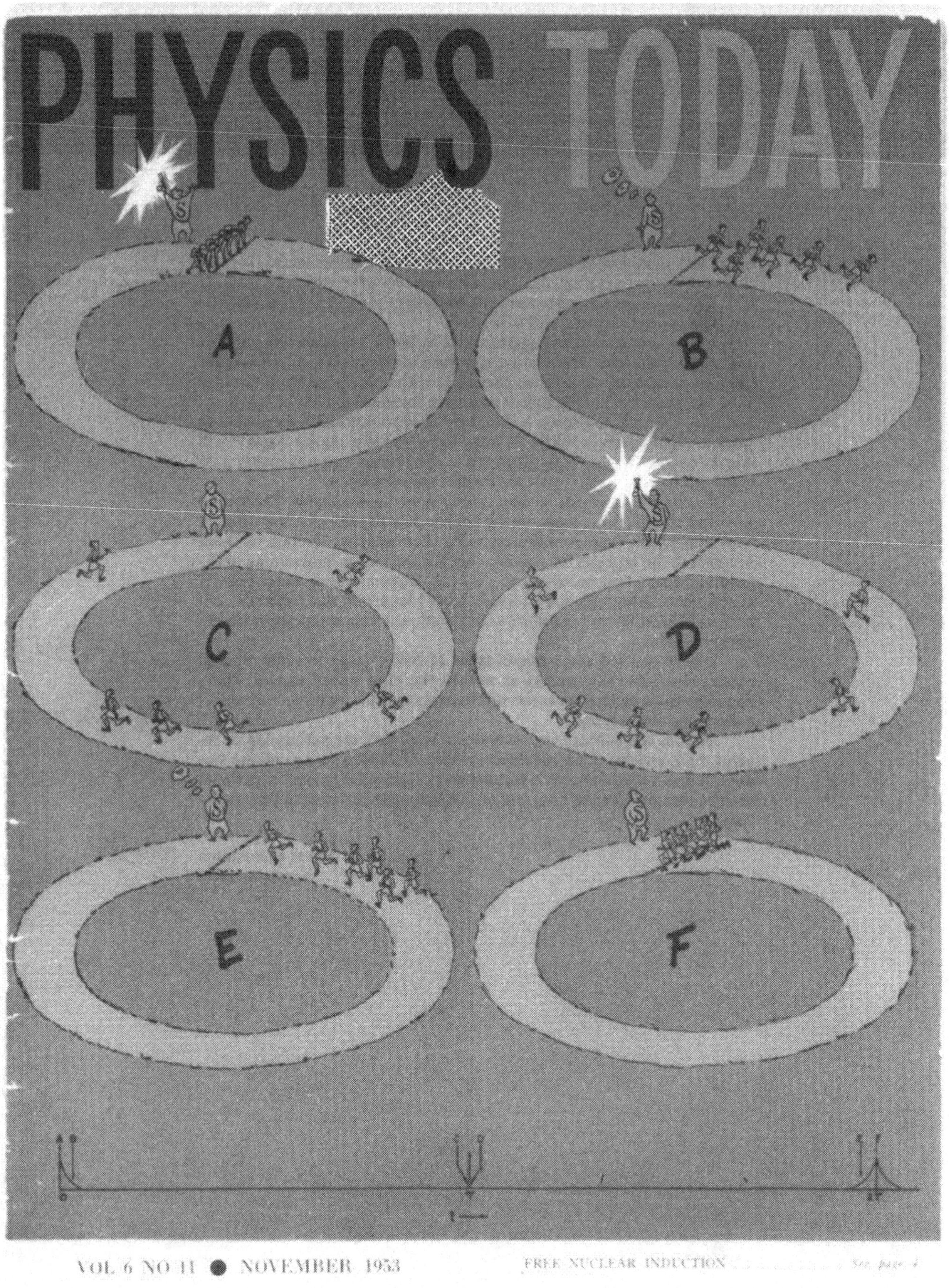

COVER: Drawing by Kay Kaszas. *A Race-track Analogy of the "Spin-echo" Effect*

See E. L. Hahn, "Free Nuclear Induction," pp. 65–70, this volume.

Nuclear Induction

F. Bloch
Stanford University, California
(Received July 19, 1946)

The magnetic moments of nuclei in normal matter will result in a nuclear paramagnetic polarization upon establishment of equilibrium in a constant magnetic field. It is shown that a radiofrequency field at right angles to the constant field causes a forced precession of the total polarization around the constant field with decreasing latitude as the Larmor frequency approaches adiabatically the frequency of the r-f field. Thus there results a component of the nuclear polarization at right angles to both the constant and the r-f field and it is shown that under normal laboratory conditions this component can induce observable voltages. In Section 3 we discuss this nuclear induction, considering the effect of external fields only, while in Section 4 those modifications are described which originate from internal fields and finite relaxation times.

1. INTRODUCTION

THE method of magnetic resonance[1] has been successfully applied to measure the magnetic moment of the neutron[2] and of various nuclei.[3] The principal feature of this method is the observation of transitions, caused by resonance of an applied radiofrequency field with the Larmor precession of the moments around a constant magnetic field. In its application to nuclear moments the deflection of molecular beams in an inhomogeneous field was used as a means of detecting the occurrence of nuclear transitions. This method of detection has proven to be very fruitful but it was clear, at the same time, that the connection between molecular beams and magnetic resonance was not of basic character. The question arose, in particular, whether nuclear transitions could not be detected by far simpler electromagnetic methods, applied to matter of ordinary density.[4]

An attempt in this direction was undertaken by Gorter and Broer[5] whose arrangement was designed to indicate magnetic resonance absorption by a slight change in frequency of an electric oscillator. The experiment was based upon considerations which apply strictly to radiofrequency fields which are so small that they cause only a slight disturbance of the spin orientation; it was carried out with LiCl and KF at low temperatures and it was suggested that the failure to find an effect was caused by the fact that the nuclei had not found the orientation, corresponding to thermal equilibrium.

The first successful experiments to detect magnetic resonance by electromagnetic effects have been carried out recently and independently in the physics laboratories of Harvard[6] and Stanford[7] Universities. The experiment of Purcell and his collaborators is very closely connected to that of Gorter and Broer, the main difference being that resonance absorption manifests itself in the change of Q of an electric oscillator rather than in a change of frequency, and it presupposes, likewise, the necessity of only slightly perturbing r-f fields.

The considerations upon which our work was based have several features in common with the two experiments, previously mentioned, but differ rather essentially in others. In the first place, the radiofrequency field is deliberately chosen large enough so as to cause at resonance a considerable change of orientation of the nuclear moments. In the second place, this change is not observed by its relatively small reaction upon the driving circuit, but by directly observing the induced electromotive force in a coil, due to the precession of the nuclear moments around the constant field and in a direction

[1] I. I. Rabi, Phys. Rev. **51**, 652 (1937).
[2] L. W. Alvarez and F. Bloch, Phys. Rev. **57**, 111 (1940).
[3] I. I. Rabi, S. Millman, P. Kusch, and J. R. Zacharias, Phys. Rev. **53**, 318 (1938); **55**, 526 (1939).
[4] The first purely magnetic experiment to find an effect due to nuclear moments by measuring the susceptibility of liquid hydrogen was published by B. G. Lasarew and L. W. Schubnikow, Sov. Phys. **11**, 445 (1937).
[5] C. J. Gorter and L. J. F. Broer, Physica **9**, 591 (1942).
[6] E. U. Purcell, H. C. Torrey, and R. V. Pound, Phys. Rev. **69**, 37 (1946).
[7] F. Bloch, W. W. Hansen, and Martin Packard, Phys. Rev. **69**, 127 (1946).

Reprinted from Physical Review **70**, 460–474 (1946);

perpendicular both to this field and the applied r-f field. This appearance of a magnetic induction at right angles to the r-f field is an effect which is of specifically nuclear origin and it is the main characteristic feature of our experiment. In essence, it is similar to the Faraday effect of rotation of the plane of polarization of light around a magnetic field, with the r-f field taking the place of the field vectors in a light wave and the observed perpendicular nuclear induction indicating a rotation of the total oscillating field around the constant magnetic field.

This effect is, of course, most outspoken at resonance (just as the Faraday effect becomes greatest in the neighborhood of a resonant frequency) and, in practice, is noticed by its sudden strong appearance under resonance conditions. It is worth while, however, to point out that the observation of nuclear induction should be possible even without any use of the magnetic resonance. Not only a weak r-f field, acting at resonance over very many Larmor periods, can produce an appreciable nuclear change of orientation, but also a strong field pulse, acting over only a few periods. Once the nuclear moments have been turned into an angle with the constant field, they will continue to precess around it and likewise cause a nuclear induction to occur at an instant when the driving pulse has already disappeared. It seems perfectly feasible to receive thus an induced nuclear signal of radiofrequency well above the thermal noise of a narrow band receiver. It is true that, due to the broadening of the Larmor frequency by internuclear fields or other causes, this signal can last only a comparatively short time, but for normal fields it will still contain many Larmor periods, i.e., it will be essentially monochromatic. The main difference between this proposed experiment and the one which we have actually carried out lies in the fact that it would observe by induction the free nuclear precession while we have studied the forced precession impressed upon the nuclei by the applied r-f field.

2. NUCLEAR PARAMAGNETISM

The existence of a resultant macroscopic moment of the nuclei within the sample under investigation is a common prerequisite for all electromagnetic experiments with nuclear moments. It is in fact a change of orientation of this macroscopic moment which causes the observed effects, and irrespective of the changes of orientation of the individual nuclei which might be induced by a r-f field, their moments would always cancel each other, if they did so initially, and thus escape observation. In the experiments with molecular beams of Stern and Rabi this necessity is avoided by separation in the beam of nuclei with different orientation.

Even in the absence of any orientation by an external magnetic field one can expect in a sample with N nuclei of magnetic moment μ to find a resultant moment of the order $(N)^{\frac{1}{2}}\mu$ because of statistically incomplete cancellation. This moment, however, would naturally be very small and in samples of normal size will be greatly increased as soon as the nuclei have found their equilibrium distribution in a field of a few thousand Gauss which, at the same time, will bring their Larmor precession into the convenient radiofrequency range.

In contrast to the familiar atomic paramagnetism which establishes itself almost immediately upon application of the polarizing field, there is no assurance for the same thing being true in the nuclear case. The time of establishment or "relaxation time" can be expected to vary anywhere between fractions of a second and many hours, depending in the most delicate manner upon the nuclear moments, the electronic structure of the atoms in the sample, their distance, and their motion. To study experimentally and theoretically this interesting relationship between nuclear relaxation time and atomic features seems to us, in fact, to be one of the fruitful fields of investigation which have now opened.[8]

It must be pointed out here that, in cases where the natural relaxation time should turn out inconveniently long, the establishment of thermal equilibrium can often be greatly accelerated by use of paramagnetic catalysts. The problem is similar to that of the conversion of

[8] A good deal of the theory of the paramagnetic relaxation time by I. Waller, Zeits. f. Physik **79**, 370 (1932), R. deL. Kronig, Physica **6**, 33, 1939. J. H. van Vleck, Phys. Rev. **57**, 426 (1940), can readily be adopted to hold also for nuclear paramagnetism. For nuclear paramagnetism and its bearing on reaching low temperatures see E. Teller and W. Heitler, Proc. Roy. Soc. **155**, 629 (1936).

ortho- and parahydrogen which was found, for example to be accelerated by paramagnetic ions in solutions and by admixture of oxygen in the gas phase.

As to the nuclear induction effect, there exists the interesting possibility of first establishing the equilibrium in a strong magnetic field under conditions of relatively short relaxation time (by evaporation into an oxygen atmosphere, addition of a paramagnetic catalyst, heating, etc.), thereupon considerably lengthening the relaxation time (by recondensation, removal of the catalyst, cooling, etc.), and thus preserving the high field polarization for a considerable time, even when the sample is removed from the high field. A subsequent nuclear induction experiment, carried out under suitable conditions, can then exhibit either a moment, pertaining to the field in which it was originally established, or indicate that the relaxation time was comparable or short, compared to the time which has elapsed since removal. We shall come back later to some more considerations of the important role of the relaxation time, which are directly connected with our experiments.

For the following purposes we shall assume that the thermal equilibrium between the nuclear moments and their surrounding atoms has been actually established in an external magnetic field of strength H. Let T be the equilibrium temperature and n the number of nuclei per unit volume, each having a magnetic moment μ and an angular momentum $jh/2\pi$. We shall write

$$M=\chi H, \tag{1}$$

where M is the resultant nuclear moment per cc in the field direction and where χ is the nuclear paramagnetic susceptibility. It is given by the familiar Curie formula

$$\chi=[(j+1)/3j](n\mu^2/kT), \tag{2}$$

in which it is assumed that $H\mu \ll kT$. This condition will always be very well satisfied, except for extraordinarily strong fields or exceedingly low temperatures.

For a numerical example we shall take the protons in water at room temperature ($T=291°$). We have here $j=\frac{1}{2}$, i.e., $(j+1)/3j=1$, and with $\mu=1.4\times10^{-23}$ c.g.s., $n=6.9\times10^{22}$ cm^{-3}, we obtain from (2) for the nuclear susceptibility

$$\chi=3.4\times10^{-10} \tag{3}$$

and with a field $H=10{,}000G$ from (1)

$$M=3.4\times10^{-6}\text{ gauss} \tag{4}$$

for the nuclear polarization.

While this value is small and would be difficult to observe directly, it is actually not M but the rate of change of the rapidly varying nuclear induction $B=4\pi M$, which is observed in our experiment and which we will show to be easily detectable.

3. PRINCIPLE OF THE NUCLEAR INDUCTION

We shall investigate the behavior of the great number of nuclei contained in a macroscopic sample of matter and acted upon by two external fields: a strong constant field and at right angles to it, a comparatively weak radiofrequency field. In order to simplify the explanation of the principle we shall, for the moment, omit some of the actually present complicating factors, and we shall assume:

(1) That the changes of orientation of each nucleus are solely due to the presence of the external fields;
(2) That the external fields are uniform throughout the sample.

The second assumption will evidently be justified with sufficient perfection of the experimental arrangement. It is the first assumption which is more serious; normally it will be far from being justified, and rather essential corrections will have to be introduced later to account for the actual conditions. The following conditions should be satisfied for its acceptance:

(1a) The atomic electrons do not cause appreciable fields to act upon the nuclei.
(1b) The interaction between neighboring nuclei can be neglected.
(1c) The thermal agitation does not essentially affect the nuclei, i.e., the relaxation time is long compared to the considered time intervals.

We shall come back to these three points in Section 4; accepting in this section the above assumptions, the discussion becomes comparatively simple: Let the z-direction be that of the constant field with strength H_0 and the x-direction that of the r-f field with circular frequency ω

and amplitude $2H_1$, so that the total external field vector **H** has the components

$$H_x = 2H_1 \cos \omega t; \quad H_y = 0; \quad H_z = H_0. \qquad (5)$$

We shall further denote by **M** the vector, representing the nuclear polarization, i.e., the resultant nuclear moment per unit volume; it is the variation with time of this vector in which we are primarily interested.

To obtain this variation does not require the solution of the Schroedinger equation. It is enough to remember the general fact that the quantum-mechanical expectation value of any quantity follows in its time dependence exactly the classical equations of motion and that the magnetic and angular momenta of each nucleus are parallel to each other.

The resultant angular momentum vector **A** of all the nuclei, contained in a unit volume will, therefore, satisfy the classical equation

$$d\mathbf{A}/dt = \mathbf{T}, \qquad (6)$$

where **T** is the total torque, acting upon the nuclei and it is

$$\mathbf{T} = [\mathbf{M} \times \mathbf{H}], \qquad (7)$$

where the vector **M** represents the resultant nuclear magnetic moment per unit volume. The parallelity between the magnetic moment $\boldsymbol{\mu}$ and the angular momentum **a** for each nucleus implies

$$\boldsymbol{\mu} = \gamma \mathbf{a} \qquad (8)$$

with the gyromagnetic ratio

$$\gamma = \mu/a \qquad (9)$$

and we have, therefore, also for the resultant quantities **M** and **A**

$$\mathbf{M} = \gamma \mathbf{A}, \qquad (10)$$

with γ likewise given by (9).[9]

Combining Eqs. (6), (7), and (10), we have, therefore, for the variation of the polarization vector **M**

$$d\mathbf{M}/dt = \gamma [\mathbf{M} \times \mathbf{H}]. \qquad (11)$$

For our purposes we are interested in a special solution of this equation which can be obtained if, for the fields H_0 and H_1 of Eq. (5) we have $H_1 \ll H_0$, and if both are positive and constant. We shall further assume that the circular frequency ω of the r-f field is in the neighborhood of the resonance frequency ω_0, given by

$$\omega_0 = \gamma H_0 \qquad (12)$$

i.e., that we have

$$|\omega - \omega_0| \ll \omega_0. \qquad (13)$$

The actual oscillating field in the x-direction can then be effectively replaced[10] by a field

$$H_x = H_1 \cos \omega t; \quad H_y = \mp H_1 \sin \omega t \qquad (14)$$

rotating around the z-direction with the sign of H_y and, therefore, the sense of rotation being negative or positive, depending upon whether the sign of γ is positive or negative.

It follows immediately from (11) that the magnitude M of the polarization is constant. Besides, there is a special solution for which its z-component M_z is likewise constant. Introducing the polar angle θ and writing

$$M_x = M \sin \theta \cos \omega t; \quad M_y = \mp M \sin \theta \sin \omega t; \quad M_z = M \cos \theta, \qquad (15)$$

one can indeed verify immediately that (11) is satisfied if θ is a constant and chosen such that

$$tg\theta = \gamma H_1/(\gamma H_0 \mp \omega), \qquad (16)$$

with the minus or plus sign before ω, depending upon whether γ is positive or negative. If we let

$$H^* = \omega/|\gamma| \qquad (17)$$

denote the "resonance field at frequency ω" i.e., that field H for which the Larmor frequency $\omega_L = \gamma H$ is equal to the frequency ω of the oscillating field, we can write (16) in the form

$$tg\theta = H_1/(H_0 - H^*). \qquad (18)$$

Equations (15) represent a solution for which the polarization rotates around the z-direction, i.e., around the strong field H_0 and in such a way that it lies at any instant in the common plane of this field and the effective rotating field (14).

The angle θ between H_0 and the polarization follows from (18) to be small, as long as H_0 is appreciably larger than the resonant field H^*.

[9] Our treatment includes evidently both cases of $\boldsymbol{\mu}$ and **a** being parallel (i.e., having the same relative orientation as for a positive rotating charge) and opposite. Both are taken into account by assigning to the quantity γ of Eq. (9) a positive or negative value respectively. We shall see below that the actual sign of γ reveals itself in the phase of the induced voltage signal.

[10] F. Bloch and A. Siegert, Phys. Rev. **57**, 522 (1940).

The direction of the polarization starts to deviate noticeably from the z-direction as the difference $H_0 - H^*$ becomes comparable or small compared to the magnitude H_1 of the effective rotating field. It is perpendicular to the z-direction for $H_0 = H^*$ and for still further decreasing values of H_0 turns toward the negative z-direction, finally pointing in a direction opposite to H_0 for $H^* - H_0 \ll H_1$.

Formulas (15) and (18) can be conveniently rewritten by introducing the difference

$$\delta = (H_0 - H^*)/H_1 = \text{cotg } \theta \qquad (19)$$

between the actual z-field H_0 and its resonance value H^* in units of the magnitude H_1 of the effective rotating field (14) or the half amplitude of the actual oscillating field (5) in the x-direction. We have then

$$M_x = M\frac{\cos \omega t}{(1+\delta^2)^{\frac{1}{2}}}; \quad M_y = \mp M\frac{\sin \omega t}{(1+\delta^2)^{\frac{1}{2}}}; \qquad (20)$$
$$M_z = M\frac{\delta}{(1+\delta_z)^{\frac{1}{2}}}.$$

These formulas show clearly the increase of the rotating component of **M** upon approach to resonance.

The solution (20) differs essentially from those previously[1,2] considered, where the nuclear moment can be assumed to be originally in one of the stationary states in which its angle with the z-direction is given by $\cos \theta = m/j (m = -j \cdots +j)$ and where it becomes suddenly subjected to the action of an oscillating field, thereby undergoing transitions in which θ changes through a change of m by ± 1. The solution (20) corresponds, in the language of quantum mechanics, not to a single stationary state, but to a "mixture" or linear combination of all stationary states with different values of m and with their amplitudes and phases so adjusted that the expectation values of the components of the angular or magnetic moment are proportional to the values (20). Particularly the z-component is not quantized, but has an expectation value which varies continuously with variation of δ. In order to obtain a persistent rotating component of the expectation value of **M**, as expressed by (20), it is essential, from this point of view, that we are dealing with a "coherent mixture" of states, i.e., that the relative phases of the wave functions, corresponding to the different states, do not undergo any changes. It can be expected, and will be shown later, that any cause which tends to destroy the phase relation, such as the interaction between neighboring nuclei, will diminish the actual observable value of the rotating component.

Our special interest in the particular solution (20) is based upon the fact that, while it has been derived under the assumption that ω, H_0, and therefore, δ are constant, it can be shown to be equally valid, provided that these quantities vary adiabatically, i.e., slowly enough so that

$$|d\delta/dt| \ll |\gamma H_1|. \qquad (21)$$

For constant H_1, i.e., for constant amplitude of the oscillating field, this variation of δ and, thereby, of the components (20) of the polarization can, according to (19), take place through two different procedures. Either the field H_0 in the z-direction is kept constant and the frequency ω of the oscillating field is slowly varied, thereby slowly varying the value H^* of the resonance field, given by (17); or ω and therefore H^* are kept constant and H_0 is varied slowly. Both procedures are recommendable in practice, depending upon the circumstances; we shall here assume that the latter procedure is chosen, i.e., that H_0 varies adiabatically with constant ω.

Whether a variation dH_0/dt of H_0 can be considered as adiabatic or not, depends, according to (19) and (21) upon the half amplitude H_1 of the r-f field, any given variation being the more adiabatic, the larger H_1. The condition (21) for adiabatic variation can also be expressed by stating that the z-field H_0 has to pass through an interval, comparable to the "resonance width" H_1 during a time which is long compared to

$$t_1 = 1/|\gamma H_1|. \qquad (22)$$

As a numerical example, we shall again consider protons with $\gamma = 2.66 \times 10^4$ c.g.s.[11] Choosing an amplitude $2H_1$ of 10 gauss, for the r-f field, i.e., with $H_1 = 5$ gauss, we have

$$t_1 = 7.5 \times 10^{-6} \text{ sec.}, \qquad (23)$$

[11] J. M. B. Kellogg, I. I. Rabi, N. F. Ramsey, Jr., and J. R. Zacharias, Phys. Rev. **56**, 728 (1939).

i.e., the variation is adiabatic, if H_0 varies by an amount comparable to the "resonance width" of 5 gauss in a time long compared to 7.5 microseconds. It is clear that with the assumed value of H_1, normal variations of H_0 can be considered as perfectly adiabatic and that H_1 has to be chosen very much smaller before this condition is violated.

We shall now make use of the preceding considerations in order to introduce the nuclear induction effect, i.e., the essential features of an experimental arrangement by which a rotating component of the nuclear polarization can be observed through an induced voltage signal.

A sample, containing among others the nuclei under consideration, shall occupy a relatively small volume between the poles of a magnet so that the field H_0 of the magnet can be considered homogeneous over the extension of the sample, its direction being chosen as the z-direction of a right-handed Cartesian coordinate system. An oscillating field with amplitude $2H_1$ and circular frequency ω shall be produced by an r-f current, passing through a wire which is wound such that the r-f field has an essentially constant amplitude over the sample region and oscillates in the x-direction. Immediately surrounding the sample there shall finally be wound a coil with its axis parallel to the y-axis, so that an r-f flux in the y-direction will manifest itself by an induced r-f voltage signal across its terminals. It is such a signal, produced by the rotating component of the nuclear polarization, in which we are primarily interested, and we shall show that one can expect it to be of easily detectable magnitude, with moderate sizes of the sample and under normal laboratory conditions.

In order to estimate the induced r-f voltage we have to consider the expression for M_y of Eq. (20). The corresponding value of the induction is

$$B_y = 4\pi M_y, \qquad (24)$$

and if N turns of the receiver coil surround a cross-sectional area A of the sample, we obtain for the effective flux through the coil

$$F = 4\pi NAM_y = \mp 4\pi NAM\frac{\sin \omega t}{(1+\delta^2)^{\frac{1}{2}}}, \qquad (25)$$

and for the induced voltage V across the terminals of the coil

$$V = -\frac{1}{c}\frac{dF}{dt} = \pm\frac{4\pi}{c}NAM\omega\frac{\cos \omega t}{(1+\delta^2)^{\frac{1}{2}}}, \qquad (26)$$

where the variation of δ has been considered slow enough so that its time derivative can be neglected compared to that of cos ωt.

The amplitude of the signal voltage V reaches evidently a maximum at resonance, i.e., for $\delta = 0$, so that here the z-field H_0 has, according to (17) and (19), the value $H_0 = \omega/|\gamma| = H^*$. We shall now assume that the sample has been for a sufficiently long time in a field $H_0 = H > H^*$, so that M has reached its thermal equilibrium value, corresponding to that field, and is given by (1) to be $M = \chi H$. If now H_0 starts to decrease and if our previous assumptions, and particularly that of the constancy of M, remain valid, we can substitute this value into (26), thus obtaining

$$V = \pm\frac{4\pi}{c}NA\chi H\omega\frac{\cos \omega t}{(1+\delta^2)^{\frac{1}{2}}}. \qquad (27)$$

It has to be observed that it is sufficient for the "equilibrium field" H to be larger than the resonance field H^* only by a small percentage, i.e., by several times the relatively weak field H_1, in order that thermal equilibrium can be established under non-resonant conditions. If we assume this to be actually the case, we can substitute in (27) for H the resonance value $H^* = \omega/|\gamma|$. We shall further rewrite the formula (2) for the nuclear susceptibility, using (9) and writing $a = jh/2\pi$ for the angular momentum, so that

$$\chi = n\frac{j(j+1)}{3kT}\left(\frac{\gamma h}{2\pi}\right)^2,$$

and finally

$$V = \frac{1}{\pi c}NAn\frac{j(j+1)}{3kT}h^2\gamma\omega^2\frac{\cos \omega t}{(1+\delta^2)^{\frac{1}{2}}}. \qquad (28)$$

The plus or minus sign in (27), referring to positive or negative γ-values, respectively, does not appear any more in this formula, provided that γ is here taken algebraically, i.e., positive or negative, depending upon the nuclei under consideration.

We shall now compute from (27) the voltage amplitude $a_r = (4\pi/c)NA\chi H\omega$ at resonance (i.e., for $\delta=0$) of the signal voltage for protons in water, using the same conditions ($H=10{,}000G$, $T=291°$), which led to the value (4) for $\chi H=M=3.4\times10^{-6}G$. The corresponding circular frequency is given by $\omega=|\gamma|H=2.66\times10^4\times10{,}000=2.66\times10^8$ sec.$^{-1}$. ($\nu=42$ megacycles.) Assuming the cross-sectional area A of the sample to be 1 cm^2 and the number of turns on the receiver coil $N=10$, we obtain the resonance amplitude in volts

$$a_r=10^{-8}\times4\pi\times10\times3.4\times10^{-6}\times2.66\times10^{-8}=1.1\times10^{-3}\text{ volt.}$$

We see thus that even under the moderate conditions, here assumed, with linear dimensions of the sample of the order of 1 cm, one can expect from these considerations an r-f signal voltage of the order of a full millivolt, giving a considerable margin in the limits of observation.

Such a margin is desirable, not only because it allows a convenient reduction in the applied values of H and ω, but also because the voltage, given by (27) or (28) represents actually an overestimate. These formulas omit the influence of internuclear and thermal interaction, and we shall see later that they can cause an appreciable reduction from the above estimate.

Not only the magnitude of the induced voltage signal is of interest, but also its sign, i.e., its relative phase with respect to the "driving field" H_x of Eq. (5). It is evident that the sign is partly determined by that of γ, but its determination in Eq. (28) depends also upon the special assumptions under which it was derived, particularly the one expressed in formula (20), that M_z has the positive value M for large positive values of δ, i.e., according to (19) for z-fields H_0 appreciably *above* the resonance field H^*. This assumption is of course justified, if thermal equilibrium has been established, previous to the passage through resonance, in fields *higher* than the resonance field. If, on the other hand, the thermal equilibrium was previously established in fields *below* the resonance field, one must obviously demand that the opposite initial condition is fulfilled, i.e., that M_z is positive for negative values of δ. This requires changes in the considerations, leading to Eq. (20), which can immediately be derived from the starting Eq. (11). This equation is linear and homogeneous in the components of the polarization vector **M** so that a change of sign of all components leads necessarily from one solution to another solution. In order to satisfy the condition for positive M_z below resonance, it is therefore only necessary to change the sign in the formula (20). The subsequent considerations will all remain valid, except for a change of sign of M_y and V, leading thus, instead of (28), to the more general formula

$$V=\pm NAn\frac{j(j+1)}{3kT}h^2\gamma\omega^2\frac{\cos\omega t}{(1+\delta^2)^{\frac{1}{2}}},\tag{29}$$

with the $\pm$ sign referring to the case of the polarization parallel to the z-field far above or below resonance, respectively. In using the phase of the induced voltage signal with respect to the driving field for a determination of the sign of γ, or of the nuclear moment, it is therefore necessary to keep also the relative magnitude of equilibrium field and resonance field in mind, since it likewise affects the phase.

The fact that our formulas express an "indefinite memory" of the nuclear polarization, as to the conditions under which it was created, rests of course upon the initial simplifying assumptions of this section, and particularly upon assumption (1c), that thermal agitation does not affect the nuclei during the time of observation. It is clear from the previous considerations that finite, although possibly rather long, relaxation times will play an important role in the actual behavior of the observable phenomena, and they will be discussed in the following section.

4. INFLUENCE OF THERMAL AGITATION AND INTERNUCLEAR ACTION

The considerations of the preceding section can be regarded only as qualitative, since they are based upon the assumption (1) that all changes of nuclear orientation are due to the external fields. It implies the omission of three major internal actions, mentioned under (1a), (1b), and (1c), of Section 3, which are likewise responsible for changes of orientation.

The first action is that of atomic moments upon the nuclei. Its importance depends evidently upon the substance under consideration, i.e., upon whether such moments are actually present or not. There are indeed many substances (e.g., water) where it is safe to assume their absence, i.e., where the electronic spin moments will be paired off and where orbital moments which may be present in the free atoms or molecules are quenched because of intermolecular action. While assumption (1a) is justified under these circumstances, one has to introduce major changes in the presence of permanent atomic moments. The fields due to these moments and acting upon the nuclei will generally be considerably stronger than the external fields, and we do not have to deal in first approximation with independent nuclei, but with nuclear moments which are strongly coupled to the atomic frame. The situation is analogous to that investigated by Rabi and his collaborators[12] for free molecules, where the r-f field causes transitions between hyperfine structure levels. Except for a broadening of these levels, due to interatomic forces, similar phenomena can be expected in liquids and solids (e.g., in the salts of the rare earths), but we shall not enter here upon their discussion and shall restrict ourselves to the case where permanent atomic moments are absent.

There remain then the internal fields due to thermal agitation and internuclear action which have to be considered. Although both these fields are usually considerably weaker than the external applied fields, they are of importance because of their cumulative effects over longer periods of time. We shall not attempt here to give a rigorous quantum-mechanical theory of these effects; an excellent start for such a theory has been made in the papers mentioned in reference 8. Instead, we shall restrict ourselves to a semi-macroscopic description, trying to introduce into Eq. (11) for the macroscopic polarization such modifications as are necessary to account for the principal features of these effects.

To arrive at these modifications we shall consider a finite polarization **M** to exist at a certain moment and shall separately investigate the changes which it will undergo due to thermal agitation and internuclear action. Although there is a certain similarity insofar as both represent random actions upon the individual nuclei, there is this essential difference, that only thermal perturbations can change the energy of the total spin system, while internuclear interactions leave this energy unchanged.

The dominant part of the total spin energy E is caused to the strong field H_0 in the z-direction and can be written in the form

$$E = -H_0 M_z. \tag{30}$$

Major changes of the total energy are therefore necessarily due to a change of the z-component of the polarization and it will be the thermal perturbations which will be responsible for these changes. The equilibrium value which M_z will approach under the influence of thermal perturbations is given by

$$M_0 = \chi H_0. \tag{31}$$

If at any time $M_z \neq M_0$, it will approach this value exponentially with a characteristic time constant T_1, which we shall call the "thermal" or "longitudinal" relaxation time, and we can describe the rate of change of M_z, due to thermal perturbations alone, by the differential equation

$$\dot{M}_z = -(M_z - M_0)/T_1 \tag{32}$$

with the stationary solution $M_z = M_0$.

The actual value of T_1 is very difficult to predict for a given substance; it depends delicately, not only upon the thermal motion of the atoms which is quite different in gases, liquids, and solids, but also upon their electronic structure and its modification, due to interatomic forces. Rough estimates can easily lead to relaxation times of many seconds or even hours. We shall see below that such long relaxation times can be inconvenient for the observation of the induction effect. It is recommendable, in this case, to add to the substance a certain percentage of paramagnetic atoms or molecules. They will essentially act as catalysts, with the relatively strong fields of their permanent moments greatly shortening the relaxation time T_1, even if they are present in a small percentage and do not otherwise affect the nuclei under consideration.

The fields which are due to neighboring nuclei, also contribute to the establishment of the equilibrium because of their thermal agitation. These fields are so small that they alone would normally lead to extraordinarily long thermal relaxation times with their influence upon the actual value T_1 being negligible. Internuclear actions can, however, be of importance for the changes of the other two components M_x and M_y of the polarization in which we are equally interested. The fact that the nuclei with their moments participate in the thermal agitation is indeed of minor importance and represents only a small correction in their effect upon these components, since changes of M_x and M_y do not affect the energy (30) of the total spin system in the field. These changes can, therefore, take place without the necessity of transferring part of the spin energy E into kinetic energy of the atoms and it is permissible, in this respect, to consider the nuclei at rest and to neglect their comparatively slow motion.

Processes in which the total energy of the spin system does not change and which therefore affect only the components of the polarization which are transversal to the field are not necessarily due to internuclear forces alone. Small and irregular inhomogeneities of the z-field H_0 and the presence of other moments, such as those of paramagnetic ions in solution, will cause similar effects. It will in fact be permissible, for a qualitative discussion, to describe all these effects, including internuclear actions, by an "effective irregularity" of the z-field of order H' and through this field by a "transversal relaxation time"

$$T_2 = 1/|\gamma| H', \tag{33}$$

which it takes for M_x and M_y to be appreciably affected.

The magnitude of T_2 can be easily estimated if H' is due to neighboring nuclei whose motion can be neglected. With μ being the magnetic moment of a neighboring nucleus and r being its distance one will expect

$$H' \cong \mu / r^3. \tag{34}$$

Choosing $\mu = 10^{-23}$ c.g.s. and $r = 2 \times 10^{-8}$ cm this leads to

$$H' = 1 \text{ gauss} \tag{35}$$

and with $\gamma = 10^4$ c.g.s. through (33) to

$$T_2 = 10^{-4} \text{ sec.} \tag{36}$$

To give a reliable estimate of T_2 in more general cases requires a more detailed investigation of the mechanisms involved and will not be attempted here. The special case leading to (36) will merely be considered as an illustration for the fact that the transversal relaxation time T_2 can be many orders of magnitude smaller than the longitudinal time T_1 and that serious errors may be committed by assuming $T_1 = T_2$. There are, on the other hand, also cases where this equality is justified, particularly those where both relaxation times are due to impacts which last during a time short compared to the Larmor period, so that the distinction between collisions which change the spin energy and those which leave it unaltered becomes immaterial.

In order to obtain a qualitative description of the total change of the nuclear polarization $\mathbf{M}$ with time we shall now introduce such terms for its rate of change which contain the essential features of a longitudinal and transversal relaxation time and which, at the same time, are chosen so as to complicate the analysis as little as possible. For this purpose we shall assume that in analogy to the change (32) of the longitudinal component, the change of M_x and M_y will likewise be of an exponential character, governed by the equations

$$\begin{aligned} \dot{M}_x &= -(1/T_2) M_x, \\ \dot{M}_y &= -(1/T_2) M_y \end{aligned} \tag{37}$$

and that the total rate of change of $\mathbf{M}$ is obtained by adding to the expression (11), which takes into account the action of external fields only, the changes (32) and (37) of its components due to internal actions. The appearance of the same time constant T_2 in both Eqs. (37) is justified if one considers the substance to be isotropic. We obtain then the following differential equations for the three components of $\mathbf{M}$:

$$\dot{M}_x - \gamma(M_y H_z - M_z H_y) + \frac{1}{T_2} M_x = 0, \tag{38a}$$

$$\dot{M}_y - \gamma(M_z H_x - M_x H_z) + \frac{1}{T_2} M_y = 0, \tag{38b}$$

$$\dot{M}_z - \gamma(M_x H_y - M_y H_x) + \frac{1}{T_1} M_z = \frac{1}{T_1} M_0. \tag{38c}$$

The components of the external field are given by Eq. (5) and we shall replace, as in Section 3, the actual oscillating x-component by the effective rotating component (14), so that we have

$$H_x = H_1 \cos \omega t; \quad H_y = \mp H_1 \sin \omega t; \quad H_z = H_0 \quad (39)$$

with the two signs of H_y referring to positive and negative values of γ, respectively.

It is convenient to introduce instead of M_x and M_y two new variables u and v through

$$M_x = u \cos \omega t - v \sin \omega t, \qquad (40a)$$

$$M_y = \mp(u \sin \omega t + v \cos \omega t). \qquad (40b)$$

We shall further choose the time scale in units of $1/|\gamma|H_1$ by introducing the dimensionless quantities:

$$\tau = |\gamma|H_1 t; \quad \alpha = \frac{1}{|\gamma|H_1T_1}; \quad \beta = \frac{1}{|\gamma|H_1T_2}; \quad \delta = \frac{|\gamma|H_0 - \omega}{|\gamma|H_1} \qquad (41)$$

where $|\gamma|$ is the absolute value of γ. Using (39)–(41), one can then write the Eqs. (38) in the form:

$$du/d\tau + \beta u + \delta v = 0, \qquad (42a)$$

$$dv/d\tau + \beta v - \delta u + M_z = 0, \qquad (42b)$$

$$dM_z/d\tau + \alpha M_z - v = \alpha M_0. \qquad (42c)$$

The Eqs. (38) differ from the three component Eqs. (11) of Section 3, insofar as they contain on their left sides the "damping terms" inversely proportional to T_1 and T_2 and that on the right side of (38c) there appears an inhomogeneous term, proportional to the equilibrium polarization $M_0 = \chi H_0$. Their solution offers no difficulty, especially if δ is a slowly varying quantity so that they appear in the form (42) as a system of linear differential equations with almost constant coefficients; a more detailed discussion will be reserved for a later occasion. We shall here give directly a particular solution which is of special interest for our present purposes; its validity can be verified, provided that the variation of δ is adiabatic in the sense of Section 3 and that both quantities α and β of (41) are assumed to be small compared to unity. The first condition implies $|d\delta/d\tau| \ll 1$; it is identical with Eq. (21) and will be the more closely satisfied for any given variation with time of δ, the larger the amplitude of the oscillating field. In order to have $\alpha \ll 1$, $\beta \ll 1$, it is necessary, according to (41), that either the relaxation times T_1 and T_2 are sufficiently large or that the amplitude $2H_1$ of the oscillating field is sufficiently large.

One can normally expect the transverse relaxation time T_2 to be smaller than the longitudinal T_1 or $\alpha < \beta$. The assumption of "large amplitudes" implies then, according to (41), that

$$H_1 \gg \frac{1}{|\gamma|T_2}$$

or, according to (33),

$$H_1 \gg H'. \qquad (43)$$

Using for a numerical example the value (35) of 1 gauss for H', we see therefore that an amplitude of $2H_1 = 10$ gauss can already be considered as "large" although it will normally be still very small compared to the strong field H_0 (of the order of several thousand gauss) in the z-direction. It is in this case $\beta = 0.2$ and no serious errors are committed in neglecting correction terms in α and β.

With the three quantities $|d\delta/d\tau|$, α and β small compared to unity our particular solution of (38) can be written in the convenient form

$$M_x = \frac{M}{(1+\delta^2)^{\frac{1}{2}}} \cos \omega t, \qquad (44a)$$

$$M_y = \mp \frac{M}{(1+\delta^2)^{\frac{1}{2}}} \sin \omega t, \qquad (44b)$$

$$M_z = \frac{M\delta}{(1+\delta^2)^{\frac{1}{2}}}, \qquad (44c)$$

with

$$\delta = \delta(t) = \frac{H_0(t) - H^*}{H_1}; \quad H^* = \frac{\omega}{|\gamma|}, \qquad (45)$$

$$M = M(t) = \frac{1}{T_1} \times \int_{-\infty}^{t} \frac{\delta(t') \exp\{-[\theta(t) - \theta(t')]\} M_0(t')}{[1+\delta^2(t')]^{\frac{1}{2}}} dt', \qquad (46)$$

$$M_0(t') = \chi H_0(t'), \qquad (47)$$

$$\theta(t) - \theta(t') = \frac{1}{T_1} \int_{t'}^{t} \frac{\delta^2(t'') + T_1/T_2}{1+\delta^2(t'')} dt''. \qquad (48)$$

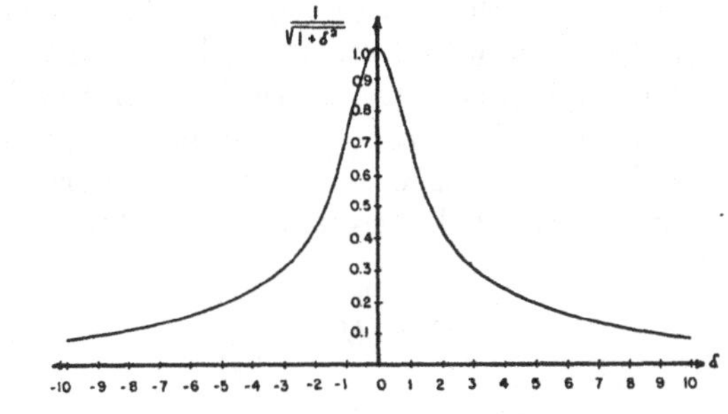

FIG. 1. Schematic representation of the voltage amplitude in the case of rapid passage. The abscissa δ is the deviation of the actual field from the resonance field in units of the half-amplitude of the oscillating field. The ordinate is proportional to the amplitude of the induced voltage and the y-component of the nuclear polarization.

The general solution of (38) is of course obtained by adding to (44) arbitrary multiples of the homogeneous solutions, their factors being determined by the initial conditions. These solutions can be seen to decrease exponentially with time so that it is permissible to omit them after a sufficient time has elapsed. In practice it is only the solution (44) which has to be considered, since it corresponds to the situation where the equilibrium polarization M_0 was zero before a sufficient time in the past, i.e., where the actual conditions have been obtained by adiabatic change, starting from an originally unpolarized sample.

The Eqs. (44) have the same form as the Eqs. (20) of Section 3, with the difference that M is not an arbitrary constant but is generally a function of time, given by (46). It has to be observed that while its absolute value $|M|$ still represents the instantaneous magnitude of the polarization, the quantity M itself is not necessarily positive but can have both signs, depending upon the positive or negative values which δ in the integrand of (46) has assumed in the past. We are primarily interested in the behavior of the polarization near resonance and it is interesting to observe that M may here be considered as essentially constant or appreciably variable, depending upon the speed of variation of δ, i.e., of either H_0 or ω in comparison to the relaxation times.

Formula (46) shows in fact that the value of M at any time is determined by the past history and that the more remote a past is of importance, the slower the decrease of the exponential. The functional dependence of M upon time and particularly its behavior near resonance will be primarily determined by the relative change of δ and of the exponential in the integrand of (46).

We shall speak of the limiting case of "rapid passage" if δ varies near resonance by an amount of order unity during a time short compared to the variation of the exponential. This case is the more approximately realized the more rapid the change of δ or, according to (48), the longer the relaxation times T_1 and T_2. It can be easily seen with the use of Eqs. (41) that because of the assumed smallness of α and β such a relatively rapid change of δ is compatible with the adiabatic condition $|d\delta/dt| \ll 1$. In the neighborhood of resonance M can be considered in this case to be essentially constant, its actual value depending mostly upon the values which M_0 and δ have assumed an appreciable time before approach of resonance conditions.

The amplitude of M_y is then proportional to $1/(1+\delta^2)^{\frac{1}{2}}$ and is schematically represented in Fig. 1. The simplified situation, considered in Section 3, is in fact a special case of "rapid passage" or long relaxation time. It is assumed here that H_0 has been held fixed for a long time at a value H, far from resonance, so that $|\delta| \gg 1$ with a subsequent establishment of resonance conditions during a time, very short compared to T_1. The main contribution to (46) arises then from past times t', where M_0 had the constant value χH and δ^2 had a value large compared to unity. If we assume, besides, for this value $\delta^2 \gg T_1/T_2$ or if we assume, irrespective of δ that $T_1 = T_2$ we obtain from (48)

$$\theta(t) - \theta(t') = \frac{t-t'}{T_1}$$

and from (46)

$$M = \pm \frac{\chi H}{T_1} \int_{-\infty}^{t} \exp\{-t-t'/T_1\} dt' = \pm \chi H. \quad (49)$$

The magnitude of M corresponds thus, as was to be expected, to the field H, to which the sample was sufficiently long subjected. The plus or minus sign in (49) has to be taken, depending on whether this field was stronger or weaker than the resonance field, i.e., whether during action of this field, far from resonance, we had $\delta/(1+\delta^2)^{\frac{1}{2}} = +1$

or $\delta/(1+\delta^2)^{\frac{1}{2}}=-1$. This duplicity of the resulting sign of M_y upon approach of resonance from "above" or "below" was discussed in Section 3 and must not be confused with the other duplicity, expressed in (44b) and depending upon the sign of γ.

The opposite limiting case, to be considered, is that of "slow passage" through resonance or short relaxation times, where the main variation in the integrand of (46) is caused by the exponential. It can be treated by writing

$$\frac{\partial}{\partial t'}\exp-[\theta(t)-\theta(t')]=\frac{\partial\theta(t')}{\partial t'}\exp-[\theta(t)-\theta(t')]$$

or using (48)

$$\exp-[\theta(t)-\theta(t')]$$
$$=T_1\frac{1+\delta^2(t')}{\delta^2(t')+(T_1/T_2)}\frac{\partial}{\partial t'}\exp-[\theta(t)-\theta(t')]$$

so that by partial integration, we obtain from (48)

$$M(t)=\frac{1+\delta^2(t)}{\delta^2(t)+(T_1/T_2)}\frac{\delta(t)M_0(t)}{[1+\delta^2(t)]^{\frac{1}{2}}}$$
$$-\int_{-\infty}^{t}\exp-[\theta(t)-\theta(t')]\frac{d}{dt'}$$
$$\times\left[\frac{1+\delta^2(t')}{\delta^2(t')+(T_1/T_2)}\frac{\delta(t')M_0(t')}{[1+\delta^2(t')]^{\frac{1}{2}}}\right]dt'. \quad (50)$$

The last integral becomes evidently negligible if the variation of δ and M_0 with time is sufficiently slow. Keeping only the dominant first term in (50) and substituting into (44), we find thus:

$$M_x(t)=\frac{M_0(t)\delta(t)}{\delta^2(t)+T_1/T_2}\cos\omega t, \quad (51a)$$

$$M_y(t)=\mp\frac{M_0(t)\delta(t)}{\delta^2(t)+T_1/T_2}\sin\omega t, \quad (51b)$$

$$M_z(t)=\frac{M_0(t)\delta^2(t)}{\delta^2(t)+T_1/T_2}. \quad (51c)$$

In contrast to the case of rapid passage, where M can be considered essentially constant and where the amplitudes of M_x and M_y reach, according to (44) their maximum at resonance, i.e., for $\delta=0$, it is seen from (51) that in the case of slow passage all three components of the polarization vanish at resonance. Since M_0 varies only little in the neighborhood of resonance, it is permissible to replace it by the equilibrium polarization M^* at the resonance field H^*, i.e., to write

$$M_0=M^*=\chi H^*=\chi\frac{\omega}{|\gamma|}.$$

The amplitudes of M_x and M_y are then given by

$$a_{x,y}=\chi\frac{\omega}{|\gamma|}f(\delta) \quad (52)$$

where

$$f(\delta)=\frac{\delta}{\delta^2+T_1/T_2}. \quad (53)$$

We have plotted this function in Fig. 2. It assumes its extremum values

$$f(\delta)=\pm\tfrac{1}{2}(T_2/T_1)^{\frac{1}{2}} \quad (54)$$

for

$$\delta=\pm(T_1/T_2)^{\frac{1}{2}} \quad (55)$$

and shows qualitatively different behavior from the simple maximum, to be expected in the case of rapid passage and represented in Fig. 1.

The case of "slow passage" can also be treated directly without the restriction $\alpha\ll1$, $\beta\ll1$ which led to the solution (44) and (51). With δ and M_0 so slowly varying that they can be considered as practically constant one can in this case obtain a solution of the Eqs. (42) by assuming u, v, and M_z likewise as practically constant, i.e., by neglecting their derivatives. With $du/d\tau=dv/d\tau$

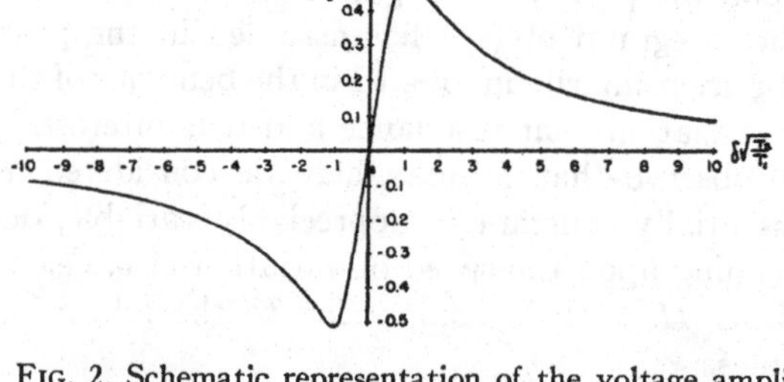

FIG. 2. Schematic representation of the voltage amplitude in the case of slow passage. T_1 and T_2 are the longitudinal and transversal relaxation times, respectively, and the scale is chosen such as to make the plot independent of their values. The significance of abscissa and ordinate is otherwise the same as in Fig. 1.

$=dM_z/d\tau=0$ one obtains directly from the Eqs. (42)

$$u=\frac{\delta}{\beta^2+\delta^2+\beta/\alpha}M_0, \quad (56a)$$

$$v=-\frac{\beta}{\beta^2+\delta^2+\beta/\alpha}M_0, \quad (56b)$$

$$M_z=\frac{\beta^2+\delta^2}{\beta^2+\delta^2+\beta/\alpha}M_0. \quad (56c)$$

Using $\beta/\alpha=T_1/T_2$ and neglecting for small values of β, v and the term β^2 in the numerators this is through (40) identical with the solution (51).

It is interesting, however, to investigate more closely this solution for arbitrary β, particularly since it allows in the case of slow passage a direct comparison between the condition, favorable for nuclear induction on one side and resonance absorption on the other. Using the Eqs. (41) and writing $|\gamma|H_0-\omega=\Delta\omega$ for the deviation of the resonance frequency from ω we can also write

$$u=\frac{|\gamma|H_1T_2^2\Delta\omega}{1+(T_2\Delta\omega)^2+(\gamma H_1)^2T_1T_2}M_0, \quad (57a)$$

$$v=-\frac{|\gamma|H_1T_2}{1+(T_2\Delta\omega)^2+(\gamma H_1)^2T_1T_2}M_0, \quad (57b)$$

$$M_z=\frac{1+(T_2\Delta\omega)^2}{1+(T_2\Delta\omega)^2+(\gamma H_1)^2T_1T_2}M_0. \quad (57c)$$

For nuclear induction it is evidently recommendable to have u as large as possible. Its maximum is obtained for

$$\Delta\omega=\frac{1}{T_2}[1+(\gamma H_1)^2T_1T_2]^{\frac{1}{2}} \quad (58)$$

and has here the value

$$u_{\max}=\frac{|\gamma|H_1T_2}{[1+(\gamma H_1)^2T_1T_2]^{\frac{1}{2}}}M_0. \quad (59)$$

This value again increases monotonically with H_1 and for

$$H_1\gg\frac{1}{|\gamma|(T_1T_2)^{\frac{1}{2}}}$$

becomes

$$u_{\max\,\max}=(T_2/T_1)^{\frac{1}{2}}M_0. \quad (60)$$

The condition that the r-f field amplitude H_1 is large enough to obtain the highest possible u is according to (41) equivalent with $\alpha\beta\ll1$ and also actually assumed in the derivation of (51) since it is valid only for $\alpha\ll1$, $\beta\ll1$.

To obtain maximum absorption, it is on the other hand necessary to make v as large as possible, since it is this quantity which, through (40a) determines the out of phase part of M_x. v has its maximum

$$v_{\max}=-\frac{|\gamma|H_1T_2}{1+(\gamma H_1)^2T_1T_2}M_0 \quad (61)$$

for $\Delta\omega=0$. Unlike $u_{\max}$ this quantity does not increase monotonically for increasing H_1, but decreases for large values of H_1. As observed by Purcell, Torrey, and Pound[6] it is therefore not advisable for absorption experiments to use too intensive r-f fields. The best possible choice is

$$H_1=\frac{1}{|\gamma|(T_1T_2)^{\frac{1}{2}}} \quad (62)$$

and yields

$$v_{\max\,\max}=(T_2/T_1)^{\frac{1}{2}}M_0. \quad (63)$$

It seems satisfactory that the maximum values (60) and (63) of the two decisive quantities u and v is thus the same, although to obtain the one, necessary for induction, requires the r-f field amplitude to be large compared to the "critical amplitude" $1/|\gamma|(T_1T_2)^{\frac{1}{2}}$ whereas the other, necessary for absorption requires it to be equal to this quantity.

It is evident from (44b) and (46) that the magnitude of the signal, induced by the component M_y of the nuclear polarization, depends not only upon the susceptibility χ but also, in a rather complicated way, upon the relaxation times and upon the magnitude and speed of variation of δ. The expression (28) for the induced voltage which was derived in Section 3 under simplifying assumptions will usually not correspond to the observed values but represents merely an estimate to be approached under favorable conditions. The special case of "rapid" passage, considered before and leading to the expression (49) of M represents such a favorable condition, and it is only in this limiting case that

(28) represents the actually observable value of the induced voltage.

The maximum value of M and thereby of the induced voltage signal at resonance must be expected normally to be smaller than (49) and in fact becomes the smaller for given T_2, and under otherwise equal conditions, the longer the relaxation time T_1. In the limit of very large T_1 one obtains from (48)

$$\theta(t)-\theta(t')=\frac{1}{T_2}\int_{t'}^{t}\frac{dt''}{1+\delta^2(t'')}, \qquad (64)$$

which is independent of T_1, so that, according to (46) M becomes inversely proportional to T_1. One may say, in this sense, that too long relaxation times T_1 are unfavorable for the observation of the effect unless the variation of δ with time is changed so as to re-establish favorable conditions. It is otherwise recommendable to use samples in which T_1 is as short as possible, possibly by adding paramagnetic catalysts as previously mentioned. As T_1 becomes sufficiently short, one will approach the case of "slow passage" mentioned above and represented in Fig. 2. While it will not be possible to have actually $T_1 \ll T_2$ it s seen from (54) that the maximum signal will still increase with decreasing T_1, the optimum being reached for $T_1=T_2$.

5. CONCLUSIONS

While the methods of molecular beams and of nuclear induction have a common ground of investigation it is evident that neither one makes the other superfluous. There remains a large complex of interesting nuclear experiments to be done, for which molecular beams are essential, particularly those which call for matter of extremely low density and the least possible interference of neighboring atoms or molecules. There are, on the other hand, many problems which become accessible or which can be more conveniently solved through nuclear induction and some of these will be mentioned here:

(1) The exact comparison of the magnetic moments of the neutron, the proton, and the deuteron is at present one of the most interesting problems, concerning nuclear forces. The main difficulty in this comparison was until now the sufficiently accurate calibration of the resonance field. It can be completely avoided by repeating the experiment of Alvarez and Bloch[2] for neutrons and by observing through nuclear induction simultaneously and in the same field the resonances of protons and deuterons. The problem of comparison of their magnetic moments is thus reduced to that of their respective resonance frequencies and can be solved with high accuracy. It was indeed with this experiment in mind, and while searching for a suitable method of comparison, that the author was led to the thought of nuclear induction, and preparations are now under way at Stanford to carry out the measurement in the near future.

(2) One of the difficulties in the determination of the gyromagnetic ratios of many nuclei by molecular beams is that of finding suitable detectors. The method of nuclear induction is free from this obstacle and should be soon applied to all elements for which this determination is of interest.

(3) While even in its very initial stage, nuclear induction was observed with a sample of 100 milligrams, there are good reasons to believe that the sensitivity can still be greatly increased. This offers the possibility to observe the effect not only in liquids and solids but also in gases under no excessive pressure. With only small amounts of matter necessary for its performance, the experiment offers a convenient way of isotope analysis and particularly also for its application to radioactive nuclei.

(4) It was shown in Section 4 that the induced signals to be expected depend not only upon the nuclear susceptibility but also upon the relaxation times. By suitable choice of the variation with time of resonant field or frequency, it is thus possible to measure these quantities separately. The study of nuclear relaxation times is of interest not only as an experimental method to investigate the establishment of thermal equilibrium, but also because of its importance for reaching extremely low temperatures through the nuclear magnetocaloric effect. While even the information gained at room temperature is valuable, it is clear that it can be greatly enlarged by studying the temperature dependence of the effect and particularly its behavior at low temperatures. It is in this same respect that the effect of paramagnetic catalysts, mentioned in Sections 1 and 4, seems of considerable interest.

(5) As in comparing the moments of neutron, proton, and deuteron, nuclear induction can well be developed as a simple and practical method to calibrate and measure high magnetic fields with great accuracy, and to apply it, for example, in the construction of cyclotrons and mass spectrographs.

There are unquestionably more problems which will become tangible in further development of the new electromagnetic effects. The fact that they are simple to obtain and require only very modest equipment should make it possible for many investigators to enter this field of research.

Relaxation Effects in Nuclear Magnetic Resonance Absorption*

N. Bloembergen,** E. M. Purcell, and R. V. Pound,***
Lyman Laboratory of Physics, Harvard University, Cambridge, Massachusetts
(Received December 29, 1947)

The exchange of energy between a system of nuclear spins immersed in a strong magnetic field, and the heat reservoir consisting of the other degrees of freedom (the "lattice") of the substance containing the magnetic nuclei, serves to bring the spin system into equilibrium at a finite temperature. In this condition the system can absorb energy from an applied radiofrequency field. With the absorption of energy, however, the spin temperature tends to rise and the rate of absorption to decrease. Through this "saturation" effect, and in some cases by a more direct method, the *spin-lattice relaxation time* T_1 can be measured. The interaction *among* the magnetic nuclei, with which a characteristic time T_2' is associated, contributes to the width of the absorption line. Both interactions have been studied in a variety of substances, but with the emphasis on liquids containing hydrogen.

Magnetic resonance absorption is observed by means of a radiofrequency bridge; the magnetic field at the sample is modulated at a low frequency. A detailed analysis of the method by which T_1 is derived from saturation experiments is given. Relaxation times observed range from 10^{-4} to 10^2 seconds. In liquids T_1 ordinarily decreases with increasing viscosity, in some cases reaching a minimum value after which it increases with further increase in viscosity. The line width meanwhile increases monotonically from an extremely small value toward a value determined by the spin-spin interaction in the rigid lattice. The effect of paramagnetic ions in solution upon the proton relaxation time and line width has been investigated. The relaxation time and line width in ice have been measured at various temperatures.

The results can be explained by a theory which takes into account the effect of the thermal motion of the magnetic nuclei upon the spin-spin interaction. The local magnetic field produced at one nucleus by neighboring magnetic nuclei, or even by electronic magnetic moments of paramagnetic ions, is spread out into a spectrum extending to frequencies of the order of $1/\tau_c$, where τ_c is a correlation time associated with the local Brownian motion and closely related to the characteristic time which occurs in Debye's theory of polar liquids. If the nuclear Larmor frequency ω is much less than $1/\tau_c$, the perturbations caused by the local field nearly average out, T_1 is inversely proportional to τ_c, and the width of the resonance line, in frequency, is about $1/T_1$. A similar situation is found in hydrogen gas where τ_c is the time between collisions. In very viscous liquids and in some solids where $\omega\tau_c > 1$, a quite different behavior is predicted, and observed. Values of τ_c for ice, inferred from nuclear relaxation measurements, correlate well with dielectric dispersion data.

Formulas useful in estimating the detectability of magnetic resonance absorption in various cases are derived in the appendix.

I. INTRODUCTION

IN nuclear magnetic resonance absorption, energy is transferred from a radiofrequency circuit to a system of nuclear spins immersed in a magnetic field, H_0, as a result of transitions among the energy levels of the spin system. For each of N non-interacting spins, characterized in the usual way by $I\hbar$ and μ, the maximum z components of angular momentum and magnetic moment, respectively, there are $2I+1$

* A brief account of this work has appeared in *Nature* 160, 475 (1947).

** Present address: Kamerlingh Onnes Laboratory, University of Leiden.

*** Society of Fellows.

levels spaced in energy by $\mu H_0/I$. The state of this (non-interacting system) can be described by the magnetic quantum numbers $m_1 \cdots m_i \cdots m_N$, where $-I \leqslant m_i \leqslant I$. Upon the application of a suitable oscillating magnetic field, transitions corresponding to stimulated emission ($\Delta m_i = +1$) and to absorption ($\Delta m_i = -1$) will occur. If transitions of the latter sort are to preponderate, so that there is a net absorption of energy from the radiation field, it is essential that there be initially a surplus of spins in the lower states. This condition will be attained eventually if there is some way by which the spin system can interact with its surroundings and come to thermal equilibrium at a finite temperature. At equilibrium the population of the $2I+1$ levels will be governed by the Boltzmann factor $\exp(\mu H_0 m_i/IkT)$, and the requisite surplus in lower states will have been established.

The exposure of the system to radiation, with consequent absorption of energy, tends to upset the equilibrium state previously attained, by equalizing the population of the various levels. The new equilibrium state in the presence of the radiofrequency field represents a balance between the processes of absorption of energy by the spins, from the radiation field, and the transfer of energy to the heat reservoir comprising all other internal degrees of freedom of the substance containing the nuclei in question. The latter process involves what we shall call for short the *spin-lattice* interaction, and is described by a characteristic time, T_1, the spin-lattice relaxation time. It is this time also which measures how long one must wait, after the application of the constant field H_0, for the establishment of thermal equilibrium.

It should be evident that the above-mentioned competition between resonance absorption and spin-lattice interaction provides a way of measuring T_1. In a precisely similar way one might study the heat transfer between a wire and a surrounding bath by examining the temperature of the wire as a function of the power dissipated in the wire because of a current flowing through it. In our case the resonance absorption itself supplies a convenient thermometer, for the intensity of the absorption, depending as it does upon the surplus of spins in lower states, reflects the "temperature" of the spin system. An equally direct method, in the case of the wire, would be the measurement of the temperature of the wire, as a function of time, after switching off the heating current. The exact nuclear absorption analog of this experiment has also been carried out, and affords a striking manifestation of the relaxation process.

It may be said at once that nuclear spin-lattice relaxation times appear to range between 10^{-4} and 10^2 seconds. Perhaps it seems remarkable at first that times so long should be associated with any atomic process. On the contrary, such times were, from the theoretical point of view, unexpectedly short, for the processes which had been examined theoretically[1] prior to the first experiments in this field yielded much longer relaxation times. In the first experimental detection of nuclear resonance absorption in a solid,[2] it was shown that the relaxation time for protons in paraffin, at room temperature and 7000 gauss, was certainly less than one minute. Subsequent work has shown that T_1 for this substance, under those conditions, is in fact very much shorter. Additional information on nuclear relaxation times has come from the work of Bloch and collaborators,[3] and of Rollin[4,5] who was the first to investigate the phenomenon at very low temperatures. The work here reported began as a more or less cursory survey of typical substances and grew, as the theoretical interpretation of the results progressed, into a systematic study of certain types of structures which forms the basis for a comprehensive theory of nuclear magnetic relaxation applicable to liquids, some gases, and certain types of solids.

It is obvious that the model of non-interacting spins described above is incomplete, for there will necessarily be interaction among the spins due to the associated nuclear magnetic moments, not to mention other possible sources of magnetic interaction such as electronic states of non-vanishing magnetic moment. We shall continue

---

[1] W. Heitler and E. Teller, Proc. Roy. Soc. **155A**, 629 (1936).

[2] E. M. Purcell, H. C. Torrey, and R. V. Pound, Phys. Rev. **69**, 37 (1946).

[3] F. Bloch, W. W. Hansen, and M. Packard, Phys. Rev. **70**, 474 (1946).

[4] B. V. Rollin, Nature **158**, 669 (1946); B. V. Rollin and J. Hatton, Nature **159**, 201 (1947).

[5] B. V. Rollin, J. Hatton, A. H. Cooke, and R. J. Benzie, Nature **160**, 457 (1947).

temporarily to ignore interactions of the latter sort, as well as possible electric quadrupole effects —it will turn out that we are justified in doing so for a sufficiently important class of substances —but we must certainly take into account the nuclear dipole-dipole interaction.

The interaction energy of neighboring dipoles separated by the distance r is μ^2/r^3 in order of magnitude, and if $H_0\mu \gg \mu^2/r^3$ we may expect, since more remote neighbors are relatively ineffective, that the result of the interactions will be a broadening of each of the $2I+1$ levels mentioned earlier. From a point of view which is somewhat naive but useful within limits, we may say that the total magnetic field in the z direction at the ith nucleus is H_0 plus a "local" field H_{loc} of order of magnitude

$$|H_{loc}| \approx (\mu/I) \sum_j m_j r_{ij}^{-3}$$

and that the resonance condition for a transition $\Delta m_i = \pm 1$ is $h\nu = (H_0 + H_{loc})\mu/I$. One would thus be led to expect the width of the absorption line, expressed in gauss, to be of the order μ/r^3. Taking another point of view, a spin j causes at its neighbor, i, an oscillating field at the resonance frequency, because of its own precession about H_0. This field is capable of inducing transitions in which i and j exchange energy. The lifetime against transitions is about hr^3/μ^2, leading again to a line width, expressed in gauss, of μ/r^3. Bloch[6] introduced the characteristic time $T_2 \approx hr^3/\mu^2$ to describe the spin-spin interaction, thus recalling the second process just mentioned. We were accustomed to use the local field as a measure of the interaction, which suggests the first process. Of course, both effects are present ordinarily and are included in any complete theory of the spin-spin interaction.† In any case, the examination of the line width is seen to be important in the experimental investigation of this interaction. One of the earliest results of resonance absorption experiments was the then surprising one that the line width of the proton resonance in liquids is, in general, much less than μ/r^3 in magnitude, a result which receives quantitative explanation in the work here reported.

[6] F. Bloch, Phys. Rev. **70**, 460 (1946).

† That the two effects are not one and the same can be seen by considering a lattice of non-identical nuclei, for which the second effect is absent but the "dispersion of local z-fields" remains.

The spin-spin interaction is of course incapable of bringing about heat transfer between the spin system as a whole and the thermal reservoir, or "lattice." However, the spin-spin and spin-lattice interactions are not unrelated, for each involves the existence at a nucleus of perturbing fields. Therefore our attention has been directed to both effects. Broadly speaking, and anticipating results described further on, the perturbing fields responsible for spin-lattice relaxation originate in the thermal motion of neighboring magnetic dipoles, whereas the spin-spin interaction, although it may be modified by such motion, is not dependent upon it for its existence, and is in fact strongest when such motion is absent.

The picture which has been sketched in these introductory remarks will be recognized as an approximation, obtained by starting with isolated non-interacting spins, then introducing the interaction as a perturbation, but clinging still to a description in terms of individual m_i's. That the picture proves as useful as it does can be attributed to the fact that the spin-spin interaction is weak compared to $H_0\mu$ and of relatively short-range character.

Finally, we review briefly the phenomenological theory of magnetic resonance absorption, before describing the experimental method. The phenomenon lends itself to a variety of equivalent interpretations. One can begin with static nuclear paramagnetism and proceed to paramagnetic dispersion;[7] or one can follow Bloch's analysis, contained in his paper on nuclear induction,[6] of the dynamics of a system of spins in an oscillating field, which includes the absorption experiments as a special case. We are interested in absorption, rather than dispersion or induction, in the presence of *weak* oscillating fields, the transitions induced by which can be regarded as non-adiabatic. We therefore prefer to describe the experiment in optical terms.

Consider a substance containing, per cm^3, N_0 nuclei of spin I and magnetic moment μ, placed in a strong uniform magnetic field H_0 along the z axis, and subjected to a weak oscillating field $H_x = 2H_1 e^{2\pi i \nu t}$; $H_y = 0$. The probability of a single transition in which one of the m_i's changes to

[7] C. J. Gorter and L. J. F. Broer, Physica **9**, 591 (1942).

m_i' can be found with the aid of the standard formula for magnetic dipole transitions:

$$W_{m_i\to m_i'}=(8\pi^3/3h^2)\,|(m_i|M|m_i')|^2\rho_\nu. \quad (1)$$

M is the magnetic moment operator. Ordinarily ρ_ν represents the energy density, in unit frequency range, in the isotropic unpolarized radiation field. We have to do here with radiation of a single frequency from levels of a finite width, which we describe by the observed shape of the absorption line, $g(\nu)$. The shape function $g(\nu)$ is to be normalized so that

$$\int_0^\infty g(\nu)d\nu=1.$$

Our radiation field is also unusual in that it consists simply of an oscillating magnetic field of one polarization. Only one of the circularly polarized components (of amplitude H_1) into which this field can be decomposed is effective. The equivalent isotropic unpolarized radiation density is easily arrived at, and we have for ρ_ν

$$\rho_\nu=3H_1^2g(\nu)/4\pi. \quad (2)$$

Introducing the nuclear gyromagnetic ratio γ, which will be used frequently hereafter,

$$\gamma=2\pi\mu/Ih, \quad (3)$$

and the matrix elements required in (1), we find

$$W_{m\to m-1}=(\pi/3)\gamma^2(I+m)(I-m+1)\rho_\nu$$
$$=\tfrac{1}{4}\gamma^2H_1^2g(\nu)(I+m)(I-m+1). \quad (4)$$

Equation (4) gives the probability for a transition $m\to m-1$, involving the absorption from the radiation field of the energy $h\nu=h\gamma H_0/2\pi$. If the spin system is initially in equilibrium at the temperature T, the population of each level m exceeds that of the next higher level, $m-1$, by

$$N_m-N_{m-1}\approx\frac{N_0}{2I+1}\frac{h\nu}{kT}. \quad (5)$$

The approximation (5) is an extremely good one, for, in the cases with which we shall deal, $h\nu/kT\approx10^{-6}$. The net rate at which energy is absorbed is now

$$P_a=\frac{N_0}{2I+1}\frac{(h\nu)^2}{kT}\sum_{m=I}^{-I+1}W_{m\to m-1}$$
$$=\frac{\gamma^2H_1^2N_0(h\nu)^2I(I+1)g(\nu)}{6kT}. \quad (6)$$

The maximum energy density in the oscillating field is $(2H_1)^2/8\pi$. The apparent "Q" of the sample, at frequency ν, is therefore given by

$$(1/Q)=P_a/\nu H_1^2=\gamma^2h^2N_0I(I+1)\nu g(\nu)/6kT. \quad (7)$$

An equivalent statement is that χ'', the imaginary part of the magnetic susceptibility of the substance, is

$$\chi''=\frac{1}{4\pi Q}=\frac{\pi}{2}\left[\frac{N_0\gamma^2\hbar^2I(I+1)}{3kT}\right]\nu g(\nu), \quad (8)$$

where $\hbar=h/2\pi$. The term in brackets will be recognized as the *static* susceptibility χ_0' of the spin system. It may be noted in passing that Eq. (8) is compatible with, and could have been obtained from, the Kramers formula connecting the real and imaginary parts of the susceptibility, which for this special case is

$$\chi_0'=\frac{2}{\pi}\int_0^\infty\frac{\chi''(\nu)d\nu}{\nu}. \quad (9)$$

If our earlier estimate of the line width is correct, the maximum value of $g(\nu)$ should be of the order of magnitude hr^3/μ^2; if so, the maximum value of χ'' to be expected, according to Eq. (8), is of the order of $h\nu/kT$, since $N_0\approx1/r^3$. A more accurate estimate based on the theory to be given later leads to the formula:

$$\chi''_{\max}\approx0.2(h\nu/kT)[I(I+1)]^{\frac{1}{2}}. \quad (10)$$

The numerical factor in (10) depends somewhat on the geometrical arrangement of the nuclei. For $\nu=10$ Mc/sec., and room temperature, $h\nu/kT=1.64\times10^{-6}$. If the line is much narrower than the above estimate—as we shall find it to be in many cases—the maximum value of χ'' is correspondingly larger. The experimental detection of the absorption thus presents the problem of detecting a small change in the susceptibility of the core of a radiofrequency coil. The absorption is, of course, accompanied by dispersion, the maximum value of χ' being of the same order of magnitude as the maximum value of χ''. The dispersive properties of the medium, moreover, are those of a substance displaying the Faraday effect, except that the effect is greatly enhanced by resonance. Advantage is taken of this in the nuclear induction method, but not in our experiments.

The formula (8) applies regardless of the origin of the line broadening, even if the line width and shape are controlled, as is often the case, by accidental inhomogeneities in the supposedly uniform H_0, *provided* that the original distribution of population among the levels remains substantially unaltered. In other words (8) is correct for vanishing H_1. As H_1 is increased, the thermal contact between spin system and lattice eventually proves unable to cope with the energy absorbed by the spin system, the spin temperature rises, and the relative absorption, measured by χ'', diminishes. It is the onset of this saturation effect which has been used to measure the spin-lattice relaxation time in most of our work. The interpretation of the data is complicated by the necessity of distinguishing various cases according to the factors which control the line width, and a detailed discussion of this problem will be postponed. However, in order to introduce at an early stage our working definition of the spin-lattice relaxation time, we shall now investigate the occurrence of saturation in a simple case.

Assume that $I=\frac{1}{2}$, so that we have to deal with two levels. Let n denote the surplus population of the lower level: $n=N_{+\frac{1}{2}}-N_{-\frac{1}{2}}$, and let n_0 be the value of n corresponding to thermal equilibrium at the lattice temperature. We *assume* that, in the absence of the radiofrequency field, the tendency of the spin system to come to thermal equilibrium with its surroundings is described by an equation of the form

$$dn/dt=(1/T_1)(n_0-n). \quad (11)$$

The assumption will be justified in Section VII. The characteristic time T_1 in Eq. (11) is called the spin-lattice relaxation time. (Wherever possible our definitions and nomenclature follow those of Bloch.)[6]

The presence of the radiation field requires the addition to (11) of another term:

$$dn/dt=(1/T_1)(n_0-n)-2nW_{\frac{1}{2}\to-\frac{1}{2}}. \quad (12)$$

A steady state is reached when $dn/dt=0$, or, using Eq. (4), when

$$n/n_0=[1+\tfrac{1}{2}\gamma^2H_1^2T_1g(\nu)]^{-1}. \quad (13)$$

We shall express the maximum value of $g(\nu)$ in terms of a quantity T_2^* defined by

$$T_2^*=\tfrac{1}{2}g(\nu)_{\max}. \quad (14)$$

The maximum steady-state susceptibility in the presence of the r-f field is thus reduced, relative to its normal value, by the "saturation factor" $[1+\gamma^2H_1^2T_1T_2^*]^{-1}$. The possibility that T_2^* itself depends on H_1 is not excluded.

The quantity T_2^* defined by (14) is a measure of the inverse line width. It is strictly connected only with the area and peak value of a line-shape curve. In the case of the damped oscillator or single-tuned-circuit curve, $1/\pi T_2^*$ is the full width between the half-maximum points, on a frequency scale. We use the notation T_2^* when admitting all sources of line broadening, including inhomogeneity in the magnetic field. The symbol T_2 will be used to specify the true line width which is observable only if the field is sufficiently homogeneous.

II. APPARATUS AND EXPERIMENTAL PROCEDURE

The substance to be investigated was located within a small coil, in the field of an electromagnet. Resonance absorption in the sample caused a change in the balance of a radiofrequency bridge containing the coil. In much of the work a frequency in the neighborhood of 30 Mc/sec. was used. This is a convenient region in which to study the H^1 and F^{19} resonances which occur at 30 Mc/sec. in fields of 7050 and 7487 gauss, respectively. Some experiments were made at 14.5 Mc/sec. and at 4.5 Mc/sec. Considerations of sensitivity alone favor the highest frequency compatible with available magnetic field intensities and the gyromagnetic ratio of the nucleus in question. (See Appendix.)

A typical r-f coil consists of 12 turns of No. 18 copper wire wound in a helix of inside diameter 7 mm, and 15 mm long. The coil is sometimes supported on a thin-walled polystyrene form, but this is not always desirable for, in addition to reducing the space available for the sample, the coil form, if it contains hydrogen, gives rise to an absorption at the proton resonance.†† The

†† An excellent insulating material of negligible hydrogen content is Teflon, the Dupont fluorine-substituted polymer. We have not been able to detect a proton resonance in Teflon. Its mechanical properties, however, are not ideal for a coil form. Glass would be acceptable, as would porcelain.

coil is surrounded by a cylindrical shield of 17 mm inside diameter, one end of the coil being connected to the shield. A hole in the end of the shield can permits insertion of the sample which, if a liquid or powder, is contained in a thin-walled glass tube. The effective volume of the sample within the coil is approximately 0.5 cm³. It would be advantageous to work with larger samples; in the present case the limitation is imposed by the width of the magnet gap. For experiments requiring lowering of the temperature, copper tubing was wound around and soldered to the shield can, and acetone cooled by dry ice was allowed to flow through the cooling tube at a controllable rate. This rather clumsy arrangement was also necessitated by lack of space in the magnet gap.

The magnet is a Société Genevoise water-cooled electromagnet provided with pole pieces 4 inches in diameter at the gap. The gap width was adjustable and was ordinarily between $\frac{3}{4}$ and 1 inch. The pole faces were flat except for a raised rim 0.015 inch high and 0.31 inch wide at the periphery of each, which was designed according to the prescription of Rose[8] to compensate the inherent radial inhomogeneity of the simple gap. The residual inhomogeneities were apparently of non-geometrical origin, and doubtless arose from magnetic inhomogeneities in the cold-rolled steel pole caps. More careful attention to the pole cap material, in future work, is to be recommended. The best position in the field was found by exploring with a liquid sample. As will be seen later, the line width in water, for example, is essentially determined by the field inhomogeneity, and therefore affords a sensitive test for homogeneity. The location of the flattest spot in the field varied somewhat with field intensity, which is evidence for ferromagnetic inhomogeneity of the poles. The magnet current was supplied by storage batteries and regulated by suitable rheostats, the main rheostat consisting of some 15 feet of heavy Advance strip, with a sliding contact. It will be readily appreciated that the examination of resonance lines a few tenths of a gauss in width in fields of several thousand gauss requires an unusually stable source of current. The current could be monitored by means of a shunt and a type K potentiometer. The power required for magnet excitation at 7000 gauss with a 0.75-inch gap was about 220 watts.

A sinusoidal modulation of small amplitude, at 30 c.p.s., was applied to the magnetic field by means of auxiliary coils around the poles, or, in some cases, by means of a small pair of coils

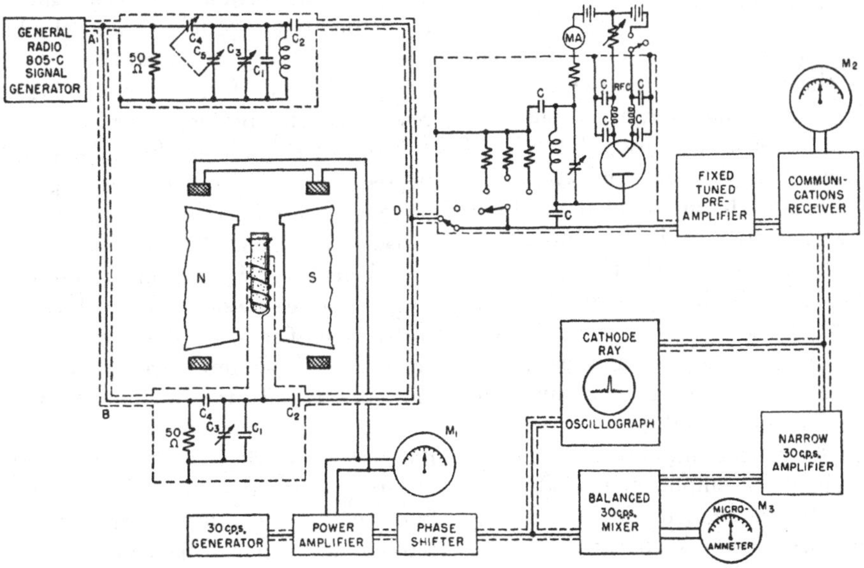

FIG. 1. Block diagram of circuit. The sample and the magnet gap are not drawn in correct proportion.

[8] M. E. Rose, Phys. Rev. 53, 715 (1938).

fitting closely against the shield surrounding the r-f coil. The 30-c.p.s. signal was derived from a small permanent-magnet alternator driven by a synchronous motor, and a 6L6 served as power amplifier to drive the modulating coils. The amplitude of modulation at the sample could be made as large as 15 gauss peak-to-peak, but was usually very much smaller.

A block diagram of the radiofrequency circuit is shown in Fig. 1. The source of the r-f signal was a General Radio 805-C signal generator. The output of this instrument covers a convenient range in power, the maximum power available being adequate to saturate most samples. The circuit elements are connected together by standard u.h.f. coaxial cable (50 ohms). Each of the two branches leading from a tee-junction near the signal generator contains a resonant circuit, one of which consists of the coil surrounding the sample and a fixed ceramic condenser shunted by a variable condenser of about 10 mmf. The dummy circuit in the other branch is similar in all respects but is not located in the field. When liquids of high dielectric constant are being investigated it is sometimes worth while to slip a dummy sample into the dummy coil, thus balancing, at least approximately, the considerable change in stray capacitance resulting from the insertion of the sample in the other coil.

An extra half-wave-length of line AB in one branch makes the point D, where the branches join to the line to the amplifier, a voltage node in the perfectly balanced condition. This is, of course, only one of many ways in which the equivalent of a balanced-pair to unbalanced line transition, essential to any bridge, might be effected, and simplicity is the only virtue claimed for it. The Q of a coil such as that described is about 150, at 30 Mc/sec., and the shunt impedance of the LC circuits at resonance is of the order of 5000 ohms. The small coupling condensers C_2 are chosen to make the impedance seen looking back from the amplifier approximate the value for which the amplifier noise figure is best. The main tuning capacity in each circuit is provided by a fixed ceramic condenser C_1.

The bridge is balanced in phase and amplitude by the adjustment of one of the trimming tuning condensers C_3 and the coupling condenser C_4.

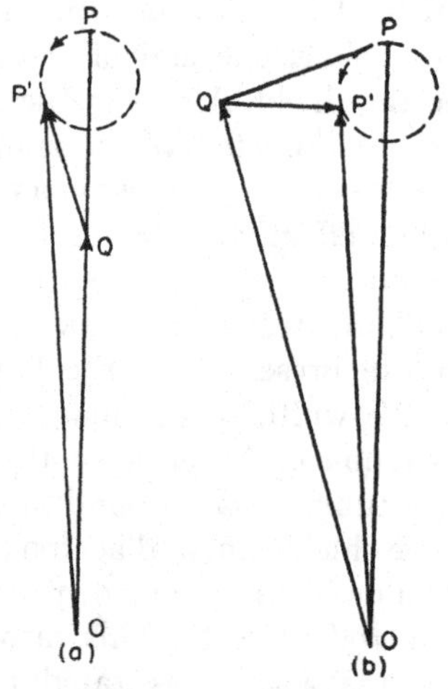

FIG. 2. Diagram illustrating the dependence of the shape of the resonance curve upon the nature of the residual unbalance of the bridge. The diagram is not to scale; actually, $PP' \ll QP \ll OP$.

The latter adjustment alone affects the phase balance as well as the amplitude balance, for the coupling condenser essentially appears in shunt with the tuned circuit. The adjustable coupling condenser C_4 therefore is ganged with a trimming tuning condenser C_5 in such a way as to leave the total tuning capacity unaltered during the change in coupling. In this way substantially orthogonal phase and amplitude adjustments are provided, and the balancing of the bridge is a simple operation.

In searching for, or examining, a nuclear resonance the bridge is *never* completely balanced, but is intentionally unbalanced to such a degree that the voltage appearing at point D is large compared to the change in this voltage caused by nuclear absorption or dispersion. The only benefits derived from the use of the bridge are (a) the reduction of the r-f level at the input to the amplifier to a value low enough to permit considerable r-f amplification before detection, and (b) the reduction in the relative magnitude of output fluctuations arising from amplitude fluctuations in the signal supplied by the signal generator. With a rather moderate degree of balance these advantages are already fully exploited, and further balance does not improve matters. We have ordinarily operated with a balance of 40 db to 60 db, by which is meant that the voltage appearing at D is between 10^{-2}

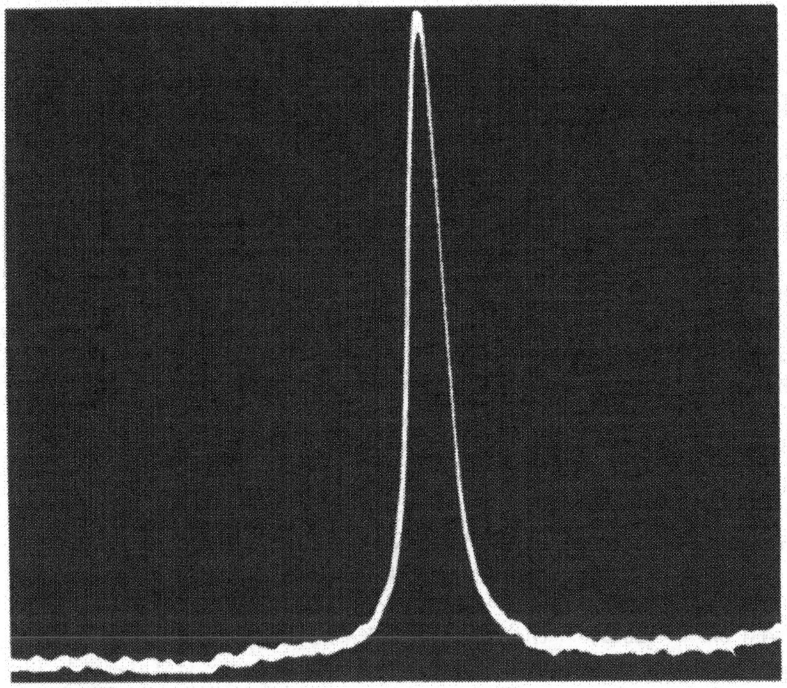

FIG. 3. Proton resonance (absorption curve) in ferric nitrate solution.

and 10^{-3} times the value it would have if one branch were removed.

It is important to be able to control the character of the residual unbalance, for upon this depends the nature of the signal observed as the magnetic field is varied through the region of resonance. The nuclear absorption has the effect of altering the shunt resistance of the tuned circuit, while the accompanying nuclear dispersion changes its resonant frequency slightly. The two effects can be described together by considering, in Fig. 2, the vector OP' as representing the signal arriving at the point D from the branch containing the LC circuit with the sample in it. As resonance is traversed the end of the vector, P', describes a small circle.††† Had the residual unbalance off resonance, QP, referred to the point D, been in amplitude only, so that the signal from the other branch is represented by the vector OP in Fig. 2, the resultant voltage applied to the amplifier, QP', would vary with magnetic field in the manner illustrated in Fig. 3 which is a trace of the receiver output after detection, *vs.* magnetic field, in the case of a proton resonance. Pure phase unbalance, on the other hand, is represented by the diagram of Fig. 2b, and the observed output variation is shown in Fig. 4. (It must be remembered that actually the resultant QP' is very much smaller than OP, and the small circle is in turn very much smaller than the unbalance voltage QP.) It is, in fact, the nuclear absorption, χ'', which is observed in Fig. 3 and the nuclear dispersion, χ', which is observed in Fig. 4. Either can be obtained as desired by first balancing the bridge to a high degree and then setting in an intentional unbalance by means of the phase knob or the amplitude knob—hence the importance of independent phase and amplitude controls. A "mixed" unbalance brings about an unwelcome mixture of the two effects recognizable by the unsymmetrical shape of the output curve. We usually preferred to work with the absorption curve (amplitude unbalance) because the apparatus is in this condition less susceptible to the frequency modulation present in the signal generator output. Indeed, absence of frequency modulation effects in the output is a sensitive criterion for pure amplitude unbalance. The trouble caused by frequency modulation is, of course, mainly due

††† The locus of P' is strictly circular only if the shape of the absorption curve is that appropriate to a damped oscillator.

to the frequency sensitivity of the half-wave line, and could be mitigated by the use of a suitably designed transformer to achieve the required 180° phase shift.

The r-f amplifier is subject to no special requirements other than the desirability of a good noise figure. We have used commercial receivers, both Hallicrafters S-36 and National HRO, preceded, when the signal frequency is near 30 Mc/sec., by a broad band preamplifier of Radiation Laboratory design[9] which has a much smaller noise figure than either of the commercial receivers. A temperature-limited diode in a circuit for noise comparison (shown in Fig. 1) has been included as a permanent fixture, and calibration by this means showed the noise figure to be 1.5 db when the preamplifier was in use.

Relatively strong signals, such as that displayed in Fig. 3, can be observed on an oscilloscope by merely connecting the detector output to the vertical deflection amplifier, and synchronizing the sweep with the 30-c.p.s. modulation of the magnetic field, the amplitude of which is adjusted to several times the line width. For most of our work, however, we have used, following the r-f amplifier and detector, a narrow band amplifier and balanced mixer, a combination variously known as a "lock-in" amplifier or a phase-sensitive detector. The magnetic field modulation is reduced to a fraction of the line width, and the resulting 30-c.p.s. modulation of the detected receiver output is amplified and converted by mixing it with a 30-c.p.s. reference signal derived from the small alternator. There is thus obtained a d.c. output proportional to the slope of the nuclear absorption curve, and this is indicated by a meter M_3. The relation between meter current and $d\chi''/dH$ is strictly linear for sufficiently small amplitudes of magnetic field modulation, because of the relatively large unbalance voltage in the bridge and the nature of the lock-in amplifier. The effective band width of the system is determined by the time constant of the final d.c. circuit containing the meter, which was usually about 1 second but could be increased to 10 seconds if desired. The amplifier was copied after one designed by R. H. Dicke.[10]

The superiority of the narrow-band-amplifier-plus-meter over the oscilloscope in the matter of detecting weak signals is not as marked as

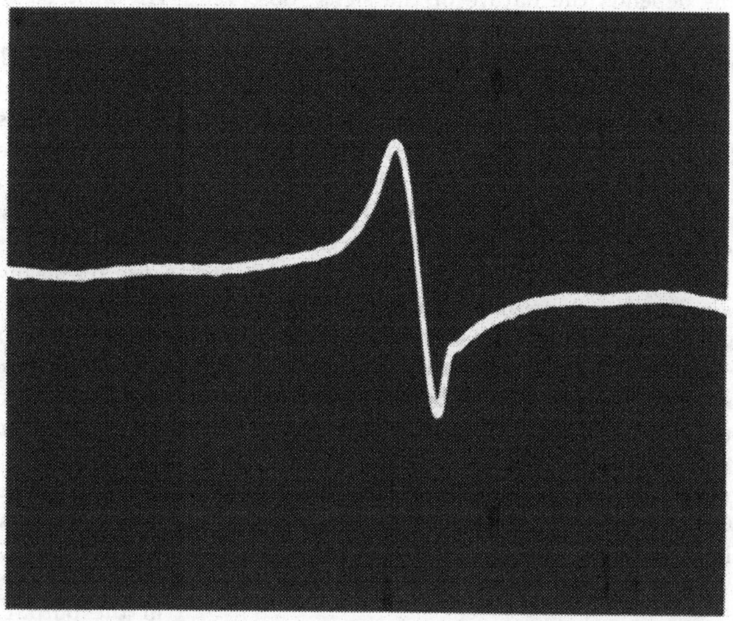

FIG. 4. Proton resonance (dispersion curve) in ferric nitrate solution

[9] G. E. Valley and H. Wallman, *Vacuum Tube Amplifiers, Radiation Laboratory Series* (McGraw-Hill Book Company, Inc., New York, in press), Vol. 18.
[10] R. H. Dicke, Rev. Sci. Inst. 17, 268 (1946).

comparison of the band widths alone would suggest, because of the new well recognized ability of the observer to integrate noise. However, for the quantitative examination of weak lines the narrow-band device is nearly indispensable.

It will be noted that the output of the lock-in amplifier, in the case of amplitude unbalance, reproduces the *derivative* of the absorption curve (which resembles qualitatively a dispersion curve), while with phase unbalance one obtains the derivative of the dispersion curve—a large peak flanked symmetrically by two dips.

III. MEASUREMENT OF LINE WIDTH

When the resonance absorption is intense enough to be displayed on the oscilloscope screen the line width can be measured directly on the trace. A sinusoidal horizontal sweep synchronous with, and properly phased with respect to, the modulation of the magnetic field, provides a horizontal scale linear in gauss. We denote by H_m the modulation amplitude, and by ω_m the angular frequency of modulation, in these experiments $2\pi\times 30$ sec.$^{-1}$. That is, $H_z = H_0 + H_m \sin \omega_m t$. To establish the scale against which the line width is measured, H_m must be known. This calibration can be effected in several ways. The signal generator frequency can be altered by a small, known amount, and the resulting shift of the resonance within the modulation interval noted, or the alternating voltage induced in a small coil of known turns×area, at the location of the sample, can be measured with a vacuum-tube voltmeter. Both methods were used and gave consistent results. The relation between H_m and the voltage applied to the modulating coils was so nearly linear that a single calibration usually sufficed for a given setting of the gap width.

The band width of the system must, of course, be adequate to permit a faithful reproduction of the line shape at the modulation frequency used. The r-f field applied to the sample must be weak enough to avoid saturation. If the line is very narrow a complication can arise from a transient effect to be described briefly later. The vestige of such an effect is responsible for the asymmetry at the base of the absorption line in Fig. 3.

For the study of broad, and hence weak, absorption lines the narrow-band amplifier is used. The modulation amplitude H_m is reduced to a fairly small fraction of the line width, and the line is explored by slowly changing the magnetic field H_0. It usually suffices to note the interval in H_0 between points giving extreme deflection of the output meter M_3, which are assumed to be the points of maximum slope of the absorption curve. A plot of the output meter readings in one case is shown in Fig. 5. In this example, the interval between extreme deflections is $\Delta H = 0.5$ gauss (curve *a*). The relation between ΔH so determined and T_2 defined by Eq. (14) would be $(1/T_2) = (\sqrt{3}/2)\gamma\Delta H$, for a damped oscillator absorption curve, or $(1/T_2) = (2\pi)^{-\frac{1}{2}}\gamma\Delta H$, for a Gaussian curve. The derivative curve (*a*) in Fig. 5 indicates an absorption line shape of intermediate character. The most conspicuous evidence for this is the ratio of the maximum negative to the maximum positive

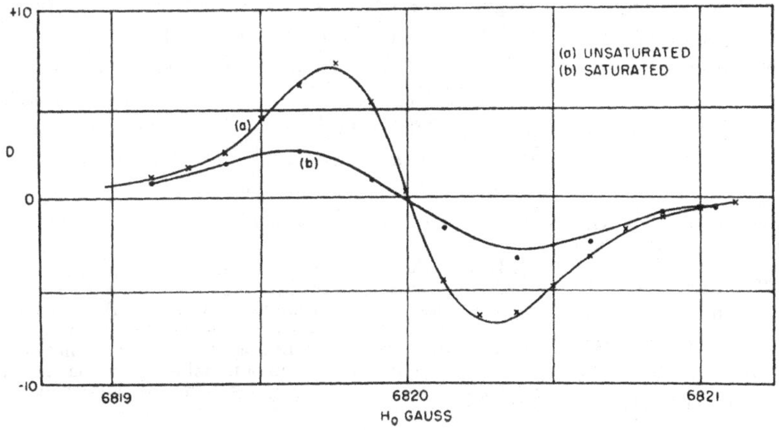

FIG. 5. Plot of the reading of the output meter M_3, for a 0.5*N* solution of ferric nitrate. Curve (*b*) was taken at a high power level and shows the effect of saturation. The value of the magnetic field at the line center was not measured, but merely computed from the frequency and the accepted value of the proton moment.

slope, which should be 4:1 for the derivative of a damped oscillator curve, 2.2:1 for the derivative of a Gaussian, and is in fact about 3:1.

The effect of saturation, evident in the curve (*b*) in Fig. 5, is to flatten and spread apart the extrema of the derivative curve. This comes about because saturation is more nearly complete at the center of the line.

IV. MEASUREMENT OF RELAXATION TIME

Two methods by which the spin-lattice relaxation time T_1, defined by Eq. (11), can be measured have been mentioned in the introduction. The direct observation of recovery from saturation was convenient only when T_1 was of the order of magnitude of one second; doubtless a wider range could be covered with suitable circuit modifications and recording instruments. The direct method has the advantage that the interpretation of the results is simple and entirely unambiguous. The saturation-curve method is applicable over a much wider range, but the analysis of the experiment is rather involved and absolute values of T_1 derived in this way are probably less reliable. The two methods complement one another, as we shall see.

Before analyzing special cases, we shall describe an instructive experiment which vividly demonstrates the saturation phenomenon and also exposes the source of some of the complications which are encountered in applying the saturation-curve method to narrow resonance lines.

The proton resonance in water was observed on the oscilloscope with a weak (non-saturating) r-f field. The line appeared as in Fig. 6(a). The modulation amplitude H_m was then reduced to zero for a few seconds, while the field H_0 was maintained at a value corresponding to the point *A*; the line was, of course, not visible on the screen during this interval. Upon turning the modulation amplitude back to its original value, and restoring the sweep, the line reappeared bearing a deep, narrow "gash" at the position *A*, as in Fig. 6(b). This hole rapidly filled up and in a few seconds was gone, as shown in the sketches (c) to (e). The explanation of the effect is this: the true width of the water resonance line was much less than the width of the original curve, which was caused entirely by

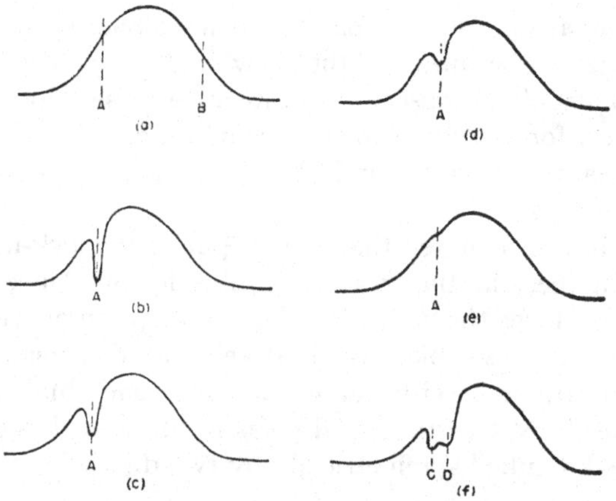

FIG. 6. The effect of local saturation upon a line whose width is caused by magnetic field inhomogeneity.

inhomogeneity of the field H_0. The absorption observed at *A* was caused by nuclei in one small part of the sample, that at *B* by those in another part. With the removal of the modulation the former group was subjected to radiation at its resonance frequency *all the time*, and was more or less thoroughly saturated. The subsequent recovery from saturation occurred when the modulation was restored. The gash in the curve betrays, quite literally, a *hot spot* (in respect to nuclear spin temperature) somewhere in the sample.

If the modulation amplitude H_m is reduced to a small but finite value, rather than zero, the r-f amplitude H_1 being meanwhile increased somewhat, subsequent observation will disclose a hole like that in Fig. 6(f), which then heals at the same rate as before. The double dip results, of course, from the sinusoidal modulation—a longer time was spent at the turning points *C* and *D* than at the middle of the hole.‡

If the attempt is made to produce the effect shown in Fig. 6 in a subsance for which the line width is determined by a strong spin-spin interaction, rather than field inhomogeneity, it is found that the whole curve saturates; it is not possible to "eat a hole" in the curve. The energy

‡ It was the observation of the effects just described which first made us aware that the true line width in water was much less than the apparent line width. By exploring the magnet gap, a spot was found where the field was considerably more homogeneous and the line correspondingly narrower. The narrowest line obtainable in our magnet at 7000 gauss was about 0.1 gauss wide; at 1100 gauss a width of 0.015 gauss was achieved in one case.

absorbed by the spins can no longer be localized, but goes to raise the temperature of the spin system as a whole. In other words, the application of an r-f field at a single frequency, with H_0 fixed, immediately (or at least in a time shorter than T_1) affects a region about $2/\gamma T_2$ wide on a magnetic field scale, or about $1/\pi T_2$ wide on a frequency scale. It is important to keep this point in mind.

In any case, it is possible to saturate all parts of the sample by modulating through the whole line, using the highest available r-f field intensity. The recovery, observed with a much weaker r-f field in the sample, appears to follow the simple law $(1-e^{-t/t_0})$, and we identify t_0 with T_1, the spin-lattice relaxation time. To make a quantitative measurement, the oscilloscope trace was photographed with a movie camera operating at a known frame speed, and the film analyzed to determine the time constant of recovery. This is the direct method referred to above. It was applied to a distilled water sample, to petroleum ether, and to a $0.002N$ solution of $CuSO_4$, yielding relaxation times of 2.3 ± 0.5 sec., 3.0 ± 0.5 sec., and 0.75 ± 0.2 sec., respectively.

The saturation curve method is based upon Eq. (13) in the Introduction. If it were possible to measure, in the steady state, the value of $\chi''_{\max}$ which is reduced from its normal unsaturated value by the factor $(1+\gamma^2 H_1^2 T_1 T_2)^{-1}$, a knowledge of H_1 and T_2 would then suffice to determine T_1. In the modulation method, however, we measure $\partial\chi''/\partial H_0$, which is equivalent to measuring $\partial\chi''/\partial\omega$. This introduces some complications which can be treated briefly only by confining the discussion to typical limiting cases.

We assume for analytical convenience that the shape of the unsaturated absorption curve is that appropriate for a damped oscillator,

$$g(\nu)=\frac{2T_2}{1+4\pi^2(\nu-\nu_0)^2T_2^2}=\frac{2T_2}{1+(\omega-\omega_0)^2T_2^2}. \quad (15)$$

In Eq. (15), ω_0, the center angular frequency, is γH_0^*. Although it is actually H_z which is varied in the course of the experiment, both by applying the modulation $H_m \sin\omega_m t$, and by slowly changing the magnet current in order to move through the resonance curve, it is convenient here to think of H_0 as fixed at the central value H_0^*, with ω as the variable. The modulation $H_m \sin\omega_m t$ is equivalent to a frequency modulation of amplitude γH_m radians/sec. If the amplitude of the modulation is small enough, $\gamma H_m < 1/T_2$, the deflection of the output meter M_3 will be nearly proportional to $\partial\chi''/\partial H_0$, or, in terms of the equivalent frequency modulation, to $\partial\chi''/\partial\omega$. The question is, what is to be regarded as constant in interpreting $\partial\chi''/\partial\omega$ in the presence of saturation.

We first consider cases in which the inhomogeneity of the magnetic field, denoted roughly by δH, has a negligible influence on the line width. That is, $\gamma\delta H \ll 1/T_2$. The apparent susceptibility of the sample in the steady state (no modulation) at the frequency ω_1, relative to its unsaturated value at ω_0, the line center, is

$$\chi''(\omega_1, H_1)/\chi''(\omega_0, 0) = (1+x^2)^{-1}[1+s/(1+x^2)]^{-1}=f(x)S(x), \quad (16)$$

evaluated at $x=x_1$, were $x=(\omega-\omega_0)T_2$ and $s=\gamma^2H_1^2T_1T_2$. The factor $[1+s/(1+x^2)]^{-1}=S(x)$ is the saturation factor given by Eq. (13). We are interested in $\partial\chi''/\partial x$, and, in particular, we ask for the extreme values which the derivative attains as x_1 is varied. These will be assumed proportional to the extreme meter deflections, D_{ex}, indicated on Fig. 5. There are two cases to be distinguished according to the relative magnitude of ω_m and $1/T_1$. If $\omega_m T_1 \ll 1$, the saturation factor $S(x)$ will vary *during* a modulation cycle. If $\omega_m T_1 \gg 1$, S cannot change during the modulation cycle and may be assigned a value appropriate to the center of the modulation swing. The meter deflection D is therefore proportional, in the two cases, to different functions of x_1, namely:

Case 1

$$(\omega_m T_1 \ll 1), \quad D(x_1) \propto \frac{\partial}{\partial x}[f(x)S(x)]\bigg|_{x=x_1};$$

Case 2

$$(\omega_m T_1 \gg 1), \quad D(x_1) \propto S(x_1)\frac{\partial}{\partial x}f(x)\bigg|_{x=x_1}.$$

D_{ex}, the extreme excursion of D, is found by an elementary computation to depend in the following way upon the saturation parameter s:

Case 1

$$D_{ex} \propto (1+s)^{-\frac{3}{2}}; \tag{17}$$

Case 2

$$D_{ex} \propto 8[2(16+16s+s^2)^{\frac{1}{2}}-2s-4]^{\frac{3}{2}} \times[8+8s-s^2+(2+s)(16+16s+s^2)^{\frac{1}{2}}]^{-1}. \tag{18}$$

The functions (17) and (18) are plotted in Fig. 7. The saturation effect makes itself evident more rapidly in *Case 1*, as was to be expected. A different line shape would, of course, lead to somewhat different results.

Suppose now that the field is not homogeneous, and, in particular, that $H_m < 1/\gamma T_2 < \delta H$, which we shall label *Case 3*. The result is then simply a ***superposition***, governed by δH, of curves of the sort analyzed in either *Case 1* or *Case 2* (depending on the magnitude of $\omega_m T_1$). If the line shape caused by the field inhomogeneity were known, the appropriately weighted superposition of curves $D(x_1)$ could be constructed. It is clear, however, that the decrease in D_{ex} with s will follow much the same course; D_{ex} will be substantially reduced when s is of the order of unity.

Case 4, specified by $1/\gamma T_2 < H_m < \delta H$, and $\omega_m T_1 > 1$, is significantly different. It is now the modulation amplitude H_m rather than $1/\gamma T_2$, which determines the width of the region momentarily saturated. The situation is, in fact, that illustrated by Fig. 6(f). An analysis of this case, which is only feasible if certain simplifying assumptions are made, shows that the quantity $\gamma H_1^2 T_1/H_m$, rather than $\gamma^2 H_1^2 T_1 T_2$, plays the role of the saturation parameter, a rapid decrease in D_{ex} occurring when $(\gamma H_1^2 T_1/H_m) \approx 1$. The indicated dependence of H_1 upon H_m, for the same degree of saturation, has been confirmed experimentally. *Case 4* is encountered in liquids with relatively long relaxation times. The measurements of D_{ex} are particularly difficult here for the line must be traversed at an extremely slow rate in order to permit a quasi-stationary state to develop at every value of x_1.

An even more complicated case is defined by $1/\gamma T_2 < H_m < \delta H$ with $\omega_m T_1 < 1$. The discussion of this case will be omitted as it was not an important one experimentally.

An experimental saturation curve is obtained by the following procedure. The modulation amplitude, H_m, is set at a suitable value and remains the same throughout a run. The output voltage of the signal generator, to which H_1 is proportional, is varied in steps by means of the attenuator in the instrument. At each step the receiver gain is adjusted to bring the rectified r-f output of the receiver to a fiducial level as indicated by the voltmeter M_2, the bridge balance remaining unaltered. This makes the constant of proportionality connecting the deflection of the final output meter with $\partial\chi''/\partial H_0$ the same for each step in H_1. At each signal generator output setting, after the receiver gain has been readjusted, the resonance curve is slowly traversed and the maximum and minimum readings of the output meter noted. The difference between the readings is then plotted against the logarithm of the signal generator voltage.

To interpret the saturation curves we assume that a given degree of saturation is associated with the same value of $\gamma^2 H_1^2 T_1 T_2$, or, if the conditions correspond to *Case 4*, of $\gamma H_1^2 T_1/H_m$. This is, of course, not strictly correct if we are comparing two saturation curves which correspond to two different cases, such as *Case 1* and *Case 2*, and allowance can be made if the precision of the data warrants.

A series of saturation curves for $Fe(NO_3)_3$ solutions of various concentrations is shown in Fig. 8. This is the ***roughest*** set of curves of any used in deriving the results reported in this

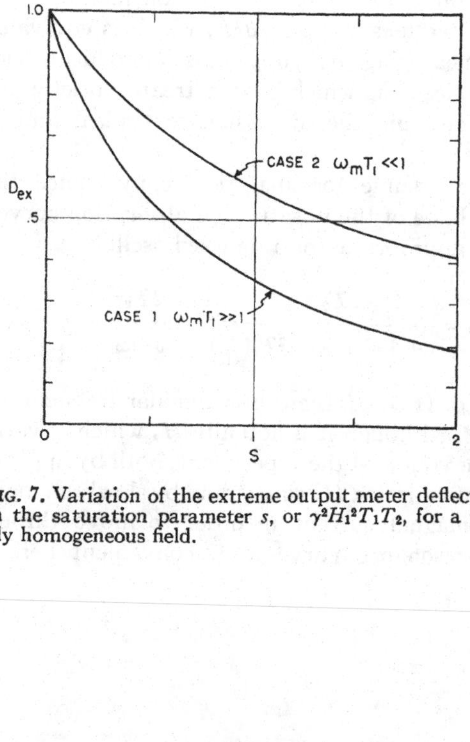

FIG. 7. Variation of the extreme output meter deflection with the saturation parameter s, or $\gamma^2 H_1^2 T_1 T_2$, for a perfectly homogeneous field.

paper, and is exhibited to indicate the experimental limitations. At the same time, it illustrates the enormous range which can be covered by the saturation-curve method. The horizontal shift of the curves is a measure of the relative change in T_1T_2. Of course, T_2 must be determined in each case by the methods already described in order to deduce the relative change in T_1. Figure 9 shows a set of saturation curves for ice at various temperatures, together with the function (18). The agreement is good enough; as a matter of fact, the difference between (17) and (18), on a logarithmic plot, is quite small, apart from a lateral shift. All saturation curves which we have observed bend downward with about the same slope.

To determine the absolute magnitude of T_1 from a saturation curve, T_2 having been measured, the relation between H_1 and the signal generator voltage must be known. The relation can be computed from the dimensions of the coil and the constants of the circuit, with which one must include the stray capacitances. We have preferred to rely on the direct measurement of T_1 for this calibration. If higher precision were required, it would be necessary to make in each case a detailed analysis like that given above for the two idealized cases *1* and *2*. The accuracy of the experimental data hardly justifies such a refinement. Fortunately, it is the relative changes in T_1, by large factors, which are most significant in the present investigation.

Discussion of the experimental results on relaxation time and line width is deferred to Sections XI to XV which follow a theoretical examination of spin-spin and spin-lattice interaction.

V. A TRANSIENT EFFECT: THE "WIGGLES"

Rather early in the course of this work a transient effect was noticed which merits a brief description if only because it can complicate the interpretation of nuclear absorption experiments in some cases. The effect is observed when a large amplitude of modulation is used to display the whole absorption curve on the oscilloscope, and is prominent only when T_1, T_2, and $1/\gamma\delta H$ are all greater than the interval spent at resonance during the modulation cycle which is roughly $(H_m\omega_m\gamma T_2)^{-1}$. An example is shown in Fig. 10, a photograph of the oscilloscope trace in the case of a water sample in a very homogeneous magnetic field. The prominent "wiggles" occur always *after* the magnetic field has passed through resonance, and therefore appear on opposite sides of the two main peaks in Fig. 10, one of which corresponds to increasing, the other to decreasing, field.

A brief explanation of the effect is most readily given in terms of the classical model of precessing

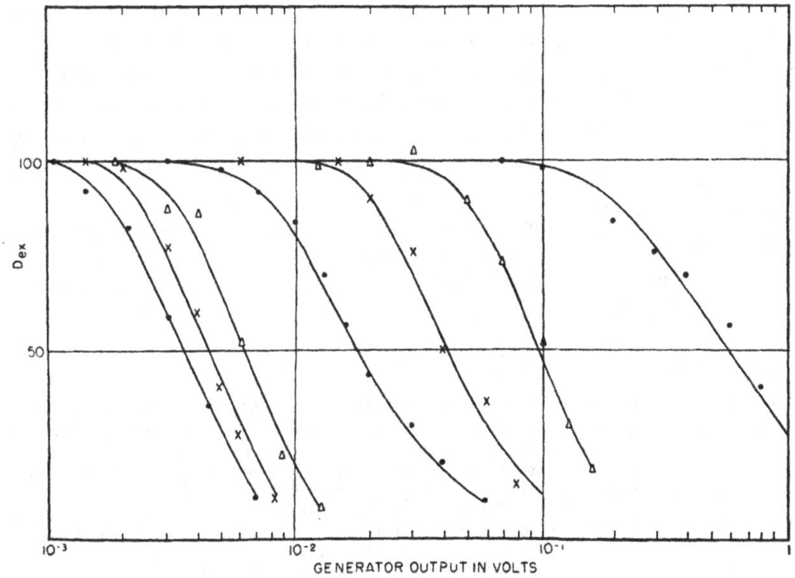

FIG. 8. Saturation curves for ferric nitrate solutions ranging in concentration from 0.0*N* to 0.6*N*. The ordinate is the extreme output meter deflection, in arbitrary units.

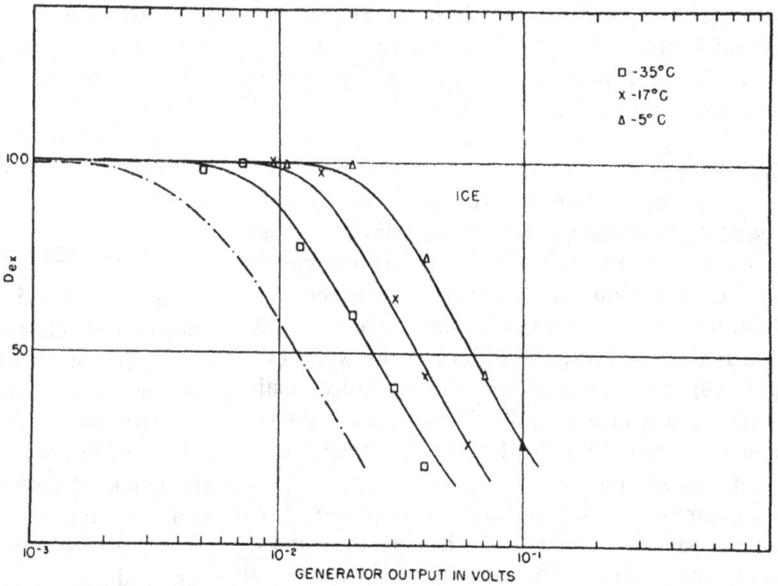

FIG. 9. Saturation curves for ice at −35°C, −17°C, and −5°C. The broken curve is a plot of Eq. (18), and may be shifted horizontally to fit any of the experimental curves.

magnetic gyroscopes. Nuclear spins which have been tipped over by the application of the r-f field at their precession frequency continue to precess in a coherent way after the removal of the r-f field, or after the radiofrequency or magnetic field has changed to a non-resonant value. In our case the magnetic field is changing, and the nuclear precession rate, following it adiabatically, comes alternately in and out of phase with the applied signal, whose frequency has remained at what was formerly the resonance value.

The duration of the "phase memory" is limited by T_2. However, if $1/\gamma\delta H$ is smaller than T_2, interference between the effects of nuclei in different parts of the sample smears out the wiggles. By slightly detuning the receiver the wiggles on the forward trace can be enhanced and those on the back trace suppressed, or *vice versa*; one set of wiggles is produced by spins precessing at a rate higher than ν_0 and the other by precession at a lower frequency. The spacing between the wiggles also varies in the manner suggested by the above explanation. A more refined treatment of the problem, which is most conveniently based on the equations of Bloch, has been made by H. C. Torrey.[11]

In our measurements of T_1 by the saturation curve method, the modulation amplitude was so small that the effect was not important. The direct measurements of T_1 were made in a field sufficiently inhomogeneous to suppress the wiggles, which would not, in any case, have interfered with the interpretation of the results.

VI. SPIN-SPIN INTERACTION

Of the processes which could conceivably bring about an exchange of energy between the spins and their surroundings, we can at once discard the interaction with the thermal radiation field. The effectiveness of this process depends upon the probability of spontaneous emission, A, which is of the order of 10^{-22} sec.$^{-1}$ for a nuclear magnetic dipole transition at the frequencies with which we are concerned. The corresponding relaxation time is $T_1 = (h\nu/kT)(1/2A) \approx 10^{15}$ seconds. Although this figure must be reduced by a large factor when the nuclear system is coupled to a resonant circuit of small dimensions,[12] the process is still inadequate to account for the relaxation effects observed and need not be considered further.

[11] H. C. Torrey (private communication).

[12] E. M. Purcell, Phys. Rev. 69, 681 (1946). In a sense, the relaxation of nuclear spins in a metal, which was considered by Heitler and Teller, reference 1, is a limiting case of the coupling of the spins to an electrical circuit, and as such requires an exception to the above dismissal of radiation damping effects. Although no experimental evidence on nuclear relaxation in a metal has appeared, it is to be expected that the mechanism treated by Heitler and Teller will play an important role.

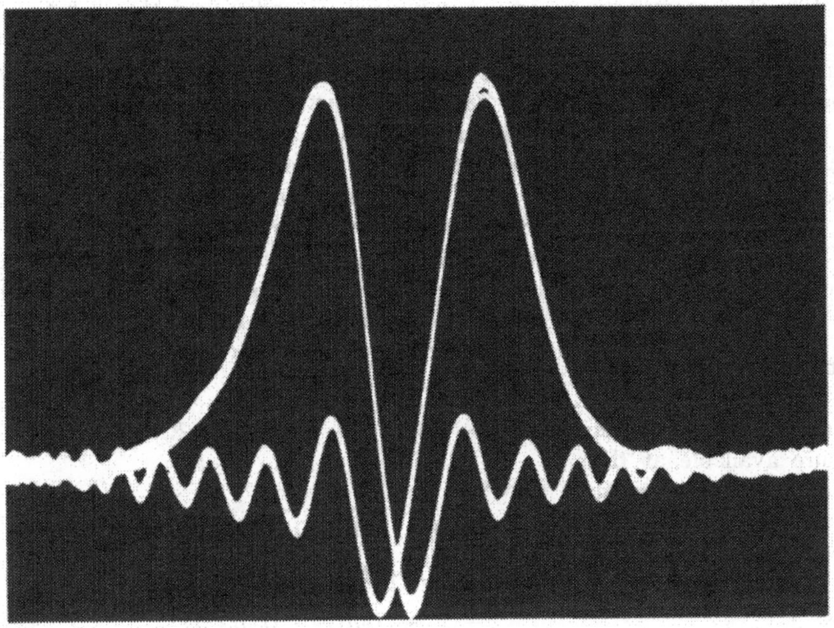

FIG. 10. An example of the transient effect described in Section IV, observed in water at 29 Mc/sec. The field modulation and the oscilloscope sweep are both sinusoidal, but there is a small phase difference between them. The wiggles occur after the magnetic field has passed through resonance.

Electric forces which act during atomic collisions directly perturb the nuclear spins only by virtue of an accompanying variation in the gradient of the electric field at the nucleus and the existence of a nuclear electric quadrupole moment. Direct electrical interactions are thus reduced to a rather minor role, quite in contrast to the situation in atomic and molecular spectroscopy, and are entirely absent if the nucleus has no quadrupole moment. We shall examine one or two cases in which the nuclear quadrupole-inhomogeneous electric field interaction is the important one, but we have been concerned chiefly with nuclei of spin $\frac{1}{2}$ for which the quadrupole moment vanishes.

Magnetic interactions will occur between the nuclear magnetic moments and magnetic moments associated with electronic states, and among the nuclear magnetic moments themselves. We shall consider the nuclear dipole-dipole interaction first, and in greatest detail, not only because it proves to be the dominant effect in most diamagnetic substances, so far as spin-lattice relaxation is concerned, but because it is entirely responsible for the spin-spin relaxation phenomena. In fact, we shall begin by considering a spatially rigid lattice of dipoles—a model necessarily devoid of spin-lattice effects.

Before elaborating upon the description outlined in the introduction in which the system of spins is characterized by individual spin quantum numbers, it is instructive to examine the problem from a somewhat more general point of view by regarding the entire lattice of N spins as a single quantum-mechanical system. The total z component of angular momentum is specified by $m\hbar$. The Hamiltonian of the system is

$$\mathcal{H}=\sum_i \gamma_i\hbar\mathbf{I}_i\cdot\mathbf{H}_0+\sum_i \gamma_i\hbar\mathbf{I}_i\cdot \sum_j\left(\frac{\gamma_j\hbar\mathbf{I}_j}{r_{ij}^3}-\frac{3\gamma_j\hbar\mathbf{r}_{ij}(\mathbf{r}_{ij}\cdot\mathbf{I}_j)}{r_{ij}^5}\right). \quad (19)$$

The second term in (19) is the dipole-dipole interaction, and the sum over j in this term is what we have called the local field, H_{loc}. In the absence of the dipole-dipole interaction the system displays $2NI+1$ discrete levels, $m=-NI$, $-NI+1, \cdots +NI$, separated in energy by $\hbar\gamma H_0$. The levels are highly degenerate; for $I=\frac{1}{2}$, for example, the degeneracy is $2^N N!/(N-2m)!\times(N+2m)!$. The interaction term splits each of these levels into a group whose width in engery is of the order, $N^{\frac{1}{2}}\hbar\gamma H_{\text{loc}}$. Now $N^{\frac{1}{2}}\gamma H_{\text{loc}}\gg\gamma H_0$ in a practical case, so that the original level structure dissolves into what is essentially a continuum showing none of the previously prominent periodicity. Broer,[13] who treated in

[13] L. J. F. Broer, Physica 10, 801 (1943).

this way the general problem of paramagnetic absorption and dispersion, showed that the calculation of the absorption for such a system, assumed to have been brought somehow into thermal equilibrium at a temperature $T \gg h\nu/k$, could be reduced to the task of finding the density function of magnetic moment, $f(\nu)$, which he defined by:

$$f(\nu)\Delta\nu = 2 \sum_{\nu-\Delta\nu}^{\nu+\Delta\nu} |M_{pq}|^2 \rho_{pq},$$

where M is the magnetic moment matrix, p and q refer to two levels separated in energy by $h\nu$, and ρ_{pq} is the average occupation of these levels. The only rigorous procedure available at this point is the method of diagonal sums which Broer shows is capable of yielding a limited amount of information about the function $f(\nu)$, namely,

$$\int_0^\infty f(\nu)d\nu, \quad \int_0^\infty \nu^2 f(\nu)d\nu, \quad \text{etc.},$$

the successively higher moments being increasingly difficult to calculate. An example of the application of the diagonal sum method to nuclear resonance absorption in crystals will be given in a forthcoming paper by J. H. Van Vleck.[14] One of the results there obtained will be invoked to complete the following discussion in which we revert to a less rigorous perturbation method.

The unperturbed state of the system is specified by the quantum numbers m_{I_j}, or for brevity, m_j, of the individual spins, the total z component of spin being given by $m = \sum m_j$. We assume that these magnetic quantum numbers still characterize a state after the introduction of the perturbation $V = \sum_i \sum_{j>i} V_{ij}$, where

$$\begin{aligned} V_{ij} = \gamma^2 h^2 r_{ij}^{-3} [& I_{x_i} I_{x_j}(1-3\alpha_1^2) \\ & + I_{y_i} I_{y_j}(1-3\alpha_2^2) + I_{z_i} I_{z_j}(1-3\alpha_3^2) \\ & - 3(I_{x_i} I_{y_j} + I_{x_j} I_{y_i})\alpha_1\alpha_2 \\ & - 3(I_{x_i} I_{z_j} + I_{x_j} I_{z_i})\alpha_1\alpha_3 \\ & - 3(I_{y_i} I_{z_j} + I_{y_j} I_{z_i})\alpha_2\alpha_3]. \end{aligned} \quad (20)$$

V_{ij} is the magnetic interaction between the ith and jth spin; α_1, α_2, and α_3 are the direction cosines of the radius vector r in coordinates with the z axis parallel to H_0. We now rearrange Eq. (20) so as to distinguish terms which leave the total z component, m, unchanged, terms for which $\Delta m = \pm 1$, and terms for which $\Delta m = +2$ or -2. This distinction is important because the terms involve a time factor $e^{i\Delta m \gamma H_0 t}$. The terms with $\Delta m = 0$ are time independent and leave the energy unchanged. We introduce the polar and azimuthal angles θ_{ij} and φ_{ij} instead of the direction cosines, after which (20) can be written in the form:

$$V_{ij} = \gamma^2 h^2 r_{ij}^{-3}(A+B+C+D+E+F), \quad (21)$$

where

$$\begin{aligned} A &= I_{z_i} I_{z_j}(1-3\cos^2\theta_{ij}), \quad (\Delta m = 0), \\ B &= -\tfrac{1}{4}[(I_{x_i} - iI_{y_i})(I_{x_j} + iI_{y_j}) \\ &\quad + (I_{x_i} + iI_{y_i})(I_{x_j} - iI_{y_j})] \\ &\quad \times (1-3\cos^2\theta_{ij}), \quad (\Delta m = 0), \\ C &= -\tfrac{3}{2}[(I_{x_i} + iI_{y_i})I_{z_j} + (I_{x_j} + iI_{y_j})I_{z_i}] \\ &\quad \times \sin\theta_{ij}\cos\theta_{ij} e^{-i\varphi_{ij}} e^{i\gamma H_0 t}, \quad (\Delta m = 1), \\ D &= -\tfrac{3}{2}[(I_{x_i} - iI_{y_i})I_{z_j} + (I_{x_j} - iI_{y_j})I_{z_i}] \sin\theta_{ij} \\ &\quad \times \cos\theta_{ij} e^{+i\varphi_{ij}} e^{-i\gamma H_0 t}, \quad (\Delta m = -1), \\ E &= -\tfrac{3}{4}(I_{x_i} + iI_{y_i})(I_{x_j} + iI_{y_j}) \sin^2\theta_{ij} \\ &\quad \times e^{-2i\varphi_{ij}} e^{+2i\gamma H_0 t}, \quad (\Delta m = 2), \\ F &= -\tfrac{3}{4}(I_{x_i} - iI_{y_i})(I_{x_j} - iI_{y_j}) \sin^2\theta_{ij} \\ &\quad \times e^{+2i\varphi_{ij}} e^{-2i\gamma H_0 t}. \quad (\Delta m = -2). \end{aligned} \quad (22)$$

Terms A and B give rise to secular perturbations; terms C to F represent periodic perturbations of small amplitude. That these last four terms cannot cause appreciable changes is a consequence of the conservation of energy. For our rigid-lattice model, then, only terms A and B need be considered. Classically A corresponds to the change in the z component of the magnetic field by the z components of the neighboring magnetic moments. This local field will vary from nucleus to nucleus, slightly changing the Larmor frequency of each. Suppose that the ith nucleus is surrounded by a number of neighbors, which for simplicity we shall assume to have spin $I = \frac{1}{2}$. Each of these can be parallel or

[14] J. H. Van Vleck, Phys. Rev. (in press).

antiparallel to H_0, and every arrangement leads to some value of the static part of the local field. From the nearest neighbors alone, we will usually obtain several possible values; each of these will be further split when the next nearest neighbors are taken into account, and so on. The result will be a continuum of possible values of the local static field, centered on H_0, whose shape would have to be calculated for the particular spatial configuration of the lattice involved, and the given direction of H_0. The calculation is simplified, in all cases of interest, by the near equality of probability for either spin orientation, which also insures the symmetry of the distribution about H_0. We have carried out the calculation for a special case of a cubic lattice of spins, obtaining a roughly Gaussian shape for the distribution. In many cases an approximately Gaussian shape is to be expected. An important exception is found, however, when the nuclei lie near together in pairs; several such cases have been studied experimentally by G. E. Pake and are treated in a forthcoming paper.[15]

We can obtain an estimate of the line width by calculating the mean square contribution of each neighbor separately and taking the square root of the sum of these contributions, thereby assuming that the orientation of each spin is independent of that of another. The contribution of term A to the mean square local field is given by

$$\langle H_{\text{loc}}^2\rangle_{\text{Av}} = \tfrac{1}{3}\gamma^2\hbar^2 I(I+1) \times \sum_j (1-3\cos^2\theta_{ij})^2 r_{ij}^{-6}, \quad (23)$$

and the mean square deviation in frequency by

$$\langle \Delta\omega^2\rangle_{\text{Av}} = \gamma^2 \langle H_{\text{loc}}^2\rangle_{\text{Av}}. \quad (24)$$

To this we must add the contribution of term B, which corresponds to the simultaneous flopping of two antiparallel spins ($\Delta m_i = +1$ and $\Delta m_j = -1$, or *vice versa*). Such a process is energetically possible and is caused by the precession of the spins around H_0 with the Larmor frequency; the precession of one spin produces a rotating field at the resonance frequency at the location of another, resulting in reciprocal transitions. We may say that this process limits the lifetime of a spin in a given state and therefore broadens the spectral line. The effect is taken into account by applying the proper numerical factor to (24). The factor has been calculated by Van Vleck by the diagonal sum method.[14] In the case of a lattice of identical spins it is 9/4, the r.m.s. (angular) frequency deviation including both effects then being,

$$[\langle \Delta\omega^2\rangle_{\text{Av}}]^{\frac{1}{2}} \equiv (1/T_2'') = \tfrac{3}{2}\gamma^2\hbar[I(I+1)/3]^{\frac{1}{2}} \times [\textstyle\sum_{j\neq i}(1-3\cos^2\theta_{ij})^2 r_{ij}^{-6}]^{\frac{1}{2}}. \quad (25)$$

If the nuclei are not identical in gyromagnetic ratio, neighbors which are dissimilar to the nucleus i contribute only through the term A, and the sum over j in (25) has to be modified in an obvious way.

We have here introduced the quantity T_2'', defined above, as a specification of the line width in this *limiting* case of a rigid lattice. We shall later use T_2' to describe the effect upon the line width of spin-spin interactions as modified by lattice motion, reserving the unprimed T_2 for the specification of the line width in the general case, to which the spin-lattice interaction also contributes. If the line shape is Gaussian, the quantity $2/T_2''$ is the width between points of maximum slope, on an ω scale. In calculating the maximum absorption for such a line shape, in the rigid lattice, the T_2 defined by Eq. (14) is to be identified with $(\pi/2)^{\frac{1}{2}}T_2''$.

An accurate experimental test of (25) requires, of course, a determination of the line shape. A substance which the rigid lattice model might be expected to represent fairly well is fluorite, CaF_2. The measurements of the F^{19} resonance line in CaF_2, previously reported,[16] and subsequent unpublished work on the same substance by G. E. Pake, are in good agreement with Eq. (25) (see reference 14).

VII. SPIN-LATTICE INTERACTION

The rigid lattice model does not exhibit thermal relaxation of the spin system as a whole. We shall now show that if the position coordinates, r_{ij}, θ_{ij}, and φ_{ij} are made to vary with time, transitions are induced in which m changes, the spin system and the "lattice" thereby ex-

[15] G. E. Pake, J. Chem. Phys. (in press).

[16] E. M. Purcell, N. Bloembergen, and R. V. Pound, Phys. Rev. 70, 988 (1946).

changing energy. Not only does this provide thermal contact between the spins and an adequate heat reservoir, it also modifies, in some cases drastically, the spin-spin interaction itself.

Waller[17] in 1932 treated the interaction of spins with lattice *vibrations*; his theory was applied by Heitler and Teller[1] to estimate the effectiveness of such an interaction in bringing about thermal equilibrium of a nuclear spin system. We want to include more general types of motion, in particular, what we may loosely call Brownian motion in a liquid or non-crystalline solid. It will turn out that such internal degrees of freedom, when active, are far more effective as a source of spin-lattice interaction than are lattice vibrations, and play an important role in substances which by many standards would be regarded as crystalline.

As far as the Hamiltonian for the nuclear spin system is concerned, we regard the motion of the molecules as produced by external forces, the atomic interactions being mainly of an electrical nature. The atom or molecule is simply a vehicle by which the nucleus is conveyed from point to point. We thus neglect the reaction of the magnetic moments of the nuclei upon the motion. The factors in (22) which contain the position coordinates are now to be expanded in Fourier integrals. We want to distinguish in our complex notation between positive and negative frequencies, and we therefore define the intensities $J(\nu)$ of the Fourier spectra of the three position functions involved by

$$\langle\Sigma_j|(1-3\cos^2\theta_{ij}(t))r_{ij}^{-3}(t)|^2\rangle_{Av}=\int_{-\infty}^{\infty}J_0(\nu)d\nu,$$

$$\langle\Sigma_j|\sin\theta_{ij}(t)\cos\theta_{ij}(t)e^{i\varphi_{ij}(t)}r_{ij}^{-3}(t)|^2\rangle_{Av}=\int_{-\infty}^{\infty}J_1(\nu)d\nu, \quad (26)$$

$$\langle\Sigma_j|\sin^2\theta_{ij}e^{2i\varphi_{ij}(t)}r_{ij}^{-3}(t)|^2\rangle_{Av}=\int_{-\infty}^{\infty}J_2(\nu)d\nu.$$

Reconsidering now the terms C, D, E, and F in the expression for the perturbation (22), we observe that if $J_1(-\nu_0)$ and $J_1(+\nu_0)$, where $2\pi\nu_0=\gamma H_0$, are different from zero, C and D become secular perturbations because the time factor cancels out. Similarly, E and F become secular perturbations if $J_2(-2\nu_0)$ and $J_2(+2\nu_0)$ do not vanish. Transitions are thus made possible in which m changes by ±1 or ±2, the thermal motion providing, or absorbing, the requisite energy.

We can form a simple picture of the processes involved by recalling that the local field at one nucleus i due to a neighbor j, in the rigid-lattice model, consists of two parts, a static part depending on I_{zj} and a rotating part depending on $(I_{xj}-iI_{yj})$. The *motion* of the neighbor j now causes the field arising from its z component of magnetic moment to fluctuate; if the motion contains frequencies synchronous with the precession of the neighbor i, the nucleus i will find itself exposed to a radiofrequency field capable of inducing a transition. This is the effect described by terms C and D. The double-frequency effect comes about as follows: the field at i which arises from the precessing components of magnetic moment of j consists in general of *both* right and left circularly polarized components. In the rigid lattice one of these has a negligible effect while the one which rotates in the same sense as the precession of i causes the perturbation expressed by term B. If, however, we impart to the nucleus j a suitable *motion*, at the frequency $2\nu_0$ we can reverse the sense, at the nucleus i, of the originally ineffective circular component, thus permitting it to interact with i. An exactly similar situation will necessarily be created at the nucleus j, with regard to the field arising from the precessing components of moment of i, hence a double transition occurs. This also shows why still *higher* multiples of ν_0 do *not* enter the problem.

We can now compute the probability for a change $\Delta m=+1$ in the magnetic quantum number of the ith spin brought about by the fields associated with the "local field spectrum." By a perturbation calculation, using terms C and E, and taking an average over all values m_j of the neighbors, we find for the transition probability,

$$W_{m_i\to m_i+1}=\tfrac{3}{4}\gamma^4\hbar^2(I-m_i)(I+m_i+1)I(I+1)\times[J_1(-\nu_0)+\tfrac{1}{2}J_2(-2\nu_0)]. \quad (27)$$

[17] I. Waller, Zeits. f. Physik 79, 370 (1932).

If the values of I and γ are not the same for all nuclei, Eq. (27) can be generalized as follows:

$$W_{m_i\to m_i+1}=\tfrac{3}{4}\gamma_i^2\hbar^2(I_i-m_i)(I_i+m_i+1)$$
$$\times\sum_j\gamma_j^2I_j(I_j+1)$$
$$\times[J_{1j}(-\nu_{0i})+\tfrac{1}{2}J_{2j}(-\nu_{0j}-\nu_{0i})],\quad(28)$$

where

$$\nu_{0j}=\gamma_jH_0/2\pi.$$

The result can be extended to include the effect of perturbing fields arising from other dipole sources such as electronic spins. In such cases the quantization of these spins may be time dependent under the influence of external forces (e.g., paramagnetic relaxation phenomena involving electronic spins) and the function $[(I_j-m_j(t))(I_j+m_j(t)+1)]^{\frac{1}{2}}$ must be included with the position functions when the Fourier expansion is performed, the factor $I_j(I_j+1)$ in (28) then being dropped.

The expressions for transitions with $\Delta m_i=-1$ are, of course, the same except that the intensities have to be taken at positive frequencies in the Fourier spectrum. Since $J(\nu)$ will in all cases be an even function of ν, it would appear that the motion should produce as many transitions up as down. However, the transition probabilities must be weighted by the Boltzmann factors of the final states, as required by the fact that the system left to itself will come to thermal equilibrium and the principle of detailed balancing. That is, if N_p and N_q are the equilibrium populations of levels p and q which differ in energy by E_p-E_q, we must have, in equilibrium,

$$N_pW_{p\to q}=N_qW_{q\to p},$$

or

$$W_{p\to q}/W_{q\to p}=N_q/N_p=e^{(E_p-E_q)/kT}.\quad(29)$$

But the probability of a single transition from p to q, in a system such as we are considering, cannot depend on the population of q; it is therefore necessary to suppose that the W's are related by the Boltzmann factor of Eq. (29) even when the spins are not in equilibrium with the lattice, T being the *lattice* temperature. We are thus enabled to trace the approach to equilibrium of the spin system coupled to a reservoir which remains throughout the process at the temperature T. This last assumption is justified if $kT\gg\gamma H_0\hbar$ for the specific heat of the lattice is then very much greater than that of the spins.

Consider first the case $I=\frac{1}{2}$. Let the total number of spins be N, with N_+ and N_- denoting the population of the lower and upper state, respectively. For $W_{+\to-}$ and $W_{-\to+}$ we take W as given by Eq. (27) modified by the appropriate Holtzmann factors:

$$W_{+\to-}=We^{-\gamma H_0\hbar/2kT}\approx W[1-(\gamma\hbar H_0/2kT)];$$
$$W_{-\to+}\approx W[1+(\gamma\hbar H_0/2kT)]\quad(30)$$

with

$$W=(9/16)\gamma^4\hbar^2[J_1(\nu_0)+\tfrac{1}{2}J_2(2\nu_0)].$$

Writing n for N_+-N_-, the surplus in the lower state, we have

$$dn/dt=2N_-W_{-\to+}-2N_+W_{+\to-}$$
$$=(N+n)W_{-\to+}-(N-n)W_{+\to-}.\quad(31)$$

If we assume that $|n|\ll N$, as it will be in all cases of present interest, and denote by $n_0=N\gamma\hbar H_0/2kT$ the equilibrium value of n, we have

$$dn/dt=2W(n_0-n),\quad(32)$$

showing that equilibrium is approached according to a simple exponential law with a characteristic time

$$T_1=1/2W.\quad(33)$$

We thus make connection with, and justify, the manner in which the spin-lattice relaxation time T_1 was introduced in Eq. (11).

For $I=\frac{1}{2}$ it is always possible to define a *spin temperature* T_s through $e^{\gamma\hbar H_0/kT_s}=N_-/N_+$, and the approach to equilibrium can be described as a cooling, or warming, of the spin system as a whole. The corresponding differential equation, in which T_L is the lattice temperature, is

$$dT_s/dt=(1/T_1)(T_s/T_L)(T_L-T_s).$$

Clearly the transient behavior of the spin system is more simply described by the reciprocal temperature, or by n, than by the temperature directly.

The situation is more complicated when $I>\frac{1}{2}$. A spin temperature cannot be defined unless the populations of successive levels are related by the same factor. Corresponding to this additional arbitrariness in the initial state of the system,

Eq. (32) has to be replaced by a number of simultaneous differential equations. However, a detailed examination of this question which can be found elsewhere,[18] leads to a simple result when the initial state of the spin system happens to be describable by a temperature T_s, and when $kT_s \gg \gamma\hbar H_0 \ll kT_L$. The approach to equilibrium then resembles that of the $I=\frac{1}{2}$ system, the surplus number in each level with respect to the next above increasing or diminishing with the same characteristic time. Any intermediate state of the system can then be assigned a temperature, and (32) applies. It is found, moreover, that value of W which must then be used in (33) is

$$\tfrac{3}{4}\gamma^4\hbar^2 I(I+1)[J_1(\nu_0)+\tfrac{1}{2}J_2(2\nu_0)].$$

That is to say, the spin-lattice relaxation time for nuclei of spin I exposed to a *given local field* is the same as the relaxation time for nuclei of the same gyromagnetic ratio γ, but with $I=\frac{1}{2}$, exposed to the same field.

The above argument would apply, for example, to a spin system with $I>\frac{1}{2}$ which is allowed to come to equilibrium in a field H_0, after which H_0 is changed adiabatically to a new value. A unique T_1 can be defined which measures the rate of approach of the populations, $N_{-I}, \cdots N_I$, to their new equilibrium distribution. Similarly, if the application of an intense radiofrequency field of frequency ν_0 has completely equalized the level populations, the recovery from such saturation (infinite spin temperature) follows the same law.

In general, however, the application of a radiofrequency field alters the distribution to one which is not precisely a Boltzmann distribution—the spin system is itself no longer in equilibrium. The most powerful agency tending to restore *internal* equilibrium among the spins is in many cases the spin-spin interaction. By means of the process associated with the term B in Eq. (22), in which $\Delta m_i=+1$, $\Delta m_j=-1$, or *vice versa*, the level populations can be readjusted. If $I=1$, for example, transitions of the type $-1\rightarrow 0$ (for m_i) accompanied by $+1\rightarrow 0$ (for m_j) augment the population of the central level, while the reverse double transition depletes it. If N_{-1}, N_0, N_{+1} are the populations of the levels, the rate at which transitions of the first type occur is proportional to $N_{-1}\cdot N_{+1}$, while the rate at which the second type occur is proportional to N_0^2, but otherwise identical. A stationary state is therefore reached only when $N_0^2=N_{-1}\cdot N_{+1}$, which is precisely the description of a Boltzmann distribution over three equidistant levels.

With the above reservations concerning the case $I>\frac{1}{2}$, we now write the general formula for the spin-lattice relaxation due to dipole-dipole interaction in the case of identical nuclei.‡‡

$$(1/T_1)=\tfrac{3}{2}\gamma^4\hbar^2 I(I+1)[J_1(\nu_0)+\tfrac{1}{2}J_2(2\nu_0)]. \quad (34)$$

VIII. EFFECT OF THE MOTION UPON THE LINE WIDTH

The terms A and B of Eq. (22) must now be reconsidered to see what effect the motion of the nuclei has upon the line width. These terms will still represent secular perturbations if we take the components near zero frequency in $J_0(\nu)$, the intensity of the Fourier spectrum of $\sum_j(1-3\cos^2\theta_{ij}(t))r_{ij}^{-3}(t)$. The question is, which frequencies are to be considered as "near zero." We may say that the perturbation is secular up to the frequency ν for which $h\nu$ is of the same order as the actual splitting of the energy levels by the perturbation. The resulting line width will be specified by $\Delta\omega=2\pi\Delta\nu=1/T_2'$, and with the application of the above criterion, T_2' is to be found from the following modification of Eq. (34):

$$1/T_2'=\tfrac{3}{2}\gamma^2\hbar[I(I+1)/3]^{\frac{1}{2}} \times\left[\int_{-1/\pi T_2'}^{+1/\pi T_2'} J_0(\nu)d\nu\right]^{\frac{1}{2}}. \quad (35)$$

Again an exception must be noted in the case of non-identical nuclei. If $\gamma_i\neq\gamma_j$, the secular perturbation from term B depends on the intensity of the J_0 spectrum in the neighborhood of $\nu=(\gamma_i-\gamma_j)H_0/2\pi$ rather than $\nu=0$.

The observed line width is not determined by T_2' alone because the spin-lattice relaxation effected by the terms C to F limits the lifetime of the nucleus in a given state. For $I=\frac{1}{2}$, the mean lifetime in each of the two levels is $1/W$, or $2T_1$. The resulting *line width*, for two levels

[18] L. J. F. Broer, *Thesis* (Amsterdam 1945), p. 56 ff.

‡‡ The case of non-identical nuclei calls for a rather obvious generalization which we omit.

broadened in this way, is, by a well-known argument[19] twice the width of one level, the line shape being given to a high degree of approximation, ($T_1 \gg 1/\nu_0$) by

$$g_1(\nu) = \frac{4T_1}{1+16\pi^2 T_1^2(\nu-\nu_0)^2}, \qquad (36)$$

a curve whose width between half-value points is $1/2\pi T_1$ on a frequency scale. With this must be combined the distribution resulting from the perturbations A and B, which, if it had a Gaussian shape, would be

$$g_2(\nu) = T_2'(2\pi)^{-\frac{1}{2}} \exp[-2\pi^2(\nu-\nu_0)^2 T_2'^2]. \qquad (37)$$

Roughly speaking, we can say that the combination of two bell-shaped curves will yield another bell-shaped curve the width of which is about the sum of the widths of the composing curves. Thus,

$$(1/T_2) = (2/\pi)^{\frac{1}{2}}(1/T_2') + (1/2T_1). \qquad (38)$$

We cannot attach a precise meaning to T_2 in general, but the numerical factors in (38) have been chosen‡‡‡ with two limiting cases in mind: (a) $T_2' \ll T_1$, and (b) $T_2' \approx T_1$. In the former case the factor $(2/\pi)^{\frac{1}{2}}$ is consistent with Eq. (14) for the Gaussian curve to which, in this case, we may expect the line to bear some resemblance. The last term in (38), which is important in case (b), is correctly written for the damped oscillator line shape. However, in this case, an additional uncertainty arises from the arbitrariness of the choice of limits in Eq. (35).

IX. THE SPECTRUM OF THE SPACE COORDINATES

In order to apply this general theory of relaxation time and line width to a particular substance, it is necessary to develop the Fourier spectra of the functions of position coordinates,

$$F_{0j} = (1-3\cos^2\theta_{ij})r_{ij}^{-3},$$
$$F_{1j} = \sin\theta_{ij}\cos\theta_{ij}e^{i\varphi_{ij}}r_{ij}^{-3}, \qquad (39)$$
$$F_{2j} = \sin^2\theta_{ij}e^{2i\varphi_{ij}}r_{ij}^{-3},$$

[19] V. Weisskopf and E. Wigner, Zeits. f. Physik **63**, 54 (1930); **65**, 18 (1930).

‡‡‡ It seems worth while to preserve internal consistency among the definitions and theoretical formulas even at the risk of making the formulas appear less approximate than they really are.

from which to obtain the functions $J_{0j}(\nu)$, $J_{1j}(\nu)$, $J_{2j}(\nu)$ governing the intensity of the local field spectrum. By postulating thermal vibrations in an otherwise rigid lattice, for example, one is led directly to Waller's result for spin-lattice relaxation in a crystal. Our interest here is in liquids, for which the functions F will vary in a random fashion with time, as the particles containing the magnetic nuclei take part in the Brownian motion. The motion will be isotropic on the average; hence, as is at once evident from (39), the time average of each of the functions F is zero.

The statistical character of the motion justifies an assumption customary in the theory of fluctuation phenomena, that

$$\langle F(t)F^*(t+\tau)\rangle_{\text{Av}} = K(\tau). \qquad (40)$$

The right side of (40) is called the correlation function of $F(t)$. It is assumed to be an even function of τ and independent of t. It follows immediately that $K(\tau)$ is real. We may assume also that $K(\tau)$ approaches zero for sufficiently large values of τ. The relation[20] between the Fourier spectrum of a function F and the correlation function of F enables us to write for the spectral intensity,

$$J(\nu) = \int_{-\infty}^{\infty} K(\tau)e^{2\pi i\nu\tau}d\tau. \qquad (41)$$

Since $K(\tau)$ is real and even, $J(\nu)$ is real and even.

In order to simplify the following discussion we make an assumption about the form of $K(\tau)$, namely, that

$$K(\tau) = \langle F(t)F^*(t)\rangle_{\text{Av}} e^{-|\tau|/\tau_c}. \qquad (42)$$

The time τ_c is a characteristic of the random motion and the function whose correlation K measures, and is called the *correlation time.* Temporarily we shall assume that τ_c is the same for each of the functions F_0, F_1, and F_2, returning to this question later. From (41) and (42) we find, for each of the three J's,

$$J(\nu) = \langle F(t)F^*(t)\rangle_{\text{Av}} 2\tau_c(1+4\pi^2\nu^2\tau_c^2)^{-1}. \qquad (43)$$

The intensity in the spectrum is nearly constant

[20] See, for example, Ming Chen Wang and G. E. Uhlenbeck, Rev. Mod. Phys. **17**, 323 (1945).

at very low frequencies but falls off rapidly when $2\pi\nu\tau_c \gg 1$. The subscript j has been omitted above, but it will be understood that, in general, not all neighbors will have the same correlation time.

X. THE THEORY APPLIED TO WATER

The application of the preceding theory will now be illustrated by a calculation of the relaxation time for protons in water. A given proton has as its nearest neighbor the other proton in the H_2O molecule, and we consider first the effect of this exceptional neighbor. Vibration of the molecule can be neglected, because of its high frequency and small amplitude. We regard the molecule as rigid, and assume that the orientation of the vector connecting the two protons varies randomly, no direction being preferred. Only the angle coordinates are now variable in time, the functions F being

$$F_0 = (1 - 3\cos^2\theta)/b^3,$$
$$F_1 = \cos\theta \sin\theta e^{i\varphi}/b^3, \qquad (44)$$
$$F_2 = \sin^2\theta e^{2i\varphi}/b^3,$$

where b is the interproton distance. We can substitute a statistical average over the spatial coordinates for the time average specified in (43). The averages are readily found:

$$\langle F_0(t) F_0^*(t)\rangle_{Av} = 4/5b^6;$$
$$\langle F_1(t) F_1^*(t)\rangle_{Av} = 2/15b^6; \qquad (45)$$
$$\langle F_2(t) F_2^*(t)\rangle_{Av} = 8/15b^6.$$

Returning to the general formula (34) with (43) and (45) we obtain for the relaxation time (involving the nearest neighbor only)

$$(1/T_1) = (3/10)(\gamma^4\hbar^2/b^6)[\tau_c/(1+4\pi^2\nu_0^2\tau_c^2) + 2\tau_c/(1+16\pi^2\nu_0^2\tau_c^2)]. \qquad (46)$$

The determination of the correlation time τ_c is closely related to the problem encountered in the Debye theory of dielectric dispersion in polar liquids; there also it is necessary to estimate the time, τ, during which molecular orientation persists.[21] Debye assumed that in the first approximation the molecule could be treated as a sphere of radius a imbedded in a viscous liquid, and thereby obtained the relation $\tau = 4\pi\eta a^3/kT$, in which η is the viscosity of the liquid. We adopt the same procedure here, noting one difference between the two problems—a rather unimportant difference, to be sure, in view of the much over-simplified model, but a characteristic one. The function whose correlation time is required in the Debye theory is $\cos\theta$, the symmetry character of which is essentially different from that of the functions F_0, F_1, and F_2 with which we are concerned. Because of this the correlation time for a given model¶ will not be the same in our case. In fact, for the sphere in a viscous liquid it is, for each¶¶ of the functions F,

$$\tau = 4\pi\eta a^3/3kT. \qquad (47)$$

Adopting Eq. (47) and assuming that $a = 1.5 \times 10^{-8}$ cm, we find for water at 20°C ($\eta = 0.01$ poise): $\tau_c = 0.35 \times 10^{-11}$ second. For comparison we note that the measurements by Saxton[22] of the dielectric constant of water in the microwave region lead to a value 0.81×10^{-11} seconds for the Debye characteristic time, at 20°C. Although this is somewhat less than 3 times our estimated τ_c, we must remember that with further refinement of the molecular model a difference between τ_c and $1/3\tau_{Debye}$ would arise from the fact that the H–H line and the polar axis of the molecule lie in different directions.

We note at once that $2\pi\nu\tau_c \ll 1$, so that the term in brackets in Eq. (46) becomes simply $3\tau_c$. In other words, we find ourselves at the low frequency end of the local field spectrum, where the spectral intensity is independent of ν and *directly* proportional to τ_c. For the interproton

[21] P. Debye, *Polar Molecules* (Dover Publications, New York, 1945), Chapter V.

¶ Another model, in some ways a better representation of the molecule and its environment, is one in which only a limited number of orientations are permitted, the transitions from one to another taking place in a "jump." Imagine, for example, that the molecular axis—in our case the H–H line—can point toward any of the eight corners of a cube centered at the molecule, transitions taking place with the probability w between any such direction and one of the three adjacent directions only. For this model the correlation functions of F_0, F_1, and F_2 are all of the form (42) with the correlation time $3/4w$, whereas the correlation time of $\cos\theta$ is $3/2w$.

¶¶ Note that F_0, F_1, and F_2 belong to the same spherical harmonic, $Y_2(\theta, \varphi)$. It can be shown that the correlation function of $Y_l(\theta, \varphi)$ for the sphere in a viscous liquid, is $\exp[-l(l+1)kT\tau/8\pi\eta a^3]$.

[22] J. A. Saxton, Radio Research Board Report C115 (declassified), D.S.I.R., March 20, 1945. (This is a British report.)

distance we take 1.5×10^{-8} cm. Upon substituting numbers into (46), we obtain for $(1/T_1)_p$ the relaxation rate due to the proton partner only, 0.19 sec.$^{-1}$.

The effect of the neighboring molecules must now be estimated. The fluctuation in the local field arising from the neighbors is mainly caused by their translational motion, and only to a minor extent by their rotation. We regard the molecules as independent, attributing to each either a magnetic moment of $2\mu_p$ and spin $I=1$, of else zero spin and moment. Three-fourths of the neighboring molecules are of the former (*ortho*) type. We ask for $\langle F(t)F^*(t)\rangle_{\text{Av}}$ and τ_c for the molecules in a spherical shell between r and $r+dr$. The center of the shell is located at, and moves with, the molecule containing the proton i. A reasonable choice for τ_c is the time it takes the molecule j to move a distance r in any direction from its original position on the shell, for in that time the field at the proton i, due to j, will have changed considerably. The relative motion of i and j is simply diffusion, and can be described most directly by means of the diffusion coefficient D of the liquid. Applying the above criterion for τ_c, and remembering that both molecules diffuse, we find,

$$\tau_0 = r^2/12D. \qquad (48)$$

To find $\langle F(t)F^*(t)\rangle_{\text{Av}}$ the angular functions are averaged as before and then, treating the molecules as independent, we sum the effects of all neighbors by integrating over the volume from the radius of closest approacy, $r=2a$, to infinity. Here also, $2\pi\nu\tau_c$ is much smaller than one for all molecules close enough to have an appreciable effect, so that we have for the relaxation rate due to the neighbors $(1/T_1)_n$,

$$\left(\frac{1}{T_1}\right)_n = \frac{3}{5}\gamma^4\hbar^2 N_0 \int_{2a}^{\infty} \frac{3}{r^6}\cdot\frac{r^2}{12D}\cdot 4\pi r^2 dr = \frac{3\pi}{10}\frac{\gamma^4\hbar^2 N_0}{aD}. \qquad (49)$$

N_0 in Eq. (49) is the number of molecules per cm^3. We can make use of the well-known relation developed by Stokes, $D=kT/6\pi\eta a$, to write Eq. (49) as

$$(1/T_1)_n = 9\pi^2\gamma^4\hbar^2\eta N_0/5kT. \qquad (50)$$

Even when an experimental value of D is available, the use of the Stokes relation represents no real sacrifice in accuracy, for the distance of closest approach, a, can at best be estimated only roughly.

Substituting numbers into (50) we obtain for water at 20°C, $(1/T_1)_n = 0.10$ sec.$^{-1}$. The relaxation time T_1 is then found by combining both effects: $(1/T_1) = (1/T_1)_p + (1/T_1)_n = 0.19$ sec.$^{-1}$ $+0.10$ sec.$^{-1}$, or $T_1 = 3.4$ sec.

The effect of the partner thus appears to be somewhat more important than that of the neighboring molecules. We have probably overestimated the effect of the neighbors by neglecting the rotation of the nearest molecules as well as the effect upon the local field produced by them of the rotation of the molecule containing the proton i. That is, the appropriate τ_c for the nearest neighbors should be somewhat smaller than Eq. (48) would predict. Such refinements are hardly warranted in view of the obvious shortcomings of the model.

The experimentally determined value of T_1 for distilled water at 20°C is 2.3 ± 0.5 seconds, obtained by the direct method described in Section IV. The theoretical estimate of T_1 is satisfactorily close to the experimental value. Unfortunately, the possibility that the experimental value was influenced somewhat by dissolved oxygen cannot be excluded. From subsequent results with paramagnetic solutions, it is estimated that a saturation concentration of O_2 could have contributed at most 0.3 sec.$^{-1}$ to $(1/T_1)$.

XI. EXPERIMENTAL RESULTS FOR VARIOUS LIQUIDS; COMPARISON WITH THEORY

A more significant test of the theory is provided by measurements of the proton relaxation time in a series of similarly constituted liquids of widely differing viscosity. The conspicuous role of the viscosity in our description of the relaxation process will already have been noticed. Despite the crudeness of a description of the details of the molecular motion in terms of the macroscopic viscosity, it can hardly be doubted that there should be a close connection between η and τ_c. Table I lists the values of T_1 obtained for a series of hydrocarbons. These values, and all relaxation times quoted hereafter, were determined by running a saturation curve on the sample, the direct measurements of T_1 for the water sample, and for the $0.002N$ $CuSO_4$ solution (see Section IV) serving as a calibration of the circuit. The viscosity of each sample was measured by conventional methods and is listed for comparison. The wide range over which a rough proportionality exists between η and $(1/T_1)$ is striking. A more heterogeneous list of substances is given in Table II. A probable explanation of the exceptional behavior of sulfuric acid is the low density of protons in the material, or, more precisely, a relatively large value of the shortest proton-proton distance (the other constituents being non-magnetic).

A wide range in viscosity is covered by water-glycerin solutions. Figure 11 shows the variation of T_1 with viscosity for the water-glycerine system at room temperature. The viscosity values were obtained from tables in this case. Exact proportionality between $(1/T_1)$ and η is hardly to be expected for the progressive change in the character of the substance must bring with it a change in the environment of each proton and in the details of the local molecular motion. Actually a 1000-fold change in η is accompanied by a change of only 100-fold in T_1.

A better experiment is one in which the same substance is used throughout, its viscosity being varied over the widest possible range by changing the temperature. Ethyl alcohol and glycerine are convenient for this purpose; each was studied over the temperature range between 60°C and −35°C. The temperature of the substance was measured with a copper-constantan thermocouple brought into the sample a short distance away from the r-f coil. The temperature of the whole enclosure containing the coil and sample was established by circulating a suitable fluid at a controllable rate. The temperature of the sample could be held constant within 0.5° while data for a saturation curve were taken. The viscosity values were taken from Landolt-Börnstein, *Physikalish-Chemische Tabellen.*

The results obtained with ethyl alcohol at two frequencies are shown in Fig. 12, in the form of a logarithmic plot of the relaxation time T_1 against the ratio of the viscosity to the absolute temperature, η/T. According to our theory τ_c, and hence $(1/T_1)$, so long as $2\pi\nu\tau_c \ll 1$, should be proportional to η/T, as indicated by the line drawn on the graph with slope −1. Moreover, the relaxation time should be independent of ν, the radiofrequency. Both predictions are confirmed by the experimental results.

A much wider range in η/T is accessible with a pure glycerine sample. Glycerine, as is well known, ordinarily becomes super-cooled down to temperatures far below its freezing point of 18°C, and at the lowest temperatures prevailing in the experiment the liquid was nearly glass-like. The glycerine results are plotted in Fig. 13 which includes also the results of line-width measurements in the low temperature region. The characteristic time T_2 associated with the line broadening is obtained by setting $\gamma\Delta H = 2/T_2$, where ΔH is the observed interval in gauss between points of maximum slope on the absorption curve. This procedure, as explained earlier, would be correct for a Gaussian line shape.

In the region of low viscosity the T_1 values for glycerine display the behavior encountered in alcohol, namely, inverse proportionality between T_1 and η/T, and independence of T_1 and ν. The new feature is the minimum in T_1, followed by increasing T_1 as η continues to increase. The inverse line-width T_2, on the other hand, decreases monotonically in the region where it could be measured, with perhaps a slight tendency to flatten off at the highest η/T values. The curves

TABLE I. Relaxation time T_1, for protons in various hydrocarbons at 20°C and 29 Mc/sec.

	Viscosity (centipoises)	T_1 (seconds)
Petroleum ether	0.48	3.5
Ligroin	0.79	1.7
Kerosene	1.55	0.7
Light machine oil	42	0.075
Heavy machine oil	260	0.013
Mineral oil	240	0.007

TABLE II. Relaxation time T_1, for protons in various polar liquids at 20°C and 29 Mc/sec.

	Viscosity (centipoises)	T_1 (seconds)
Diethyl ether	0.25	3.8
Water	1.02	2.3
Ethyl alcohol	1.2	2.2
Acetic acid	1.2	2.4
Sulfuric acid	25	0.7
Glycerine	1000	0.023

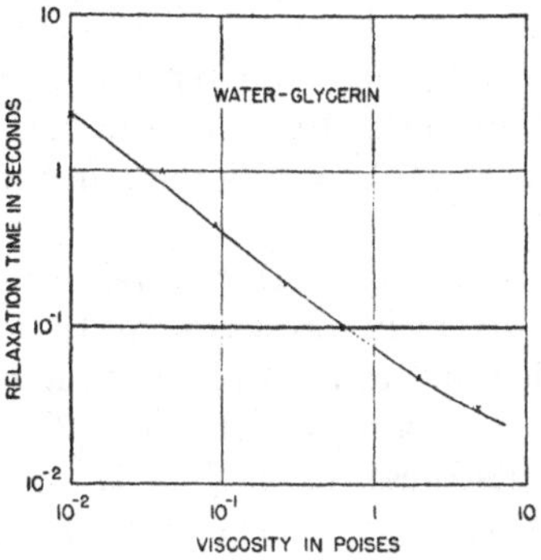

FIG. 11. Relaxation time for protons in water-glycerin solutions, measured at 29 Mc/sec.

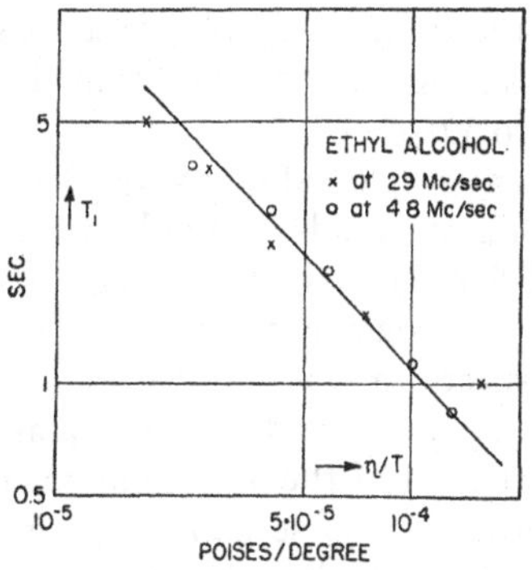

FIG. 12. Relaxation time for protons in ethyl alcohol, measured at 29 Mc/sec. and at 4.8 Mc/sec.

drawn on the graph are based on the theory, which must now be applied to this more general case, the restriction $2\pi\nu\tau_c \ll 1$ being lifted.

The discussion will be simplified by the assumption that τ_c, at a given temperature, is the same for all sources of the local field. We write Eq. (46) in the form:

$$\frac{1}{T_1} = C_1\left[\frac{\tau_c}{1+\omega^2\tau_c^2} + \frac{2\tau_c}{1+4\omega^2\tau_c^2}\right], \qquad (51)$$

where C_1 includes the factors which are independent of the temperature and frequency. (The variation of b and a with temperature is a negligible effect.) The asymptotic behavior of T_1 is simple: for $\omega\tau_c \ll 1$, $(1/T_1) \approx 3C_1\tau_c$; for $\omega\tau_c \gg 1$, $(1/T_1) \approx 3C_1/(2\omega^2\tau_c)$. In the intermediate region a minimum in T_1 occurs for $\omega\tau_c = 1/\sqrt{2}$, with $T_{1\,(\mathrm{min})} = 3\omega/(2^{\frac{3}{2}}C_1)$. If we may assume that τ_c is proportional to η/T, the shape of the curve $\log T_1$ *vs.* $\log(\eta/T)$ should be the same for all substances to which the theory applies; changes in C_1 or in the factor connecting τ_c and η/T merely shift the curve. Two such curves have been drawn on the graph in Fig. 13.

The salient features of the experimental results are well reproduced, although there are quantitative discrepancies. The horizontal shift of the minimum is somewhat less than the ratio of the frequencies, and the experimental minimum is flatter than predicted. This may be blamed in part upon experimental inaccuracy, but it must be remembered that Eq. (51) neglects the variation in τ_c among the neighbors. As we saw earlier, a proper treatment of the neighbors introduces a distribution of relaxation times, which would have the effect of flattening the minimum in the T_1 curve.

To discuss the line-width parameter, T_2, we go back to Eq. (35) which determines T_2' and insert for $J_0(\nu)$, $(4/5b^6)2\tau_c/(1+\omega^2\tau_c^2)$. We thus obtain, making use of the constant C_1 of Eq. (51),

$$(1/T_2')^2 = (3/\pi)C_1 \tan^{-1}(2\tau_c/T_2'). \qquad (52)$$

For sufficiently large values of τ_c, T_2' attains the value $(2/3C_1)^{\frac{1}{2}}$, which is identical with T_2'', the time characterizing the spin-spin coupling in the rigid lattice. When τ_c is small, we have, from (52), $(1/T_2') \approx (6/\pi)C_1\tau_c$. The dashed line in Fig. 13 has been drawn according to this relation in a region in which the contribution of $(1/T_1)$ to the line width is negligible. The agreement with experiment is satisfactory.

To the left of the minimum in T_1 we should have $(1/T_1) = 3C_1\tau_c$ so that in this region, according to Eq. (38),

$$\begin{aligned}(1/T_2) &= (2/\pi)^{\frac{1}{2}}(1/T_2') + (1/2T_1)\\ &= 3C_1\tau_c[(2/\pi)^{\frac{1}{2}} + \tfrac{1}{2}] = 1.01/T_1. \qquad (53)\end{aligned}$$

This result signifies only that T_1 and T_2 are *approximately* equal when $2\pi\nu\tau_c \ll 1$.

The general features of the simplified theory are presented in Fig. 14. The T_2 curve is located by the value of T_2'', which depends on the spin-spin interaction in the "frozen" substance, according to Eq. (25). The break between the sloping and the horizontal part of the T_2 curve occurs near $\tau_c = T_2''/\sqrt{2}$, according to (52). The minimum in the T_1 curve is located by $\omega\tau_c = 1/\sqrt{2}$, and thus both curves are fixed. That the transition regions occupy, in most cases, rather small ranges on the logarithmic plot is a consequence of the weakness of the spin-spin interaction compared to $\gamma H_0 \hbar$.

We can now understand why the true line width in many liquids is unobservably small. The rapid fluctuations in the perturbing fields very nearly average out; the intensity in the local field spectrum is no greater at $\nu=0$ than it is at $\nu=\nu_0$, resulting in the near equality of T_1 and T_2 to the left of the T_1 minimum. But if T_2 is of the order of one second, as T_1 is in many cases, the corresponding line width is of the order of 10^{-4} gauss. To observe this width, the applied field H_0 would have to be homogeneous over the sample to less than 1 part in 10^8.

As τ_c increases the line width increases, eventually becoming observable and finally attaining its asymptotic value. A striking qualitative demonstration of this effect is obtained by allowing molten paraffin to solidify in the sample tube while observing the proton resonance on the oscilloscope. If the field is reasonably homogeneous, a pronounced broadening of the line is observed when the paraffin freezes.

XII. RELAXATION TIME AND LINE WIDTH IN SOLUTIONS CONTAINING PARAMAGNETIC IONS

The addition of paramagnetic ions to water, as was first shown by Bloch, Hansen, and Packard,[3] can influence markedly the proton relaxation time. The magnetic moment of such an ion is of the order of one Bohr magneton, and the interaction energy $\mu_{\rm ion}\mu_p/r^3$ is thus some 10^3 times larger than that of the interactions previously considered. Nevertheless, the discussion of the last section can be adapted to this case without much modification. The problem is essentially that treated in estimating the effect of the neighboring molecules. The appropriate value of τ_c is determined in the same way, with one possible exception to be mentioned later, and will not be much different in magnitude, so that we again have $2\pi\nu\tau_c \ll 1$. Equation (50), adapted to this case, then reads:

$$(1/T_1) = 2\pi\gamma^2\mu_{\rm eff}^2 N_{\rm ion}/5aD$$
$$= 12\pi^2\gamma^2\eta N_{\rm ion}\mu_{\rm eff}^2/5kT. \quad (54)$$

In general, the value of $\mu_{\rm eff}^2$ can be determined from data on paramagnetic susceptibility. Stokes' relation has again been used to express aD in terms of η/kT; any difference between the mobility of an ion and that of a water molecule is thereby neglected. $N_{\rm ion}$ is the number of ions per cm^3.

According to Eq. (54) the relaxation time should be inversely proportional to the ionic concentration and to the square of the magnetic moment of the paramagnetic ion. The experimental results presented in Fig. 15 confirm the first prediction, except for a marked and unexplained deviation at very low concentrations.

The predicted proportionality between $(1/T_1)$ and $\mu_{\rm eff}^2$ is tested by the data in Table III. The values of $\mu_{\rm eff}$ in the middle column were derived from measurements of T_1 in paramagnetic solutions of known concentration by assuming arbitrarily the value 2.0 Bohr magnetons for Cu^{++}. These numbers are to be compared with the last column which contains the values of $\mu_{\rm eff}$ derived from susceptibility measurements. It is not surprising that in the cases of Ni^{++}, Co^{++}, and $Fe(CN)_6^{---}$ the $\mu_{\rm eff}$ obtained from the nuclear relaxation measurements is too small, for there are important contributions from non-diagonal elements to the magnetic moment of these ions.[23] Indeed, Co^{++} and $Fe(CN)_6^{---}$ show strong deviations from Curie's law. Non-diagonal elements would give rise to high frequency perturbations which would be ineffective in producing nuclear relaxation.

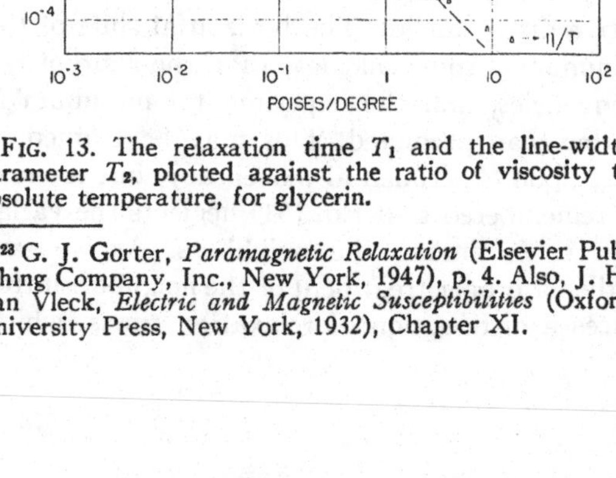

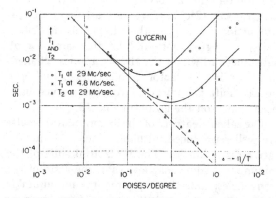

FIG. 13. The relaxation time T_1 and the line-width parameter T_2, plotted against the ratio of viscosity to absolute temperature, for glycerin.

[23] G. J. Gorter, *Paramagnetic Relaxation* (Elsevier Publishing Company, Inc., New York, 1947), p. 4. Also, J. H. Van Vleck, *Electric and Magnetic Susceptibilities* (Oxford University Press, New York, 1932), Chapter XI.

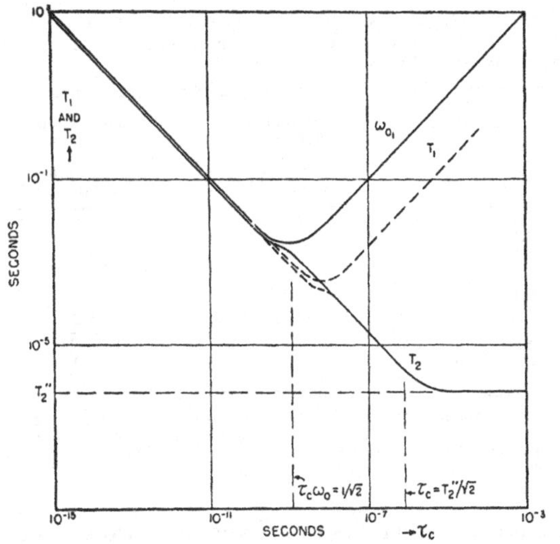

FIG. 14. Dependence of T_1 and T_2 upon τ_c, according to the simplified theory in which all interactions are assumed to have the same correlation time.

The magnitude of T_1 as predicted by Eq. (54) agrees well with experiment. In the case of Fe^{+++}, for example, Eq. (54) with $\eta=0.01$, $\mu_{\text{ion}}=5.9$ Bohr magnetons, yields $T_1=0.08$ second, for a concentration of $10^{18}/\text{cm}^3$. The value observed at this concentration is 0.06 second. Such close agreement must be regarded as accidental.

An effect which has not been taken into account in the above discussion is the relaxation of the electron spins (i.e., paramagnetic relaxation) which may in some cases limit the value of τ_c. Paramagnetic relaxation times as short as 10^{-9} second have been observed in crystals at room temperature.[24] Some measurements of paramagnetic dispersion in solutions, reported by Zavoisky,[25] indicate relaxation times of the order of 10^{-8} to 10^{-9} second. In applying Eq. (54) we have tacitly assumed that the paramagnetic relaxation time, under the conditions of our experiment was not much shorter than 10^{-11} second. It will be interesting to examine nuclear relaxation induced by paramagnetic ions under conditions for which the characteristic time τ_c due to thermal motion is quite long. The actual τ_c may then be determined by the paramagnetic relaxation effect.

The *line width* in paramagnetic solutions is of particular interest, for the extremely short spin-lattice relaxation time observed in concentrated solutions permits a test of the relation between T_1 and T_2. As we have seen, the theory predicts that T_1 and T_2 should be about equal when $2\pi\nu\tau_c \ll 1$. That this is indeed the case can be seen by comparing Fig. 15 with Fig. 16, which gives the measured line width, expressed in terms of T_2, for highly concentrated Fe^{+++} solutions.

XIII. NUCLEAR RELAXATION EXPERIMENTS INVOLVING F^{19}, Li^7, AND D^2

Several additional experiments on liquids will be reported briefly. The resonances of both H^1 and F^{19} were studied in the liquid $CHFCl_2$, a "Freon," which was condensed in a glass tube and sealed off. Both lines were narrow and the intensities were nearly equal, as would be expected. (The spin of F^{19} is $\frac{1}{2}$, and its gyromagnetic ratio only a few percent less than that of the proton.) The measured relaxation times were 3.0 sec. and 2.6 sec. for H^1 and F^{19}, respectively. According to the theory these times should be equal, for the dominant effect should be the H^1-F^{19} interaction within the molecule, and the product $\gamma_{H^1}{}^2\gamma_{F^{19}}{}^2$ occurs in the generalized form of Eq. (34) in either case. Within the experimental error this is confirmed.

The compound BeF_2 combines with water in any proportion. The behavior of both the proton and the fluorine resonances in $BeF_2 \cdot xH_2O$ was found to resemble that of the proton resonance in glycerine. With increasing viscosity (decreasing x) T_1 decreases to a minimum value of about 10^{-3} second, thereafter increasing to a final value of 0.2 second in pure BeF_2, an amorphous glassy solid. The line width meanwhile increases from an unobservably small value to approximately 10 gauss in the anhydrous solid.

The resonance absorption of Li^7 was examined at 14.5 Mc/sec. in solutions of LiCl and $LiNO_3$, with the results given in Table IV. The addition of paramagnetic ions affects the proton relaxation time more drastically than it does that of Li^7, which is probably to be explained by the mutual repulsion of the positive ions, or by the existence of a "shield" of hydration around the Li ion. In the absence of paramagnetic ions the lithium relaxation time exceeds that of the protons, but

[24] R. L. Cummerow, D. Halliday, and G. E. Moore, Phys. Rev. 72, 1233 (1947).
[25] E. Zavoisky, J. Phys. USSR 9, 211 (1945).

not by as large a factor as the simplified theory would predict. This may be attributable to the effect of nuclear quadrupole interaction, a phenomenon which is much more conspicuous in the following example.

The deuteron and proton resonances were examined, at the same frequency, in a sample of 50 percent heavy water. The relaxation times observed for D^2 and H^1 were 0.5 and 3.0 sec., respectively. Contrary to what would be expected from the ratio of the magnetic moments, the deuteron relaxation time is the shorter. However, the interaction of the electric quadrupole moment of the deuteron with a fluctuating inhomogeneous electric field can bring about thermal relaxation. We omit the analysis of this process, which parallels closely the treatment of dipole-dipole interaction, and point out only that the interaction energy, in order of magnitude, is $eQ(\partial\mathcal{E}_z/\partial z)$, where Q is the nuclear quadrupole moment in cm²; this is to be compared with μH_{loc}. If the fluctuation spectrum of $\partial\mathcal{E}_z/\partial z$ is assumed similar to that of H_{loc}, the observed value of T_1 for the deuteron in our experiment can be accounted for by assuming a reasonable value of $\partial\mathcal{E}_z/\partial z$ in the D_2O or HDO molecule.¶¶¶ The quadrupole interaction is probably less effective in the case of the lithium ion because the immediate surroundings of the ion are electrically more symmetrical.

XIV. NUCLEAR RELAXATION IN HYDROGEN GAS

We have reported earlier[26] the observation of nuclear resonance absorption in hydrogen gas at room temperature, at pressures between 10 and 30 atmospheres. The low density of nuclei severely limited the accuracy of measurement, but it was possible to ascertain that the relaxation time at 10 atmospheres was approximately 0.015 second. T_1 appeared to increase with increasing pressure. The width of the line was less than 0.15 gauss.

The interpretation of these results is closely related to our theory of relaxation in liquids. In each case the key to the problem is the observation that the local field fluctuates at a rate much higher than ν_0. There are two distinct local magnetic fields at the position of one proton in the H_2 molecule. One arises from the rotation of the molecule as a whole; the other originates, as in the water molecule, in the magnetic moment of the other proton in the molecule. The corresponding interaction energies are fairly large and have been accurately measured in molecular beam experiments[27] where the interaction causes a splitting of the proton resonance into six components, in the lowest rotational state of *ortho*-hydrogen. The interactions involve the rotational quantum number J and its spatial quantization m_J. In a gas at ordinary pressures, m_J, and, to a lesser extent, J, may be expected to change frequently as a result of molecular collisions. If m_J were to change with each colli-

¶¶¶ A treatment of relaxation by quadrupole coupling will be found in the *Thesis* to be presented by one of the authors (N.B.) at the University of Leiden.

[26] E. M. Purcell, R. V. Pound, and N. Bloembergen, Phys. Rev. 70, 986 (1946).

TABLE III. Comparison of magnetic moments of ions inferred from nuclear relaxation measurements and from static susceptibility measurements.

Ion	μ_{eff} in Bohr magnetons (from nuclear relaxation experiments)	μ_{eff} in Bohr magnetons (from susceptibility measurements)*
Er^{+++}	8.5	9.4
Fe^{+++}	5.5	5.9
Cr^{+++}	4.1	3.8
Cu^{+++}	2.0	1.9
Ni^{++}	1.8	3.2
Co^{++}	1.1	4.5–5.3
$Fe(CN)_6^{---}$	0.1	2.4

* The numbers in this column are those listed by Gorter, reference 23, pp. 14 and 15, with the exception of the value for $Fe(CN)_6^{---}$ which is based on the measurements of L. C. Jackson, Proc. Phys. Soc. 50, 707 (1938).

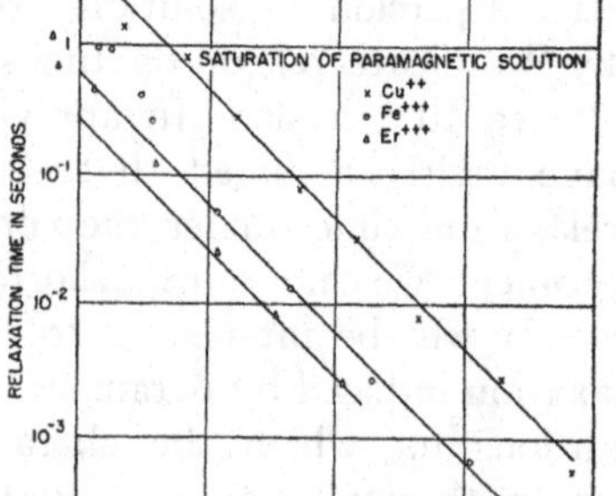

FIG. 15. Relaxation time T_1, for solutions of paramagnetic ions, measured at 29 Mc/sec.

[27] J. M. B. Kellogg, I. I. Rabi, N. F. Ramsey, and J. R. Zacharias, Phys. Rev. 57, 677 (1940).

TABLE IV.

Substance	(gram atom H / gram atom Li)	Relaxation time in sec. for Li^7	for H^1
$LiCl+H_2O$	5.8	1.75	0.4
$LiNO_3+H_2O$	14	2.7	1.1
$LiNO_3+H_2O+Fe(NO_3)_3$	8	0.11	0.0023
$LiCl+H_2O+CrCl_3$	6.5	0.24	0.0095
$LiCl+H_2O+CuSO_4$	6.6	0.18	0.013

sion the correlation time for the local fields would be the mean time between collisions, which in hydrogen at 10 atmospheres is approximately 10^{-11} second. The situation is thus similar to that in a liquid of low viscosity, in that $2\pi\nu\tau_c \ll 1$. We should therefore be prepared to find (a) that $T_1 \propto 1/\tau_c$; (b) that T_1 and T_2 are about equal, and very much larger than the reciprocal of the frequency splitting observed in the molecular beam experiment. Since we have identified τ_c with the time between collisions, T_1 should increase with increasing pressure, as observed. The implication of (b) is that the line should be very narrow, as observed, and that the line width should vary *inversely* with the pressure. This last effect would have been masked by inhomogeneities in the magnetic field, but the experimental results on closely analogous cases in liquids leave little room for doubt that "pressure narrowing" can occur.

The effect of the magnetic fields associated with neighboring molecules is negligible except at extremely high pressures, and in this respect the theoretical analysis is less uncertain than it was in the case of the liquids. We omit the derivation and present only the result, which was

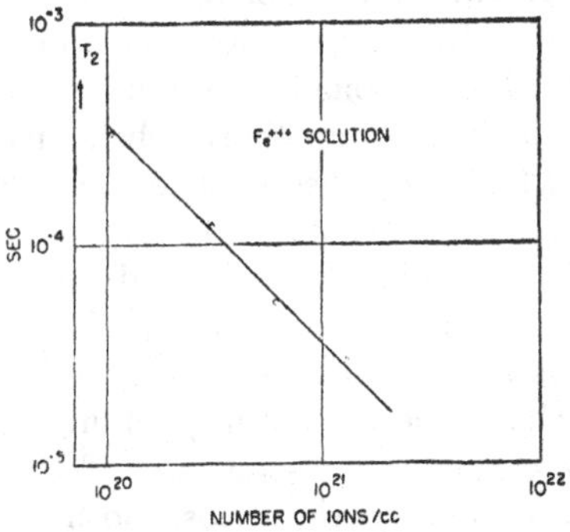

FIG. 16. The line-width parameter T_2, determined from measurements of line width in solutions containing a high concentration of ferric ions.

first obtained by J. S. Schwinger, and can be reached by a process similar to that followed in Section VII.

$$\frac{1}{T} = 2\tau_c\gamma^2\left[\frac{1}{3}H'^2J(J+1) + 3H''^2\frac{J(J+1)}{(2J-1)(2J+3)}\right]. \quad (55)$$

The field at one proton due to the molecular rotation is specified by means of the constant H' in Eq. (55), which has the value 27 gauss.[27] H'' is a measure of the field at one proton due to the other and has the value 34 gauss. At room temperature we have to do mainly with the state $J=1$; molecules with $J=0$, 2, etc., (para-hydrogen) are, of course, inert in the experiment. Identifying τ_c with the mean time between collisions calculated from gas kinetic data we find, using Eq. (55), $T_1=0.03$ seconds, at 10 atmospheres, in satisfactory agreement with experiment. The theoretical value will be *reduced* if m_J does not change in every collision, as assumed.

XV. NUCLEAR RELAXATION IN ICE

The process which we have invoked to explain nuclear relaxation in liquids has no counterpart in an ideal crystal. Nevertheless, many solids show evidence, in the form of specific heat anomalies or dielectric dispersion, of internal degrees of freedom other than vibration. We shall discuss only one example, which we have studied experimentally in some detail, that of ice.

Ice shows a marked dielectric dispersion at rather low frequencies. The residual entropy of ice also provides evidence for internal freedom which has been attributed[28] to the two available positions for the proton in the O–H–O bond. Without inquiring into the connection between these two phenomena, or into the details of the internal motion, we might assume that our parameter τ_c is proportional to the Debye characteristic time τ_D, deduced from dielectric dispersion data. The nuclear relaxation data for protons in ice have been plotted against τ_D in Figs. 17 and 18, the value of τ_D for each temper-

[28] Linus Pauling, *The Nature of the Chemical Bond* (Cornell University Press, Ithaca, New York, 1940), p. 302 ff.

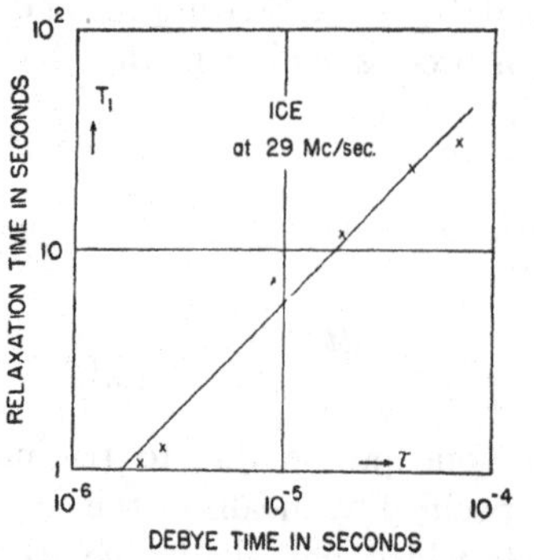

FIG. 17. Proton relaxation time in ice, plotted against the Debye relaxation time derived from dielectric dispersion data.

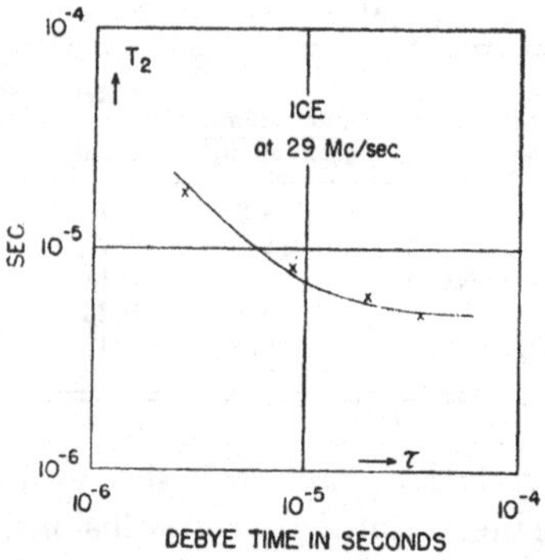

FIG. 18. T_2 for ice, plotted against the Debye relaxation time.

ature having been taken from the measurements of Wintsch.[21] Figure 17 shows T_1 for ice, in the temperature range $-2°C$ to $-40°C$. It is clear that we have to do with the ascending branch of the typical T_1 curve, and that the assumption $\tau_c \approx \tau_D$ was not a bad one. The line width increases as the temperature is lowered, as shown by the results plotted in Fig. 18. The curve in Fig. 18 was calculated by means of Eq. (52). It appears that T_2 is approaching, at the lowest temperatures, its asymptotic value. The limit approached by the line width is indicated as about 16 gauss, which is in good agreement with the value calculated from the crystal structure of ice, assuming the nuclei at rest.

It is interesting to note that although the relaxation time T_1, for water at 20°C, is about equal to that of ice at $-5°C$, the two cases represent opposite extremes of behavior. In water $2\pi\nu\tau_c \ll 1$, and the line is narrow; in ice $2\pi\nu\tau_c \gg 1$, and the line is broad.

XVI. CONCLUDING REMARKS

We have been able to account for the line widths and spin-lattice relaxation times observed in a number of liquids, in paramagnetic solutions, in hydrogen gas, and in ice. In particular, the extremely small width of the resonance line in ordinary liquids, and the accompanying inverse dependence of T_1 upon the viscosity of the liquid, appear as consequences of a situation in which all perturbing fields fluctuate in a random manner at a rate much higher than the nuclear Larmor frequency—a situation without an obvious parallel in atomic or molecular spectroscopy. The quantitative agreement is as good as we could have expected in view of the simplifying assumptions which were necessary in applying the theory to a particular substance, and the considerable uncertainty in the absolute magnitude of the experimental relaxation times. Not all of the features of the theory are thereby tested; for example, we cannot infer from the experimental data on water the relative magnitude of the inter- and intramolecular effects which were treated separately in Section X. Possibly with improvements in the accuracy of measurement of relaxation times such points could be tested. On the other hand, further refinement of the theory, in the direction of a more faithful representation of the molecule and its immediate environment, would appear to be neither easy nor, at present, very profitable, except perhaps in the case of a structure like that of ice which is well determined.

Since rotational and translational degrees of freedom play such an important role in determining line widths and relaxation times, the study of nuclear resonances provides another means for investigating those phenomena in solids which have been ascribed to the onset, or cessation, of a particular type of internal motion. Experiments of this sort have already been reported by Bitter and collaborators.[29]

An understanding of the factors controlling T_1 and T_2 should be helpful in planning experiments to detect nuclear resonances and to measure nuclear gyromagnetic ratios. Both T_1 and

[29] F. Bitter, N. L. Alpert, H. L. Poss, C. G. Lehr, and S. T. Lin, Phys. Rev. 71, 738 (1947); N. L. Alpert, Phys. Rev. 72, 637 (1947).

T_2 must be taken into account in estimating the sensitivity attainable.

Finally, we want to emphasize that our understanding of the main features of the spin-lattice relaxation process does not extend to tightly bound crystalline structures in which, presumably, only vibrational degrees of freedom are active. We have observed in a fluorite crystal at room temperature a relaxation time of about 9 seconds. This alone is surprising, for the Waller theory applied to this case predicts a relaxation time of the order of one hour. A more remarkable result is that reported by Rollin and Hatton,[5] who find that the relaxation time in lithium fluoride remains of the order of seconds down to 2°K.

ACKNOWLEDGMENT

The authors have benefited from numerous illuminating and stimulating discussions with Professor H. C. Torrey of Rutgers University, Professor J. H. Van Vleck, Professor J. S. Schwinger, and with Professor C. J. Gorter of the University of Leiden, Visiting Lecturer at Harvard in the summer of 1947.

The work described in this paper was supported by a Frederick Gardner Cottrell Special Grant-in-Aid from the Research Corporation.

APPENDIX

The intensity of the radiofrequency signal resulting from nuclear paramagnetic resonance absorption varies over a very wide range depending on the various parameters descriptive of the sample. For a large signal such as that produced by the protons in water, little attention need be given to the detection technique. Other samples often require extreme care lest a very small signal be lost in noise. To see what experimental conditions are imposed by the nature of the sample it is useful to develop a formula that expresses the signal-to-noise ratio as a function of the parameters descriptive of the apparatus and of the sample.

It might be well to point out that, though here we work entirely with the nuclear absorption, the detectability of the nuclear dispersion, observed with an apparatus of the bridge type when the residual signal is due to a phase unbalance, is almost exactly the same. As shown by Bloch,[6] the main difference between the two effects is that the signal caused by the dispersion effect increases asymptotically to a maximum as the r-f field H_1 is increased, whereas for the signal caused by the absorption there is an optimum value for H_1. As a result, one of the effects may be the more useful for some purposes and the other for others. For conducting a search for unknown lines, an apparatus using the dispersion effect might be best since the search need not be made to cover a range of values of r-f field strength and modulation amplitude. For investigating line shape and relaxation phenomena the absorption effect gives results that are probably easier to interpret.

In all experimental arrangements so far employed the sample is contained in a resonant circuit, most often in the coil of a shunt resonant coil and condenser. In Section I it was shown that a quantity, called there the Q of the absorption, was given by

$$1/Q \equiv \delta = \gamma^2 h^2 N_0 I(I+1)\nu_0 T_2^*/3kT. \qquad (56)$$

This quantity is the ratio of the energy lost per radian from the r-f field to the system of nuclear spins to the energy stored in the oscillating r-f field. Thus it is the same kind of a quantity as the ordinary Q used as a figure of merit for simple coils and condensers. The effect of the nuclear resonance on the circuit is, then, to reduce the Q of the circuit from Q_0 to a value Q_r given by

$$1/Q_r = (1/Q_0) + \zeta\delta. \qquad (57)$$

The quantity ζ is a filling factor for the circuit determined by the fraction of the total r-f magnetic energy that is actually stored in the space occupied by the sample.

To obtain the largest signal for detection with a circuit of a given Q, it is now apparent that the addition of the loss attributable to δ should produce the largest possible change in the voltage across the tuned circuit. This is achieved when the generator driving the circuit is of the constant current type, or, in other words, if the generator has a very high internal impedance compared to the impedance of the tuned circuit at resonance. To achieve this the very small coupling condensers are used on the input sides of the circuits described in Section II.

Consider the use of a modulated magnetic field. At times when the field is such that the nuclear resonance does not occur the voltage across the resonant circuit is proportional to Q_0. At the peak of the resonance, the voltage is proportional to Q_r which, since δ is very small compared to $1/Q_0$, may be written

$$Q_r = Q_0(1 - \zeta\delta Q_0).$$

The resulting modulated voltage may be expanded in Fourier series if the shape of the absorption line is known, the signal indicating the presence of nuclear absorption being contained in the sidebands. For observation of the effect on an oscillograph all of these should be taken into account but for a system using the "lock-in" amplifier, sensitive only in the immediate neighborhood of a single frequency, only a single pair of sidebands differing in frequency from the driving frequency by just the modulation frequency need be considered. The signal-to-noise ratio computed from these also gives an estimate of the usefulness of an oscillograph as an indicator, if the noise bandwidth of the system is suitably chosen.

The voltage across the tuned circuit may be considered to be given by

$$E = (2RP_0)^{\frac{1}{2}}[\cos\omega_0 t + \tfrac{1}{4}A\zeta\delta Q_0 \cos(\omega_0 + \omega_m)t + \tfrac{1}{4}A\zeta\delta Q_0 \cos(\omega_0 - \omega_m)t],$$

where R is the shunt resistance of the resonant circuit, P_0 is the power dissipated in the resonant circuit, ω_0 is 2π times the radiofrequency, ω_m is 2π times the modulation frequency, and A is a constant, of the order of unity, that depends on the exact shape of the line, the amount of the modulation swing and the shape of the modulating wave. Since this voltage is for the unloaded circuit, the power available as a signal in the sidebands is

$$P_s = P_0(A\zeta\delta Q_0)^2/32. \qquad (58)$$

It follows from the analysis, in Section IV, of the saturation effect that there is an optimum strength of the driving signal field H_1, given by

$$(\gamma H_1)^2 T_1 T_2^* = 1. \qquad (59)$$

For this magnitude of the driving signal the value of δ is reduced to one-half its value in Eq. (56) because of partial saturation. Remembering that H_1 is one-half the amplitude of the oscillating field, Eq. (59) may be used to compute the optimum value for P_0, giving

$$P_0 = \omega_0 V_c / 2\pi\gamma^2 T_1 T_2^* Q_0, \qquad (60)$$

where V_c is an effective volume for the part of the circuit in which magnetic energy is stored. For a simple solenoid this is very nearly the volume inside the coil, since the field outside is weak, and the energy is proportional to the square of the field. This equation may be put into many other forms. One useful form is

$$P_0 = H_0 H_2 V_c / 2Q_0 T_1,$$

where H_2 is the full width in gauss of the absorption line between points of half the peak absorption, assuming the damped-oscillator line shape. This shows that for lines having widths determined by inhomogeneity in the magnetic field, the input power required to produce saturation depends on *none* of the properties of the sample except T_1. Putting Eq. (60) into Eq. (58) for P_0 and using one-half the value of δ given by Eq. (56), one gets, for the total signal power available,

$$P_s = V_c Q_0 (A\zeta)^2 h^4 N_0^2 \gamma^2 \nu_0^3 T_2^* \times [I(I+1)]^2 / 1152 T_1 (kT)^2.$$

If the detection apparatus is described by an effective noise band width $2B$ (B being the band width at the indicating instrument) and an effective over-all noise figure,[30] F, the resulting signal-to-noise power ratio is

$$P_s/P_n = V_c Q_0 (A\zeta)^2 h^4 N_0^2 \gamma^2 \nu_0^3 T_2^* \times [I(I+1)]^2 / 2304 (kT)^2 (kTBF) T_1. \qquad (61)$$

The two factors containing kT are kept separate to emphasize that the noise figure of the detector is not independent of the temperature. If F is large, most of the noise originates in the apparatus and not in the circuit connected to its input terminals.

In most cases the output indicator will respond linearly to the signal and noise amplitudes and, thus, the ratio of the response to the absorption line to the r.m.s. noise fluctuation is

$$A_s/A_n = V_c^{\frac{1}{2}} Q_0^{\frac{1}{2}} A\zeta h^2 N_0 \gamma \nu_0^{\frac{3}{2}} T_2^{*\frac{1}{2}} I \times (I+1) / 48 kT (kTBF)^{\frac{1}{2}} T_1^{\frac{1}{2}}. \qquad (62)$$

---

[30] H. T. Friis, Proc. I.R.E. **32**, 419 (1944).

To demonstrate completely the dependence on the frequency and on the dimensions of the circuit an empirical relation showing the dependence of Q_0 on these quantities may be inserted. Thus

$$Q_0=0.037\nu_0^{1/2}V_c^{1/3}, \qquad (63)$$

where ν_0 is to be measured in c.p.s. and V_c in cm^3 may be used. The numerical constant is based on experimental values of Q for coils of about 1-cm^3 volume in the region of 30 Mc/sec.

Often the frequency of operation is determined by the magnetic field available. In such a case, using Eq. (63) for Q_0, Eq. (62) becomes

$$A_s/A_n\approx 1.62\times 10^{-4}V_c^{1/2}\zeta A h^2 N_0 \gamma^{11/4} H_0^{7/4} \times T_2^{*1/2} I(I+1)/kT(kTBF)^{1/2}T_1^{1/2}. \qquad (64)$$

Observe that the signal-to-noise amplitude ratio is proportional to $\gamma^{11/4}$, for a given field, concentration, volume of the circuit, filling factor for the circuit, and ratio of T_1/T_2^*. The quantity $V_c\zeta N_0$ is the total number of nuclei present. Thus Eq. (64) shows that the signal-to-noise ratio is inversely proportional to $V_c^{1/2}$ for a given number of nuclei and for all the other quantities constant. This favors as high a concentration as possible. Note, however, that if the sample is a crystal with a rigid lattice so that T_2^* is determined by the static dipole-dipole coupling, it is preferable to have the nuclei spread out. In this event, the signal-to-noise ratio is proportional to $V_c^{1/6}$, ζ and the total number of nuclei in the sample being held constant. This is true only if, as the concentration is reduced, interactions other than the spin-spin interaction among the nuclei under investigation remain negligible.

The signals produced by the nuclei in liquids, if the relaxation time T_1 can be made so short as to be an important contributant to the line width, are particularly strong. This condition can be obtained by the use of solutions containing paramagnetic ions to shorten T_1 sufficiently to make $T_2^*/T_1\approx 1$. For this case, Eq. (64) may be written as,

$$A_s/A_n\approx 1.2\times 10^{-26}NH_0^{7/4}\mu^{11/4} \times (I+1)/V_c^{1/2}B^{1/2}I^{7/4}, \qquad (65)$$

where μ is the magnetic moment in nuclear magnetons. This gives a signal-to-noise ratio of 1.4×10^6 for the protons in a 1-cm^3 water sample with a magnetic field of 15,000 gauss and a band width of 1 c.p.s. Thus even with an oscillograph and a high modulation frequency, requiring a noise band width of perhaps 5 kc/sec., a signal of the order of 2×10^4 times the r.m.s. amplitude of noise may be expected. On the other hand, a 1-molar solution of nuclei having magnetic moments of 0.1 nuclear magnetons produces a signal-to-noise ratio of only about 1.3 when a 1-cm^3 sample in a 15,000-gauss field is used with an apparatus having a 1-c.p.s. band width. For this magnetic field, band width, and volume of circuit, about 10^{18} nuclei having spins of $\frac{1}{2}$ and magnetic moments of one nuclear magneton would be required to give a signal equal to noise.

Spin Echoes*†

E. L. Hahn‡
Physics Department, University of Illinois, Urbana, Illinois
(Received May 22, 1950)

Intense radiofrequency power in the form of pulses is applied to an ensemble of spins in a liquid placed in a large static magnetic field H_0. The frequency of the pulsed r-f power satisfies the condition for nuclear magnetic resonance, and the pulses last for times which are short compared with the time in which the nutating macroscopic magnetic moment of the entire spin ensemble can decay. After removal of the pulses a non-equilibrium configuration of isochromatic macroscopic moments remains in which the moment vectors precess freely. Each moment vector has a magnitude at a given precession frequency which is determined by the distribution of Larmor frequencies imposed upon the ensemble by inhomogeneities in H_0. At times determined by pulse sequences applied in the past the constructive interference of these moment vectors gives rise to observable spontaneous nuclear induction signals. The properties and underlying principles of these spin echo signals are discussed with use of the Bloch theory. Relaxation times are measured directly and accurately from the measurement of echo amplitudes. An analysis includes the effect on relaxation measurements of the self-diffusion of liquid molecules which contain resonant nuclei. Preliminary studies are made of several effects associated with spin echoes, including the observed shifts in magnetic resonance frequency of spins due to magnetic shielding of nuclei contained in molecules.

I. INTRODUCTION

In nuclear magnetic resonance phenomena the nuclear spin systems have relaxation times varying from a few microseconds to times greater than this by several orders of magnitude. Any continuous Larmor precession of the spin ensemble which takes place in a static magnetic field is finally interrupted by field perturbations due to neighbors in the lattice. The time for which this precession maintains phase memory has been called the spin-spin or total relaxation time, and is denoted by T_2. Since T_2 is in general large compared with the short response time of radiofrequency and pulse techniques, a new method for obtaining nuclear induction becomes possible. If, at the resonance condition, the ensemble at thermal equilibrium is subjected to an intense r-f pulse which is short compared to T_2, the macroscopic magnetic moment due to the ensemble acquires a non-equilibrium orientation after the driving pulse is removed. On this basis Bloch[1] has pointed out that a transient nuclear induction signal should be observed immediately following the pulse as the macroscopic magnetic moment precesses freely in the applied static magnetic field. This effect has already been reported[2] and is closely related to another effect, given the name of "spin echoes," which is under consideration in this investigation. These echoes refer to spontaneous nuclear induction signals which are observed to appear due to the constructive interference of precessing macroscopic moment vectors after more than one r-f pulse has been applied. It is the purpose of this paper to describe and analyze these effects due to free Larmor precession in order to show that they can be applied for the measurement of nuclear magnetic resonance phenomena, particularly relaxation times, in a manner which is simple and direct.

II. FEATURES OF NUCLEAR INDUCTION METHODS

(A) Previous Resonance Techniques (Forced Motion)

The chief method for obtaining nuclear magnetic resonance has been one whereby nuclear induction signals are observed while an ensemble of nuclear spins is perturbed by a small radiofrequency magnetic field. A large d.c. magnetic field $\bar{H}_0$ establishes a net spin population at thermal equilibrium which provides a macroscopic magnetic moment $\bar{M}_0$ oriented parallel to $\bar{H}_0$. The forced motion of $\bar{M}_0$ is brought about by subjecting the spin ensemble to a rotating radiofrequency field $\bar{H}_1$ normal to $\bar{H}_0$ at the resonance condition $\omega=\omega_0=\gamma H_0$, where γ is the gyromagnetic ratio, ω is the angular radiofrequency, and ω_0 is the Larmor frequency. The techniques which obtain resonance under this condition provide for the application of a driving r-f voltage to an LC circuit tuned to the Larmor frequency. The sample containing the nuclear spins is placed in a coil which is the inductance of the tuned circuit. At resonance a small nuclear signal is induced in the coil and is superimposed upon an existing r-f carrier signal of relatively high intensity. In order to detect this small nuclear signal the r-f carrier voltage is reduced to a low level by a balancing method if the LC circuit is driven by an external oscillator,[1,3] or the LC circuit may be the tuned circuit of an oscillator which is designed to change its level of operation when nuclear resonance absorption occurs.[4,5] In general, a condition exists whereby transitions induced by H_1, which tend

* This research was supported in part by the ONR.
† Reported at the Chicago Meeting of the American Physical Society, November, 1949; Phys. Rev. 77, 746 (1950).
‡ Present address: Physics Dept. Stanford University, Stanford, California.
[1] F. Bloch, Phys. Rev. 70, 460 (1946).
[2] E. L. Hahn, Phys. Rev. 77, 297 (1950).
[3] Bloembergen, Purcell, and Pound, Phys. Rev. 73, 679 (1948).
[4] R. V. Pound, Phys. Rev. 72, 527 (1947); R. V. Pound and W. D. Knight, Rev. Sci. Inst. 21, 219 (1950).
[5] A. Roberts, Rev. Sci. Inst. 18, 845 (1947).

to upset the thermal equilibrium of the spins, are in competition with processes of emission due to lattice perturbations which tend to restore equilibrium. Spin relaxation phenomena, which are measured in terms of the relaxation times T_2 and T_1 (spin-lattice), must be distinguished simultaneously from effects due to the influence of r-f absorption. Consequently the study of resonance absorption line shapes, intensities, and transients must carefully take into account the intensity of H_1 and the manner in which resonance is obtained.

In practice, resonance takes place over a range of frequencies determined by the inhomogeneity of H_0 throughout the sample. For resonances concerning nuclei in liquids it is generally found that the natural line width given by $1/T_2$ on a frequency scale is much narrower than the spread in Larmor frequencies caused by external field inhomogeneities, whereas the converse is true in solids. Therefore steady state resonance lines due to nuclei in liquids are artificially broadened; transient signals are modified in shape and have decay times which are shorter than would otherwise be determined by T_1 and T_2.

(B) Nuclear Induction Due to Free Larmor Precession

The observation of transient nuclear induction signals due to free Larmor precession becomes possible at the resonance condition described above if the r-f power is now applied in the form of intense, short pulses. The r-f inductive coil which surrounds the sample is first excited by the applied pulses and thereafter receives spontaneous r-f signals at the Larmor frequency due to the precessing nuclei. In particular, the echo effect is brought about by subjecting the sample to two r-f pulses in succession (the simplest case) at pulse width $t_w<\tau<T_1$, T_2, where τ is the time interval between pulses. At time τ after the leading edge of the second pulse the echo signal appears. Since H_1 is absent while these signals are observed, no particular attention need be given to elaborate procedures for eliminating receiver saturation effects (as must be done in the forced motion technique) providing that T_2 is large enough to permit observation of echoes at times when the receiver has recovered from saturation due to the pulses. Because the T_2 of nuclei in liquids is generally large enough to favor this condition, the technique for obtaining echoes in this work has been largely confined to the magnetic resonance of nuclei in liquid compounds. Preliminary observations of free induction signals in solids, where T_2 becomes of the order of microseconds, indicate again, however, that procedures must be undertaken for preventing receiver saturation due to intense pulses.

For spin ensembles in liquids it will be shown that the analysis of observed echo signals yields direct information about T_1 and T_2 without requiring consideration of the effect of H_1 on the measured decay of the signal. Because of the inhomogeneity in H_0, the self diffusion of "spin-containing liquid molecules" brings about an attenuation of observed transient signals in addition to the decay due to T_1 and T_2. However, this is only serious for liquids of rather low viscosity which also have a large T_2 for the resonant nuclei concerned, whereas in conventional resonance methods (forced motion), field inhomogeneities obscure a direct measurement of T_2 in liquids over a much wider range of viscosities. The free motion technique, which will hereafter be denoted by the method of spin echoes or free nuclear induction, also reveals in a unique manner differences in resonant frequency between nuclear spins of the same species located in different parts of a single molecule or in different molecules. Such differences have been observed by previous resonance methods,[6,7] and the echo technique gives at least as good a resolution in the measurement of small shifts.

In this investigation the in-phase condition of a precessing spin ensemble is considered to be eventually destroyed because of lattice perturbations which limit the phase memory time of Larmor precession. The precession of an individual spin may be interrupted either because its energy of precession is transferred to neighboring spins in a time $\sim T_2''$ (mutual spin-spin flipping), or because this energy is transferred to the lattice as thermal energy in a time $\sim T_1$. The spread in Larmor frequencies, due to local "z magnetic field" fluctuations at the position of the nucleus caused by neighboring spins and paramagnetic ions, also serves to disturb phase memory (H_0 is in the z direction). In a formal treatment[3] this effect is considered in conjunction with the interaction giving rise to T_2'', and a general relaxation time T_2' is formulated. The inverse of the total relaxation time, $1/T_2 \sim 1/T_2' + 1/T_1$, therefore becomes the uncertainty in frequency of a precessing spin, which can then acquire an uncertainty in phase of the order of one radian in time T_2.

It will be convenient to describe the formation of free induction signals by considering the free precession of individual macroscopic moment vectors $\bar{M}_0(\omega_0)$. Each of these vectors has a magnitude at a given ω_0 which is determined by a z magnetic field distribution imposed upon the ensemble by inhomogeneities in H_0. In this spectral distribution $\bar{M}_0(\omega_0)$ can be defined as an isochromatic macroscopic moment which consists of an ensemble of nuclear moments precessing in phase at the assigned frequency ω_0. The precessional motion of any $\bar{M}_0(\omega_0)$ vector about the total magnetic field (with or without r-f pulses) can be followed regardless of what phases the individual isochromatic moments have with respect to one another throughout the entire spectrum. At the time a short r-f pulse initiates the free precession of $\bar{M}_0(\omega_0)$ from a classically non-precessing initial condition at thermal equilibrium, relaxation and possibly diffusion processes begin to diminish the magnitude of

[6] W. G. Proctor and F. C. Yu, Phys. Rev. 77, 717 (1950).
[7] W. C. Dickinson, Phys. Rev. 77, 736 (1950).

the precessing vector $\bar{M}_0(\omega_0)$ as the individual nuclear spins get out of phase with one another or return to thermal equilibrium.

The actual H_0 field which persists at the position of a precessing spin accounts for a given ω_0 and hence for a given $\bar{M}_0(\omega_0)$. In liquids this persisting field and the way it is distributed over the sample is taken to be entirely due to the magnet; any contributions to the local field at the nucleus by neighbors in the lattice average out in a time short compared to a Larmor period. Free induction signals from nuclei in solids, however, indicate that a broad distribution in H_0 exists (compared to a relatively homogeneous external field) which is determined by fixed lattice neighbors, and now this local field distribution does not average out.

The description of free induction effects is simplified by transforming to a coordinate system in which the $x'y'$ plane (Fig. 1) is rotating at some convenient reference angular frequency ω'. This frequency is usually chosen to be the center frequency of a given distribution of isochromatic moments, where the distribution is typically described by a Gaussian or Lorentz (damped oscillator) function. In the next section definite properties of the rotating coordinate representation are presented. The precessional motion as viewed in the rotating system is conveniently followed when (1) $\bar{M}_0(\Delta\omega)$ undergoes forced transient motion during the driving pulse, and (2) when the $\bar{M}_0(\Delta\omega)$ vectors precess freely, where $\Delta\omega=\omega_0-\omega'$ and $\bar{M}_0(\omega_0)=\bar{M}_0(\Delta\omega)$. The condition in (1) has already been analyzed theoretically and experimentally.[8,9,10] Although it is strictly a condition in which $\bar{M}_0(\Delta\omega)$ precesses about the total field $\bar{H}_0+\bar{H}_1$, as viewed in the laboratory system, it has been characterized by the fact that not only does $\bar{M}_0(\Delta\omega)$ appear to precess about the z axis at a high Larmor frequency, but also it appears to nutate with respect to the z axis at a much lower frequency.[11]

III. THEORY AND APPLICATIONS

(A) The Moving Coordinate Representation

Consider the torque equation, with no damping, which describes the precession of $\bar{M}$ as seen in the laboratory system:

$$d\bar{M}/dt=\gamma(\bar{M}\times\bar{H}), \tag{1}$$

where $\bar{H}$ is the total magnetic field. During the application of r-f pulses, $\bar{H}=\bar{H}_0+\bar{H}_1$; and during the free precession of $\bar{M}$ in the absence of pulses, $\bar{H}=\bar{H}_0$. During a pulse it is convenient to transform to a moving coordinate system in which $\omega'=\omega$, and $\bar{H}_1$ is chosen to lie along the x' axis. It will be pointed out, however, that, regardless of the choice of direction of $\bar{H}_1$ in the $x'y'$ plane, the description of the spin echo model presented later is not affected, except under a very special condition. If $D\bar{M}/dt$ is the observed torque in the moving coordinate system, then by a well-known transformation,

$$d\bar{M}/dt=D\bar{M}/dt+\bar{\omega}\times\bar{M}, \tag{2}$$

where $\bar{M}\equiv M(u, v, M_z)$ and $\bar{H}\equiv H(H_1, 0, H_0)$. Combining (1) and (2) we obtain

$$D\bar{M}/dt=\bar{M}\times(\Delta\bar{\omega}+\bar{\omega}_1) \tag{3}$$

as the torque in the moving system during a pulse. The vector $\bar{M}$ is identified with the isochromatic moment $\bar{M}_0(\Delta\omega)$ which appears to precess about the effective field vector $(\Delta\bar{\omega}+\bar{\omega}_1)/\gamma$. Let $(\Delta\omega)_{\frac{1}{2}}$ be the width at half-maximum of an assumed function which describes the distribution of $\bar{M}_0(\Delta\omega)$ over the inhomogeneous external field, and let ω' be the center frequency of this distribution. If, during a pulse, the inequality $1/t_w$, $\omega_1\gg(\Delta\omega)_{\frac{1}{2}}$ applies at resonance $(\omega\approx\omega')$, then the precession of any $\bar{M}_0(\Delta\omega)$ vector will appear to take place practically about the $\bar{\omega}_1$ vector in the moving system. This precessional frequency is given by $\omega_1=\gamma H_1$ (of the order of kilocycles) which appears in the laboratory system as a frequency of nutation superimposed upon a high Larmor precession frequency (~30 Mc). In the rotating system any $\bar{M}_0(\Delta\omega)$ vector will precess in a cone whose axis is in the direction of $\bar{H}_1$ and whose angle is determined by the angle between $\bar{M}_0(\Delta\omega)$ and $\bar{H}_1$ at the time $\bar{H}_1$ is suddenly applied. When $\bar{H}_1$ is suddenly removed, the vector $\bar{M}_0(\Delta\omega)$ is oriented at a fixed angle θ with respect to the z axis, and precesses freely at angular frequency $\Delta\omega$ about the effective magnetic field $\Delta\bar{\omega}/\gamma$ along the z axis. The angle θ will be determined by $\omega_1 t_w$ and the initial conditions established by successive pulses applied in the past.

(B) Simple Vector Model of the Spin Echo

For spin ensembles in liquids a simple vector model will account for the manner in which two applied r-f pulses establish a given spectral distribution of moment components in the $x'y'$ plane, where the axis of the inductive coil is oriented. This distribution then freely precesses to form, by constructive interference, a resultant "echo" in the $x'y'$ plane. This is formulated by integrating a general expression for the $x'y'$ component of the isochromatic moment over all frequencies $\Delta\omega$ imposed by H_0 field inhomogeneities. Purcell[12] has suggested a three-dimensional model of the echo, Fig. 1, which arises in a special case. At $t=0$, when $\bar{H}_1$ is suddenly applied, $\bar{M}_0(\Delta\omega)$ is at thermal equilibrium, aligned parallel to $\bar{H}_0$ along the z axis. During time t_w

[8] N. Bloembergen, *Nuclear Magnetic Relaxation* (Martinus Nijhoff, The Hague, 1948).

[9] H. C. Torrey, Phys. Rev. 76, 1059 (1949).

[10] E. L. Hahn, Phys. Rev. 76, 461 (1949).

[11] This is observed to come about in the laboratory system as the resonance absorption mode becomes modulated at the low nutation frequency. Classically speaking, the term nutation is applied only to the physical top, in which the presence of angular momentum about an axis other than the spin axis is responsible for the nutation. Although a nuclear spin possesses extremely negligible angular momentum about any axis other than its spin axis, the term nutation is convenient to retain here in order to refer to the tipping motion of $\bar{M}_0(\Delta\omega)$ with respect to the z axis.

[12] Private communication.

of the first pulse, let $\bar{M}_0(\Delta\omega)$ precess an angle $\omega_1 t_w = \pi/2$ about $\bar{H}_1$, so that all moment vectors in the spectrum will have nutated into the $x'y'$ plane. Let $\tau \gg 1/(\Delta\omega)_{1/2}$, $T_1 = T_2 = \infty$, and assume a rectangular spectrum over $\Delta\omega$, i.e., $g(\Delta\omega) = \text{const.}$, where $g[(\Delta\omega)_{1/2}] = 0$. During time $t_w \leqslant t \leqslant \tau$, the various isochromatic moment pairs $M_0(+|\Delta\omega|)$, $M_0(-|\Delta\omega|)$, will precess at frequency $\Delta\omega$, maintaining a symmetry about the y' axis but rotating in opposite directions. These precessing moments will attain an isotropic distribution in a time $\gtrsim 2\pi/(\Delta\omega)_{1/2}$ prior to which a free induction decay is observed.[2] At time τ the second r-f pulse, identical with the first one, will rotate the moment pairs from angular positions $\varphi = 3\pi/2 \pm |\Delta\omega|\tau$, $\theta = \pi/2$ to $\varphi = (0, \pi)$, $\theta = \pi - |\Delta\omega|\tau$ in spherical coordinates. During the time interval $\tau + t_w \leqslant t < 2\tau$ all moment vectors interfere destructively with one another and distribute themselves isotropically over a unit sphere until the time $t \approx 2\tau$ when they interfere constructively. At time 2τ all of the moment vectors will have again precessed angles $\Delta\omega\tau$ respectively from their positions at $t = \tau + t_w$ so that they terminate in a figure eight pattern whose equation is $\theta = \varphi$. Free induction for $t \geqslant \tau + t_w$ will be obtained from the linearly polarized component of magnetization

$$v(\Delta\omega, t) = M_0 \sin\Delta\omega\tau \sin\Delta\omega(t-\tau) \tag{4}$$

along the y' axis. The observed induction voltage will be due to the integrated precessing moment

$$V(t) = \int_{-(\Delta\omega)_{1/2}}^{(\Delta\omega)_{1/2}} g(\Delta\omega) v(\Delta\omega, t) d(\Delta\omega), \tag{5}$$

where

$$\int_{-(\Delta\omega)_{1/2}}^{(\Delta\omega)_{1/2}} g(\Delta\omega) d(\Delta\omega) = 1.$$

Therefore, from (4) and (5) we obtain

$$V(t) = \frac{M_0}{2}\left[\frac{\sin(\Delta\omega)_{1/2}(t-2\tau)}{(\Delta\omega)_{1/2}(t-2\tau)} - \frac{\sin(\Delta\omega)_{1/2} t}{(\Delta\omega)_{1/2} t}\right]. \tag{6}$$

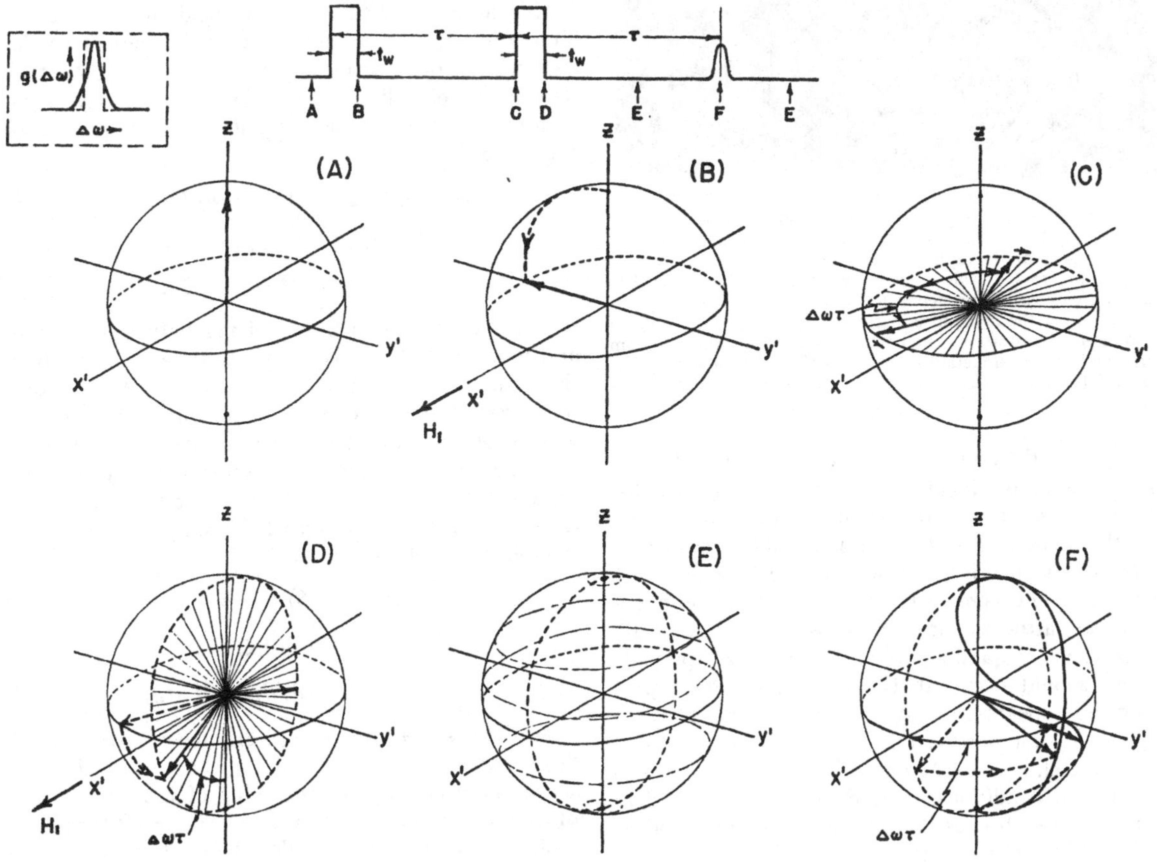

$\omega_1 \gg (\Delta\omega)_{1/2}$, $t_w \ll \tau < T_1, T_2$, $\omega_1 t_w = \frac{\pi}{2}$

FIG. 1. For the pulse condition $\omega_1 t_w = \pi/2$, the formation of the eight-ball echo pattern is shown in the coordinate system rotating at angular frequency ω. The moment vector monochromats are allowed to ravel completely in a time $\tau \gg 1/(\Delta\omega)_{1/2}$ before the second pulse is applied. The echo gives maximum available amplitude at $\omega_1 t_w = 2\pi/3$.

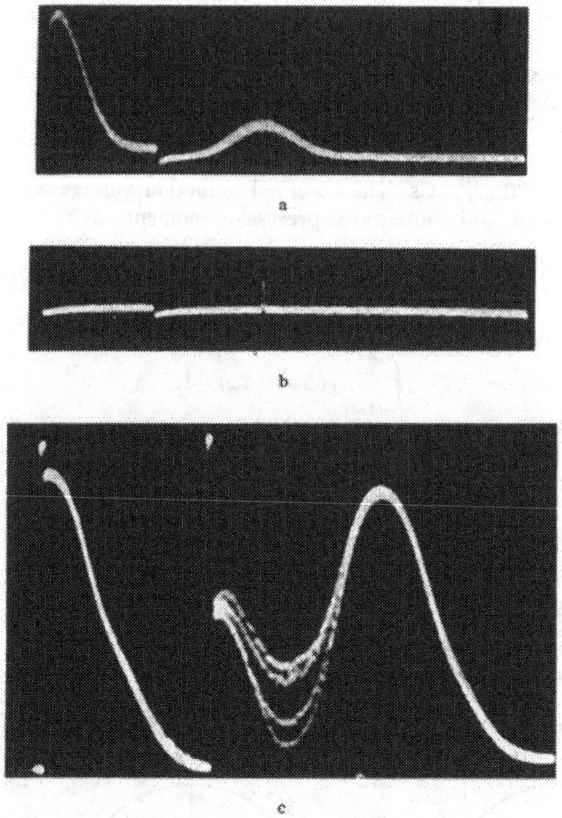

FIG. 2. Oscillographic traces for proton echoes in glycerine. The two upper photographs indicate broad and narrow signals corresponding to H_0 fields of good and poor homogeneity. The pulses, scarcely visible, are separated by 0.0005 sec. The induction decay following the first pulse in the top trace has an initial dip due to receiver saturation. The bottom photograph shows random interference of the induction decay with the echo for several exposures. The two r-f pulses are phase incoherent relative to one another.

According to the first term on the right side of (6) the echo maximum occurs at $t=2\tau$, and the signal lasts for $\sim 4\pi/(\Delta\omega)_{\frac{1}{2}}$ seconds. No free induction is predicted after the second pulse for this particular case, which is illustrated in Fig. 2 (top). For extremely large $(\Delta\omega)_{\frac{1}{2}}$ the echo becomes very sharp and the free induction decay after the first pulse becomes practically unobservable. Equation (6) indicates that periodic maxima should occur at times $2\pi/(\Delta\omega)_{\frac{1}{2}}$ sec. apart during the appearance of free induction signals. These maxima are not observed in general for this reason, because the choice of $g(\Delta\omega)$ here does not correspond to experimental conditions. A modulation is observed in particular cases because of an entirely different effect which will be discussed later.

(C) General Analysis

The echo effect will now be treated in a general way, after which some of the simplifying assumptions outlined for the very simple case just described will be applied. By making use of Bloch's equations[1] and choosing $g(\Delta\omega)$ to approximate the actual distribution of spins over H_0, the decay of echo signals due to T_1, T_2, and self-diffusion (in the case of some liquids) can be accounted for. As in the case illustrated above, $\bar{H}_1$ will be chosen to lie along the x' axis in the rotating system for both pulses. Actually $\bar{H}_1$ may appear in any possible position in the $x'y'$ plane during the second pulse since the r-f is not necessarily coherent for both pulses. However, free induction signals will be independent of this random condition as long as $\tau \gg 1/(\Delta\omega)_{\frac{1}{2}}$.[13] This signifies that free induction decay following a single pulse does not interfere with the echo (see Fig. 2, bottom, where this interference effect is shown). Ordinarily the scalar differential equations obtained from (3) are written to include additional torque terms due to relaxation according to Bloch. In the case of echo phenomena it is found that nuclear signals due to precessing nuclear moments contained in liquid molecules (particularly of low viscosity) are not only attenuated by the influence of T_1 and T_2, but also suffer a decay due to self-diffusion of the molecules into differing local fields established by external field inhomogeneities. Consequently, the phase memory of Larmor precession can be destroyed artificially to an appreciable extent. The effect of self-diffusion will be qualitatively accounted for by using Bloch's equations with a diffusion term added:

$$du/dt+[\Delta\omega+\delta(t)]v=-u/T_2 \qquad (7\text{-A})$$

$$dv/dt-[\Delta\omega+\delta(t)]u+\omega_1 M_z=-v/T_2 \qquad (7\text{-B})$$

$$dM_z/dt-\omega_1 v=-(M_z-M_0)/T_1. \qquad (7\text{-C})$$

u and v are the components of magnetization parallel and normal to $\bar{H}_1$ respectively. As time increases from the point where the first pulse is applied, $\delta(t)$ is taken to represent, due to diffusion, the shift in Larmor frequency of the u and v components away from the initial value of $\Delta\omega$. If the decay terms during a pulse are neglected, since t_w is very short compared to all decay time constants, the motion will be simply described by the following solutions of (7):

$$u(t)=(\Delta\omega/\beta)AQ+u(t_i) \qquad (8\text{-A})$$

$$v(t)=A\sin(\beta t+\xi) \qquad (8\text{-B})$$

$$M_z(t)=-(\omega_1 A/\beta)Q+M_z(t_i), \qquad (8\text{-C})$$

where $Q=\cos(\beta t+\xi)-\cos(\beta t_i+\xi)$ and $\beta=[(\Delta\omega)^2+\omega_1{}^2]^{\frac{1}{2}}$. The constants A, ξ, $u(t_i)$, and $M_z(t_i)$ are determined by initial conditions at the beginning of the pulse ($t=t_i$) and the assumption that $M_z(t_i)^2+u(t_i)^2+v(t_i)^2=M_z(t)^2+u(t)^2+v(t)^2$ during the pulse. When the r-f pulse is removed at $t=t_i'$, then $\omega_1=0$, and Eqs. (8-A) and

[13] In the calculation which follows, $\bar{H}_1$ during the second pulse could be assumed to have an arbitrary angle α with respect to $\bar{H}_1$ which existed during the first pulse. It can easily be shown that all nuclear signals are independent of α and that the direction of the echo resultant will be at an angle $\alpha+\pi/2$ with respect to the direction of $\bar{H}_1$ which was applied during the second pulse.

(8-B) combine to give a solution

$$F(t)=F(t_i')\exp\{-(t-t_i')/T_2+i[\Delta\omega(t-t_i')+\int_{t_i'}^{t}\delta(t'')dt'']\},\quad (9)$$

where $F=u+iv$ and $t\geqslant t_i'$.

A constant field gradient, $(dH_0/gl)_{Av}=G$, shall be assumed to exist throughout the sample, where l is any direction in which the field gradient has the given average value G. The actual direction of $\bar{H}_0$ must vary in the sample. Any precessing moment which experiences a change in the magnitude of $\bar{H}_0$ due to diffusion will adiabatically follow a corresponding change in Larmor frequency of precession which will take place about the new direction of $\bar{H}_0$. Therefore $\bar{H}_1$ will not have the same magnitude during both pulses for a particular spin because the component of the applied r-f field perpendicular to the different directions of $\bar{H}_0$ will differ. Free induction signals will suffer negligible distortion because of this as compared to the distortion caused by variation in direction and magnitude of $\bar{H}_1$ throughout the sample due to coil geometry. For purposes of simplicity, the analysis will not attempt to take into account any sort of inhomogeneity of the $\bar{H}_1$ field.

In (9) let $\delta(t'')=\gamma Gl(t'')$ and

$$\int_{t_i'}^{t}\delta(t'')dt''=\Phi(t)-\Phi(t_i'),\quad (10)$$

where $\Phi(t)-\Phi(t_i')$ is the total phase shift accumulated in a time $t-t_i'$ by a precessing spin due to diffusion. The solution (9) must be averaged over all Φ, using a phase probability function $P(\Phi, t)$, by considering in particular the integral

$$\int_{-\infty}^{\infty}\{\exp i[\Phi(t)-\Phi(t_i')]\}P(\Phi,t)d\Phi=\exp\left[-\frac{kt^3}{3}-i\Phi(t_i')\right],\quad (11\text{-A})$$

where $k=(\gamma G)^2D$, and D is the self-diffusion coefficient of the spin-containing molecule. It can be shown[14] that

$$P(\Phi,t)=(4\pi kt^3/3)^{-\frac{1}{2}}\exp[-\Phi^2/(4\pi kt^3/3)].\quad (11\text{-B})$$

From Eq. (7-C) the solution for $M_z[\Delta\omega+\delta(t)]$ must be averaged over the probability that the moment vector corresponding to it is precessing at frequency $\Delta\omega+\delta(t)$ at time t. The ordinary diffusion law will be assumed to apply as regards the distance of diffusion l which corresponds to frequency shift δ. General solutions of (7) representing free motion can therefore be written as follows:

$$F(t)=F(t_i')\exp\{-(t-t_i')/T_2-\tfrac{1}{3}kt^3+i[\Delta\omega(t-t_i')-\Phi(t_i')]\}\quad (12)$$

$$M_z(t)=M_0\left[1+\frac{M_i-M_0}{M_0}\exp-(t-t_i')/T_1\right],\quad (13)$$

where

$$M_i=\int_{-\infty}^{\infty}M_z(\Delta\omega+\delta,t_i')P(\delta,t)d\delta\quad (14)$$

and[15]

$$P(\delta,t)=\frac{1}{[4\pi k(t-t_i')]^{\frac{1}{2}}}\exp[-\delta^2/4k(t-t_i')].$$

For the case in which twin pulses are applied, we have at $t=0$, $M_z=M_0$ and $u=v=0$. At $t=t_w$ the moments in the rotating system are obtained from (8). At time τ the r-f pulse is again applied and removed at $t=\tau+t_w$. After the second pulse the initial values of the magnetization components which undergo free motion are as follows:

$$u(\tau+t_w)=\frac{\Delta\omega}{\omega_1M_0}[v(\tau)v(t_w)-u(\tau)u(t_w)]+\frac{u(t_w)M_z(\tau)}{M_0}+u(\tau)\quad (15\text{-A})$$

$$v(\tau+t_w)=\frac{v(t_w)}{M_0}\left[M_z(\tau)-\frac{\Delta\omega}{\omega_1}u(\tau)\right]+v(\tau)\cos\beta t_w\quad (15\text{-B})$$

$$M_z(\tau+t_w)=\frac{1}{M_0}[M_z(\tau)M_z(t_w)+u(\tau)u(t_w)-v(\tau)v(t_w)].\quad (15\text{-C})$$

The v component, which is an even function in $\Delta\omega$, provides the free induction voltage, whereas the u component is an odd function in $\Delta\omega$ and does not contribute to the integral which will be applied in (18). Imposing the condition $\omega_1\gg\Delta\omega$ and $\tau\gg t_w$ we obtain:

(a) $(t\leqslant\tau)$:

$$v(t,\Delta\omega)\approx-M_0\sin\omega_1t_w\cos\Delta\omega t\exp(-t/T_2-\tfrac{1}{3}kt^3)\quad (16)$$

(b) $(t\geqslant\tau)$:

$$v(t,\Delta\omega)\approx M_0\sin\omega_1t_w[\sin^2\tfrac{1}{2}\omega_1t_w\cos\Delta\omega(t-2\tau)-\cos^2\tfrac{1}{2}\omega_1t_w\cos\Delta\omega t]\exp(-t/T_2-\tfrac{1}{3}kt^3)-M_z(\tau)\sin\omega_1t_w\cos\Delta\omega(t-\tau)\exp[-(t-\tau)/T_2-\tfrac{1}{3}k(t-\tau)^3].\quad (17)$$

[14] First one must take into account all possible paths (essentially all possible areas expressed by the integral in (10)) which the diffusing molecule may take in the l, t plane so that the total phase shift accumulated by the precessing spin which the molecule carries with it has a given value which is the same regardless of path length and final position of the molecule. The ordinary diffusion law is assumed to apply in expressing the probability of a given path under the constraint that a certain $\Phi(t)$ be accumulated. The distribution function (11-B) over all phases then follows by applying a standard deviation theorem (see James V. Uspensky, *Introduction to Mathematical Probability* (McGraw-Hill Book Company Inc., New York, 1937), p. 270). The author is indebted to Dr. C. P. Slichter for this derivation.

[15] E. H. Kennard, *Kinetic Theory of Gases* (McGraw-Hill Book Company, Inc., New York, 1938), p. 283.

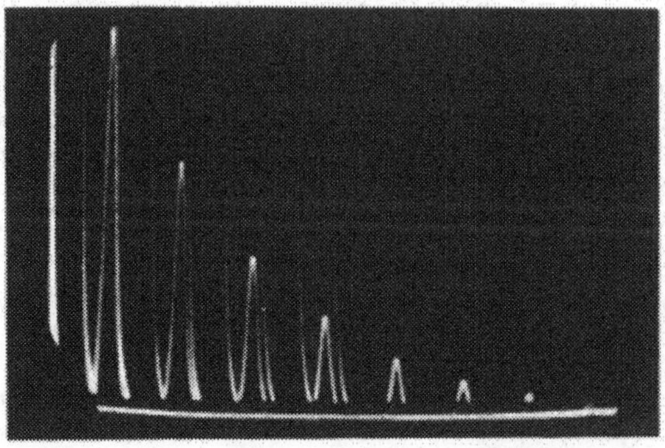

FIG. 3. Multiple exposures of proton echoes in a water solution of $Fe(NO_3)_3$ (2.5×10^{18} Fe^{+++} ions/cc). The faint vertical traces indicate paired pulses which are applied at time intervals $\gg T_2$, with the first pulse of each pair occurring at the same initial position on the sweep. For each pulse pair the interval τ is increased by 1/300 sec. The echoes are spaced 2/300 sec. apart and the measured decay time constant of the echo envelope gives $T_2=0.014$ sec.

The measured signal will be due to the integral

$$V(t)=\int_{-\infty}^{\infty} g(\Delta\omega)v(t,\Delta\omega)d(\Delta\omega). \tag{18}$$

For convenience $g(\Delta\omega)$ is chosen to be a Gaussian distribution:

$$g(\Delta\omega)=(2\pi)^{-\frac{1}{2}}T_2^*\exp[-(\Delta\omega T_2^*)^2/2],\quad T_2^*=(2\ln 2)^{\frac{1}{2}}/(\Delta\omega)_{\frac{1}{2}}, \tag{19}$$

where the integral of $g(\Delta\omega)$ over all $\Delta\omega$ is equal to unity. Integration of (16) and (17) according to (18) gives the following:

(a) $(t\leq\tau)$:

$$V(t)\approx -M_0\sin\omega_1 t_w\exp\left[-\left(\frac{t^2}{2T_2^{*2}}+t/T_2+\frac{kt^3}{3}\right)\right], \tag{20}$$

(b) $(t\geq\tau)$:

$$V(t)\approx M_0\sin\omega_1 t_w\left\{\left(\sin^2\frac{\omega_1 t_w}{2}\right)\exp\left[-\frac{(t-2\tau)^2}{2T_2^{*2}}\right]-\left(\cos^2\frac{\omega_1 t_w}{2}\right)\exp\left(-\frac{t^2}{2T_2^{*2}}\right)\right\}\times\exp\left(-\frac{kt^3}{3}-t/T_2\right)-M_z(\tau)\sin\omega_1 t_w\times\exp-\left[\frac{1}{2}\left(\frac{t-\tau}{T_2^*}\right)^2+\frac{t-\tau}{T_2}+\frac{k}{3}(t-\tau)^3\right]. \tag{21}$$

The echo at $t=2\tau$ is accounted for by the first term in (21) and has a width of $\sim T_2^*$ seconds. The remaining terms in (20) and (21) predict the occurrence of free induction decay signals immediately following the removal of the pulses. Actual shapes of all induction signals are determined mainly by what shape $g(\Delta\omega)$ happens to have due to external field inhomogeneities over the magnet. T_2 will play a significant role in affecting the shape only if $T_2\gtrsim T_2^*$. Signal amplitudes are independent of T_2^* as long as $\omega_1\gg 1/T_2^*$. In practice $g(\Delta\omega)$ is roughly a function which is some compromise between the Gaussian distribution given above and the Lorentz damped oscillator function given by

$$g(\Delta\omega)=\frac{2T_2^*}{1+(\Delta\omega T_2^*)^2}\quad\text{where}\quad T_2^*=\frac{1}{(\Delta\omega)_{\frac{1}{2}}}.$$

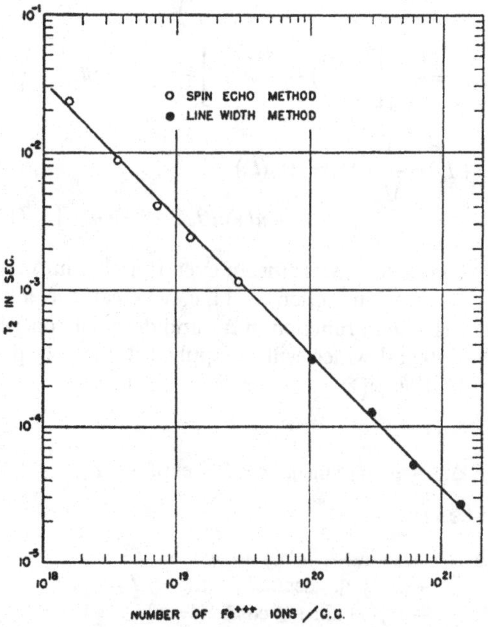

FIG. 4. T_2 measurements from the envelope decay of proton echoes are obtained for given concentrations of $Fe(NO_3)_3$ in H_2O. The plot compares with measurements made by the line width method (see reference 3).

(D) Measurement of T_2

If $\frac{1}{3}kt^3\ll t/T_2$ and $T_2^*\ll\tau<T_2$, T_1, then T_2 can be measured directly by plotting the logarithm of the maximum echo amplitude at $t=2\tau$ *versus* arbitrary values of 2τ. Figure 3 illustrates photographs of echoes on the oscilloscope for protons in a water solution of Fe^{+++} ions under these conditions. Figure 4 indicates how the measured T_2 for various concentrations of Fe^{+++} ions agrees with results obtained by Bloembergen, *et al.*[3] using the line width method. The law $C\propto 1/T_2$ is obeyed where C is the number of Fe^{+++} ions/cc for a given sample.

(E) Secondary Spin Echoes

If a third r-f pulse (identical to pulses producing the primary echo) is applied to the sample at a time T with respect to $t=0$, where $2\tau<T<T_2$, additional echoes occur at the following times: $T+\tau$, $2T-2\tau$, $2T-\tau$, $2T$. For $\tau<T<2\tau$ the signal at $2T-2\tau$ is absent but the others remain (see Fig. 5). These additional echoes can be readily predicted by rewriting Eq. (15) such that $\tau+t_w\rightarrow T+t_w$, $\tau\rightarrow T$, $t_w\rightarrow\tau+t_w$ ($\cos\beta t_w$ remains un-

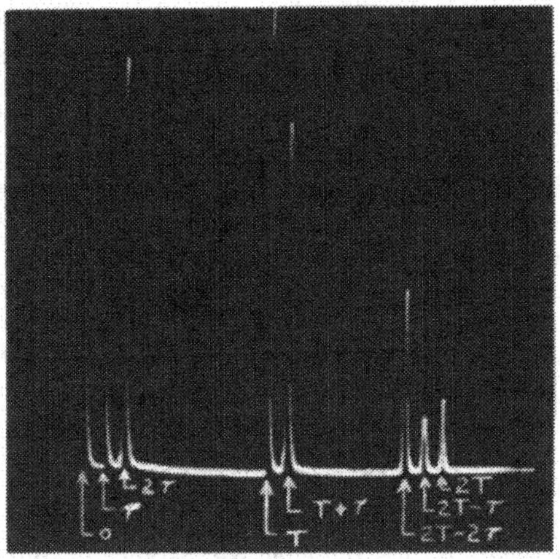

a

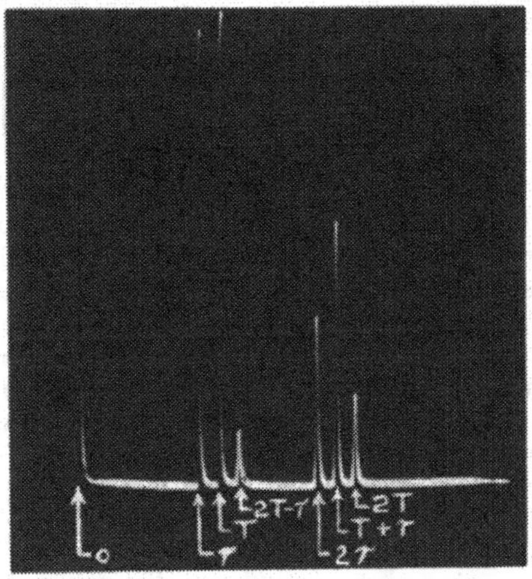

b

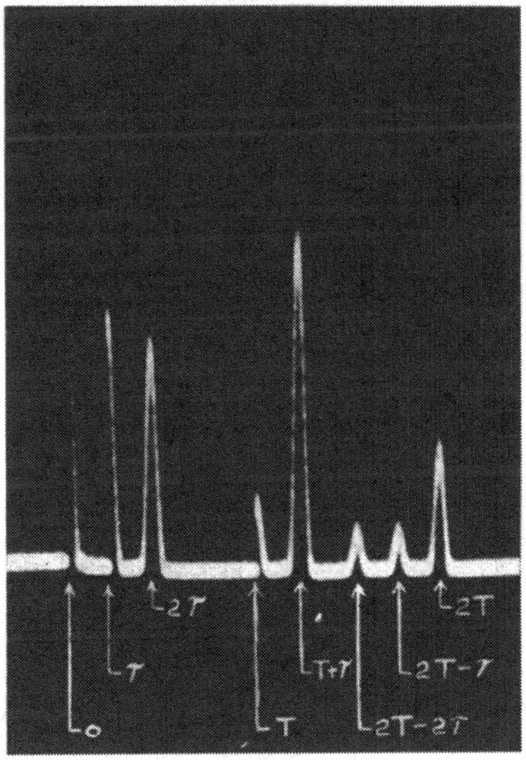

c

FIG. 5. Proton echo patterns in H_2O resulting from three applied r-f pulses. The pulses are visible in the upper two traces, and have a width $t_w \sim 0.5$ msec. In the upper trace $\tau = 0.008$ sec., $T = 0.067$ sec., and for the second trace $\tau = 0.046$ sec. and $T = 0.054$ sec. The bottom photograph shows a similar pattern for the case $T > 2\tau$ where induction decay signals can be seen following very short invisible r-f pulses. Saturation of a narrow band communications receiver, used in the case of the upper two traces, prevents the observation of these signals, whereas a wide band i.f. amplifier makes this observation possible in the bottom photograph.

changed in (15-B)) and applying the resulting expressions as initial conditions in (12), (13), and (14). In this manner, by successive application of accumulating initial conditions, the echo pattern resulting from any number and sequence of r-f pulses can be predicted. After integrating $v(t)$ over $\Delta\omega$ for $t \geqslant T > 2\tau$, using $g(\Delta\omega)$ according to (19), the following expression for $V(t)$ is obtained (terms due to induction decay directly following the pulse are omitted and assumed not to interfere with the echoes since $\tau \gg T_2{}^*$):

$$V(t) \approx \frac{M_0}{2} (\sin^3 \omega_1 t_w) \times \exp\left\{ -(T-\tau)\left(\frac{1}{T_1} - \frac{1}{T_2}\right) - t/T_2 - \frac{k}{3}[\tau^3 + (t-T)^3] - k\tau^2(T-\tau) - \frac{[t-(T+\tau)]^2}{2T_2{}^{*2}} \right\} \quad \text{(22-A)}$$

$$-M_0\left(\sin\omega_1 t_w \sin^4 \frac{\omega_1 t_w}{2}\right) \times \exp\left\{ -\frac{[t-(2T-2\tau)]^2}{2T_2{}^{*2}} - t/T_2 - \frac{kt^3}{3} \right\} \quad \text{(22-B)}$$

$$+M_z(\tau)\left(\sin\omega_1 t_w \sin^2 \frac{\omega_1 t_w}{2}\right) \times \exp\left\{ -\frac{[t-(2T-\tau)]^2}{2T_2{}^{*2}} - (t-\tau)/T_2 - k(t-\tau)^3/3 \right\} \quad \text{(22-C)}$$

$$+\frac{M_0}{4} (\sin^3 \omega_1 t_w) \times \exp\left\{ -\frac{(t-2T)^2}{2T_2{}^{*2}} - t/T_2 - \frac{kt^3}{3} \right\}. \quad \text{(22-D)}$$

Term (22-A) provides a "stimulated echo" signal at $T+\tau$. The signal at $2T-2\tau$ (22-B) can be expected qualitatively by considering the "eight ball" alignment in Fig. 1 as equivalent to an initial orientation of moments in a given direction by an imaginary r-f pulse at 2τ. Therefore, it follows that the stimulating pulse at T causes an "image echo" to occur at $2T-2\tau$. The signals at $2T-\tau$ (22-C) and $2T$ (22-D) are essentially primary echoes corresponding to twin pulses at τ, T and 0, T respectively. The signal at $2T$ is modified by the presence of the second r-f pulse at τ so that it does not have the same trigonometric dependence on $\omega_1 t_w$ as do the primary echoes at τ and $2T-\tau$.[16] Experimentally the various echo signals are observed to go through maxima and minima in general agreement with their respective trigonometric dependences on $\omega_1 t_w$ as this quantity is varied. The stimulated echo at $T+\tau$ is particularly interesting and useful in view of the fact that if τ is sufficiently small so that all terms in the exponent of (22-A) are negligible except T/T_1, the signal survives as long as T_1 permits. The remaining echo signals in liquids of low viscosity have maxima which attenuate in a time much shorter than T_1 as T is arbitrarily increased for a given τ. This is due to the diffusion factor $\frac{1}{3}kl^3$ which occurs in the exponents of (22-B), (22-C), and (22-D), but occurs only as $k\tau^2 T$ in (22-A). This property of the stimulated echo is schematically indicated in Fig. 6. The constructive interference at $T+\tau$ is due to moment vectors which previously existed as $M_z(\Delta\omega)$ components distributed in a spectrum approximately as $\cos\Delta\omega\tau$ during the time interval $\tau+t_w\rightarrow T$. This can be seen by noting that $M_z(\tau+t_w)$ has a term $v(\tau)$ proportional to $\cos(\Delta\omega+\delta)\tau$ from (15-C). This cosine distribution becomes smeared due to diffusion and must be averaged over all δ by applying the integral in (14). However, the self-diffusion of spin-containing molecules will not seriously upset this frequency pattern providing $1/\tau \gg \gamma(dH/dl)l(T)$ (let $T \gg \tau$), where $l(T)$ is the effective distance of diffusion in time T over which a shift in Larmor frequency can occur. The attenuation effect of diffusion upon echoes

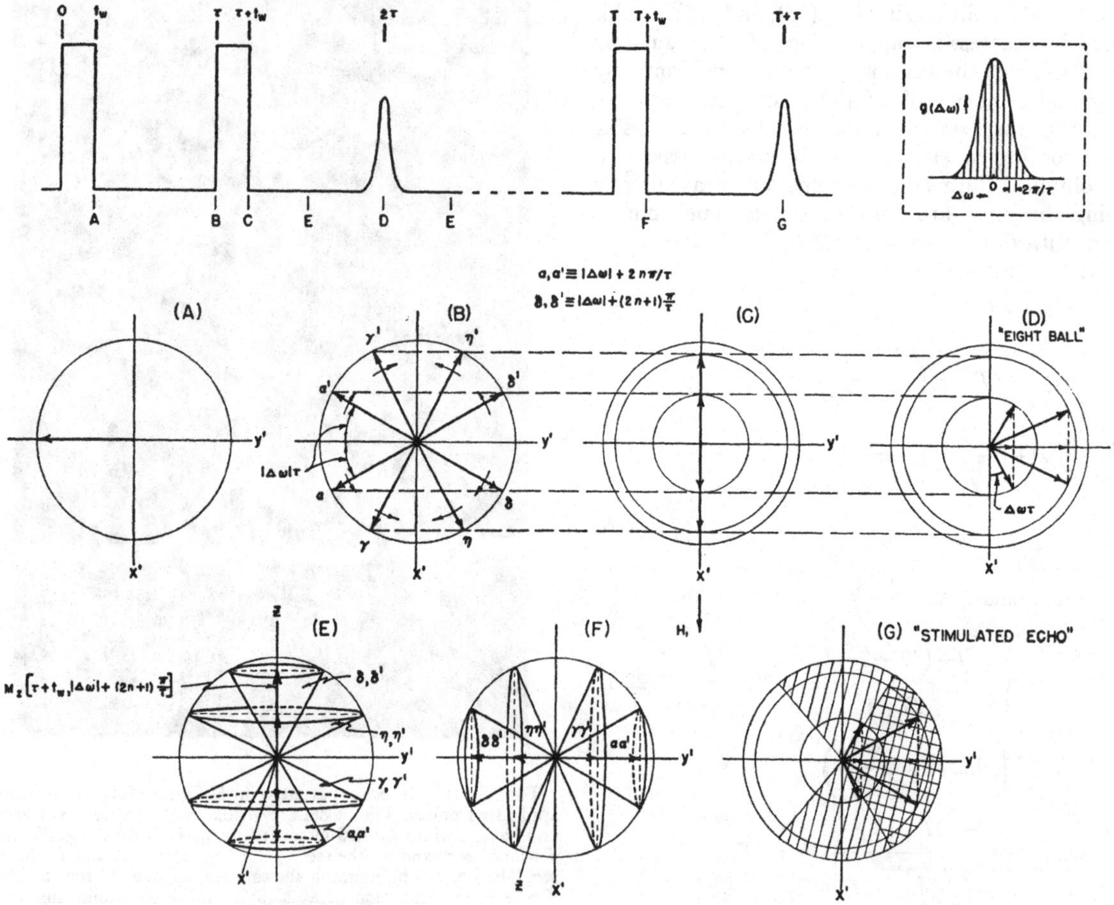

FIG. 6. A vector representation which accounts for the stimulated echo at $t=T+\tau$ is shown under conditions of the special case for the primary echo model in Fig. 1. For a given $|\Delta\omega|$, the symbols α, α' and δ, δ' denote those moments which have Larmor frequencies such that they precess angles $|\Delta\omega|\tau+2n\pi$ and $|\Delta\omega|\tau+(2n+1)\pi$ respectively in time $t=\tau$. n is any integer which applies to frequencies within the spectrum which will lie in a pair of cones corresponding to a specific $|\Delta\omega|$. These cones provide M_z components (after the pulse at τ) which are available for stimulated echo formation after the pulse at T. The shaded area in G indicates the density of moment vectors. The absence of vectors on the $-y'$ side leaves a dimple on the unit sphere.

[16] For $\tau<T<2\tau$ the echo at 2τ is modified and has the coefficient $M_0/4\ \sin^3\omega_1 t_w$ instead of the one given by (21).

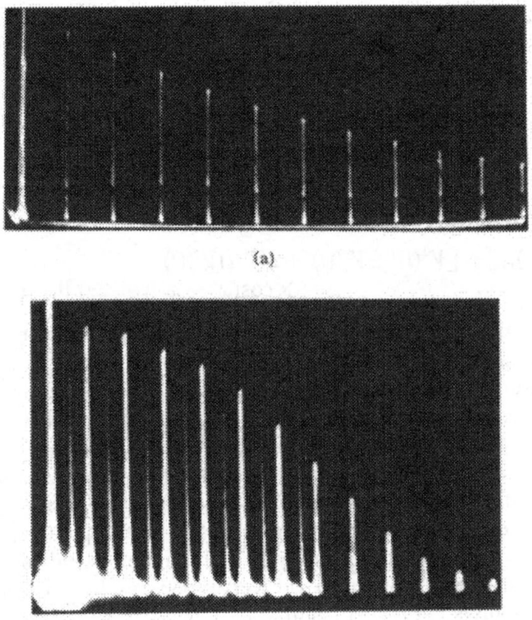

(a)

(b)

FIG. 7. A typical exponential plot of stimulated echo amplitudes is shown in the top photograph for protons in H_2O. This is obtained in a manner described for Fig. 3, except that T for the third pulse here is increased by 16/60 sec. intervals while τ is fixed at 0.0039 sec. The measured decay of the envelope is 1.89 sec. which serves as a point on the graph in Fig. 8. The apparent break in intensity in each of the stimulated echoes (seen as vertical traces because of the slow sweep speed) is due to a condition where the echo follows so soon after the stimulating pulse that it superimposes upon the voltage recovery of the receiver detector RC filter.

The bottom photograph indicates approximately an $\exp(-kt^3/3)$ decay law for the primary echo envelope in H_2O. The separation between echoes is 1/60 sec.

whose configuration depends purely upon phase and not frequency is much greater due to the exponential factor $\frac{1}{3}kt^3$ rather than $k\tau^2T$ which occurs only for the stimulated echo.

(F) Measurement of T_1; Qualitative Confirmation of the Diffusion Effect

If the condition $k\tau^2T \ll T/T_1$ is maintained by choosing τ very small, a plot of the logarithm of the stimulated echo maximum amplitude *versus* arbitrary values of T gives a straight line whose slope provides an approximate measure of T_1. In this manner glycerine is found to have a $T_1=0.034$ sec. The self-diffusion coefficient of glycerine is apparently sufficiently small so that T_1 can be measured directly as well as T_2, according to the discussion in III-D. A measured value of $T_2=0.023$ sec. is obtained, which is in substantial agreement with previous measurements.[3,9] The data for T_1 is obtained from oscillographic traces, an example of which is shown in Fig. 7 (top) for protons in distilled water. All relaxation measurements are made at room temperature, at $\omega=30$ Mc. A better value of T_1 is obtainable by plotting $1/T_m$ against τ^2 where T_m is the measured envelope decay time for the stimulated echo which decays as e^{-T/T_m}. A straight line is obtained which has the equation $1/T_m=1/T_1+k\tau^2$, which is seen from the exponential factor in (22-A) where $t=T+\tau$ and $\tau \ll T$. Such a plot is given in Fig. 8 for protons in distilled water (not in vacuum) where the reciprocal of the ordinate intercept gives $T_1=2.3\pm0.1$ sec., in agreement within experimental error with previous measurements.[3,17] Using the value of $D=2\times10^{-5}$ cm²/sec. for the water molecule,[18] the field gradient, G, calculated from the measured slope is 0.9 gauss/cm, which correlates roughly with the actual gradient over the sample. The gradient is expressed as $G\approx(\Delta H)_{\frac{1}{2}}/d$ where $(\Delta H)_{\frac{1}{2}}$ $\sim$0.2 gauss is measured directly from the resonance absorption line width (or echo width) and $d\sim3$ to 4 mm is the average thickness of the cylindrical sample. In Fig. 7 (bottom) the echo envelope for protons in distilled H_2O is reduced to $1/e$ of its maximum amplitude at $t=(3/k)^{\frac{1}{3}}$, since we neglect the decay due to T_2, which is negligible compared to diffusion. The calculated $(\Delta H)_{\frac{1}{2}}$ here is also in rough agreement with the actual field inhomogeneity present. This agreement with the predicted diffusion law confirms the existence at least of a smooth gradient in H_0 over the sample. If the sample is slightly rotated while r-f pulses are applied to obtain echoes, the echo amplitude is markedly reduced as the spin ensemble rotates into varying field inhomogeneity patterns.

(G) The Echo Beat and Envelope Modulation Effects

It has been found that the exact magnetic resonance frequency of nuclear moments of a given species depends upon the type of molecule in which it is contained. It is apparent that the local magnetic field at the position of the nucleus is shifted from the value of the applied external field by an amount which is too large to be accounted for by the normal diamagnetic correc-

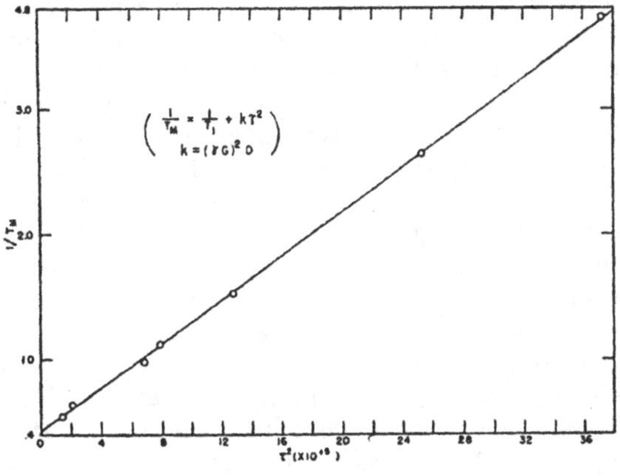

FIG. 8. Stimulated echoe measurement of spin-lattice relaxation time (T_1) of protons in H_2O.

---

[17] E. L. Hahn, Phys. Rev. 76, 145 (1949).
[18] W. J. C. Orr and J. A. V. Butler, J. Chem. Soc. 1273 (1935).

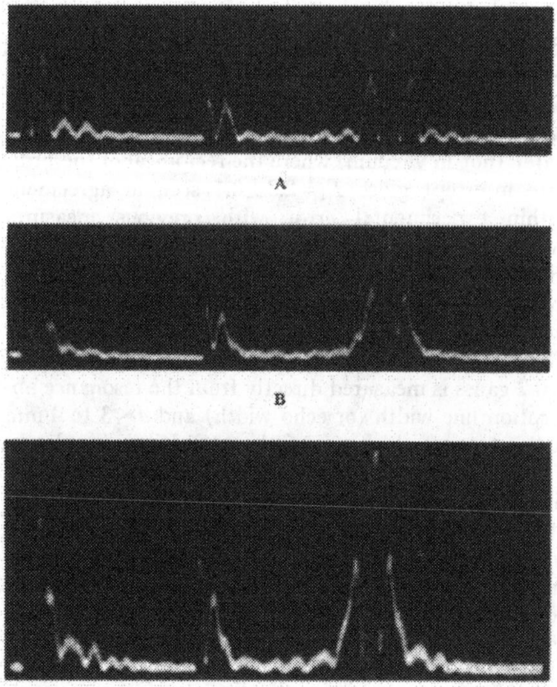

FIG. 9–A–B–C. Heterodyne beat signals for different F^{19} resonance frequencies due to the chemical Larmor shift effect.
(A) $CF_3CCl{=}CCl_2$ and 1,4 difluoro-benzene ($C_6H_4F_2$) mixture
(B) $CF_3CCl{=}CCl_2$ and 1,2,4 trifluoro-benzene ($C_6H_3F_3$) mixture
(C) 1, trifluoro-methyl 2,3,6 trifluoro-benzene ($C_6H_2F_3CF_3$)

tion.[19] Ramsey has shown that there exists the possibility of a much stronger field shift[20] due to second-order paramagnetism arising from the type of molecule which contains the nuclear spin. The echo technique reveals simultaneously the presence of two or more groups of resonant nuclei having slightly different Larmor frequencies due to such possible shifts in the local field at the nucleus, providing they are of the order of $(\Delta\omega)_{1/2}/\gamma$ gauss in magnitude. Echoes and free induction decay signals are modulated by beat patterns (Fig. 9) due to the fact that two or more spin groups of one species are contained in the same molecule or different molecules and have non-equivalent molecular environments in the same sample. For example, let ω' and ω'' denote respectively the Larmor frequencies at which the rotating coordinate systems of two spin ensembles may precess, and allow symmetric distribution functions $g'(\Delta\omega')$ and $g''(\Delta\omega'')$ to be a maximum for $\Delta\omega'=\omega_0-\omega'=0$, $\Delta\omega''=\omega_0-\omega''=0$. Therefore, identical echo configurations will result in two frames of reference, each rotating with frequencies ω' and ω'' respectively. The r-f induction is due to the magnetization component $v(\Delta\omega, t)\sin\omega t$ for an individual spin group where $v(\Delta\omega, t)$ is described as in (16) and (17). Integration over all frequencies leading to (20) and (21) provides the following total induction:

$$V(t)=V'(t)\sin\omega'(t-t_1')+V''(t)\sin\omega''(t-t_1'). \quad (23)$$

$V'(t)$ and $V''(t)$ signify the free induction signals due to each of the spin groups alone. The envelope of the echo signal (Fig. 9-A) is given by

$$V(t)_{max}=[V'(t)^2+V''(t)^2+2V'(t)V''(t) \times\cos(\omega''-\omega')(t-2\tau)]^{1/2}. \quad (24)$$

As typical examples of this effect it has been found that the signals due to F^{19} nuclei in certain organic compounds yield modulation patterns which obey the heterodyne law expressed by (24). In order to observe this effect the condition $2\pi/(\omega''-\omega')\lesssim T_2{}^*$ must be attained in order to observe at least one period of the modulation within the lifetime $T_2{}^*$ of an echo or induction decay signal following a pulse. Consequently, a high degree of homogeneity in the magnetic field must be attained in order to get very good resolution; i.e., to resolve very small shifts in Larmor frequency. It appears that this approach to the determination of very small Larmor shifts has a resolution no better than ordinary magnetic resonance absorption methods[1,3] in which the limitation is also due to external field inhomogeneities. However, the echo method is fast and lends itself more conveniently to search purposes in finding these shifts.[21] Somewhat higher resolution than that available by the normal method can be attained beyond the limitation imposed by field inhomogeneities by introducing into the receiver an r-f signal at a frequency somewhere near the Larmor frequencies present. An audio beat modulation appears having an envelope which is modulated in turn by the Larmor shift beat note. These beats can then be more easily distinguished from noise for the condition $2\pi/(\omega''-\omega')\gtrsim T_2{}^*$ in favorable cases in which the induction signals are sufficiently intense. Periods of the order of $3, 4T_2{}^*$ may possibly be observed, in which case Larmor shifts as small as 0.01 gauss, of the order of normal diamagnetic shifts, may be detected, assuming a $(\Delta H)_{1/2}\sim 0.05$ gauss is available out of a total field of 7000 gauss. It can be seen from Fig. 9 that the modulation on the echo and decay signals (following r-f pulses) correlate in pattern. It is significant that the pattern on the echo is always symmetric regardless of the spacing τ between the two r-f pulses. This is understandable in view of the fact that two rotating frames of reference, for example, increase in phase difference by $(\omega''-\omega')\tau$ radians between the pulses. The second pulse produces an initial condition such that the two frames

[19] W. E. Lamb, Phys. Rev. **60**, 817 (1941).
[20] This is treated theoretically in a paper in Phys. Rev. **78**, 699 (1950), kindly forwarded to the author in advance of publication by Professor N. F. Ramsey.

[21] One must be careful that the observed modulation is not due instead to a condition where the H_0 magnetic field inhomogeneity pattern over the sample has two or more discrete bumps in it. The modulation will again be symmetric on the echo and can only be distinguished from a true beat effect by moving the sample to a different part of the field in the magnet gap and noting whether or not the modulation disappears or varies in frequency.

of reference now rotate into one another by the same amount and coincide at the time 2τ of the echo maximum. This principle is inherent in the echo effect itself: the phase differences of all moment vectors (with respect to the initial orientation established by the first pulse) are effectively cancelled at the time of the echo maximum. This cancellation is made possible by the second pulse. If no further pulses are applied, the echo at 2τ can never repeat itself, as might be expected, because the "eight-ball" configuration is essentially only a single recurrence of the initial in-phase condition of the moment vectors at $t=t_w$, though not quite the same due to a spread in Larmor frequencies.[22]

Fluorine nuclei in the compounds[23] $CF_3CCl=CCl_2$ and $C_6H_4F_2$ (1,4 difluoro-benzene) give induction signals in separate samples in which no significant beat patterns appear. Weak beats may appear due to other fluorine compound impurities used in the synthesis of these compounds. Figure 9-A indicates the beat which results when these two molecules are mixed in liquid form in a single sample such that two fluorine spin groups are in a one-to-one ratio in concentration. The separate molecules contain fluorine atoms located in equivalent positions and therefore cannot give rise to a beat among themselves. A mixture of the two molecules, having fluorine nuclei which are relatively non-equivalent in molecular environment, now reveals a separation of 1.9 kc in Larmor frequency for the two groups in a field of 7500 gauss. According to (24) the modulation pattern goes to a complete null at this frequency since the mixture is adjusted so that $V'(t) = V''(t)$. By observing the normal resonance absorption signal of this mixture on the oscilloscope, using 30 cycle field modulation, two distinct absorption lines are observed, separated by 1.9 kc on a frequency scale. By using a mixture in which the concentration of one molecule exceeds that of the other, the relative difference in intensity of the absorption lines indicates that the fluorine resonance frequency in $C_6H_4F_2$ lies on the high side relative to that in $CF_3CCl=CCl_2$. It is reasonable to expect this if the charge density of the electronic configuration about the fluorine nuclei in $C_6H_4F_2$, being less than that in $CF_3CCl=CCl_2$, can be correlated with a correspondingly smaller negative magnetic shielding correction. This property appears to exist in all mixtures and single molecules so far investigated in which a distinction between spin groups has been made. Within experimental error, the Larmor frequency shifts observed here appear to be proportional to the applied field, based on measurements made at 7070 and 3760 gauss. Figure 9-B shows the beat pattern due to approximately a one-to-one mixture (in terms of fluorine nuclei) of the compounds 1, 2, 4 trifluoro-benzene and $CF_3CCl=CCl_2$. More than one beat modulation frequency is evident, due obviously to the presence of more than two fundamental spin groups. Figure 9-C shows how a similar complexity in beat pattern arises from a sample of 1-trifluoro-methyl 2, 3, 6 trifluoro-benzene. All observable beat frequencies are of the order of a few kilocycles.

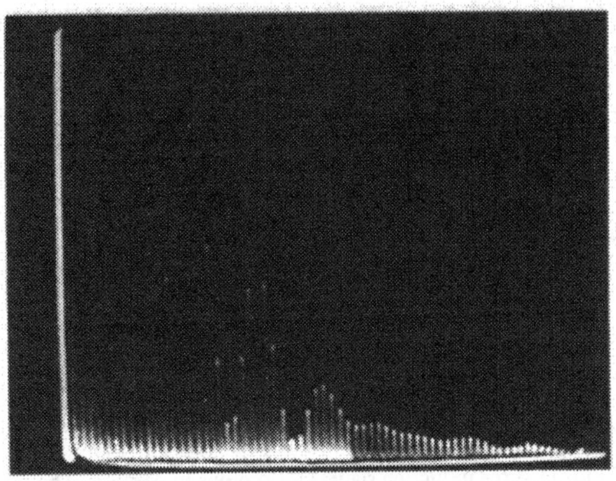

FIG. 10. The echo envelope modulation effect for protons in C_2H_5OH. Paired pulses are applied in the usual manner for obtaining multiple exposures. The echo separation is 1/300 sec. The first echo at the left follows so closely after the r-f pulses that it is not at normal amplitude because of receiver saturation.

Preliminary studies have been made of another effect which is shown in Fig. 10. The envelope of the normal echo maximum envelope plot is modulated by a beat pattern which is in violation of the normal decay due to self-diffusion and T_2. The envelope shown for C_2H_5OH (period$\approx$0.027 sec.) is an example which is typical for protons in various organic compounds. If the effect is present in the particular substance investigated it is readily observable only if the period of the modulation is shorter than the normal decay time of the echo envelope upon which it is superimposed. Several organic compounds studied so far have been observed to have characteristic periods of the order of 0.1 to 0.01 sec. Modulation patterns in many cases do not contain a single frequency but perhaps several as it appears in C_2H_5OH. The period of the modulation is found in general to be greater than T_2^*, the echo lifetime. This modulation effect cannot be attributed to an interference between several spin groups because the observed echo maximum is always due to the sum of the echo maxima contributed by each of the spin groups alone. This is true regardless of the number of different spin groups present, and therefore the beat frequencies due to such Larmor shifts cannot show up in the envelope of the echo maxima. Within experimental error

[22] It is interesting to note that the configuration at $t=t_w$, namely, $M_{xy}=M_0$, can in principle be exactly repeated at $t=2\tau$ by doubling the second r-f pulse width with respect to the first one which is at the pulse condition $\omega_1 t_w=\pi/2$ (see Fig. 1). Actual experiment indicates that the inhomogeneity in H_1 throughout the sample prevents this from exactly taking place, but shows an increase in the available echo amplitude beyond the optimum amplitude at $\omega_1 t_w=2\pi/3$ (Eq. 21). The stimulated echo at $t=T+\tau$ then nearly disappears.

[23] The fluorine compounds used were kindly provided by Dr. G. C. Finger of the Fluorspar Research Section of the Illinois State Geological Survey, where they were synthesized.

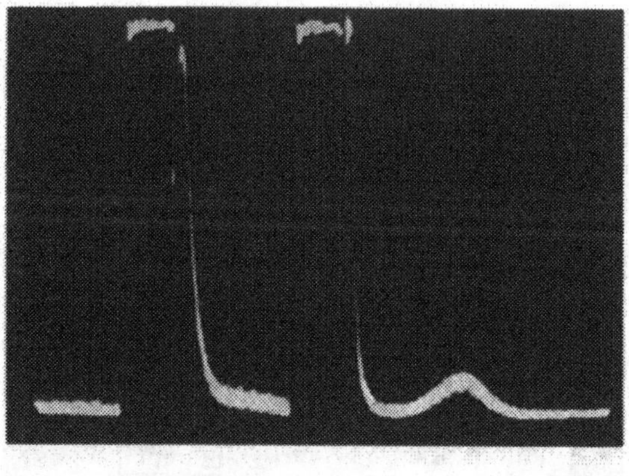

FIG. 11. Free induction signals for protons in paraffin. The echo lasts for $\sim 1.4\times 10^{-5}$ sec. The r-f pulses, about 25 μsec. wide, cause some blocking of the i.f. amplifier. The echo envelope decay time is also of the order of the single echo lifetime.

(five to ten percent) the period of the envelope modulation is found to be inversely proportional to the applied magnetic field. It is possible that an interaction between the nuclear spin and the molecule which contains it causes a periodic reduction of the echo amplitude by a modification of the echo constructive interference pattern. This effect will be treated in a later paper in greater detail.

(H) Free Nuclear Induction in Solids

It has been established in the case of liquids that, if one excludes the effect of diffusion, the lifetime of the nuclear induction decay following a pulse is given by $T_m = T_2T_2^*/(T_2+T_2^*)$, and the single echo lifetime is given by T_2^*. Qualitative observations of free nuclear induction signals due to nuclei in solids, however, indicate that the role played by T_2^* is no longer significant. The ensemble instead precesses in a magnetic field distribution described by a function $G(\delta\omega)$, where $\delta\omega = \omega_0' - \omega'$ and $\omega_0' = \gamma(H_{\text{local}} + H_{\text{magnet}})$. The local field H_{local}, due to lattice neighbors, is superimposed on the externally applied field at the positions of the precessing nuclei. This local field is spread over a width much greater than the width due to the magnet. In one case echoes have been observed for protons in paraffin (Fig. 11) where it appears that the echo and induction decay lifetimes are now given by $\sim 1/(\delta\omega)_{\frac{1}{2}}$ seconds. Extremely intense r-f power is required in order to excite all of the spins over a broad spectrum of Larmor frequencies in a pulse time $t_w \gtrsim 1/(\delta\omega)_{\frac{1}{2}}$ seconds, and therefore the condition $1/t_w,\ \omega_1 \gg (\delta\omega)_{\frac{1}{2}}$ must apply. A striking indication of the predominance of either T_2^* or $1/(\delta\omega)_{\frac{1}{2}}$ is shown by observing how the broad free induction signals from protons in liquid paraffin become very narrow as the paraffin cools and solidifies. It appears that echoes in solids can be observed in principle in a time $\lesssim T_2 \sim 1/(\delta\omega)_{\frac{1}{2}}$, because a given distribution in H_{local} determined by $G(\delta\omega)$ (which now plays the role of $g(\Delta\omega)$ in the case of liquids) is able to last roughly for a time T_2. The local z magnetic field due to neighboring magnetic moments (nuclear spins, paramagnetic ions and impurities) therefore not only depends upon the particular location of these moments with respect to the precessing nucleus, but also upon a time T_2. This time determines how long a given parallel and antiparallel configuration of these neighboring moments can exist with respect to the externally applied field. It follows, therefore, in the case of paraffin, that the stimulated echo, which depends upon frequency memory of the spin distribution, cannot be observed out to times $T+\tau \sim T_1 = 0.01$ sec., where $T_1 \gg T_2$.

Although the Bloch theory is highly successful in accounting for the echo effect in liquids where T_1 and T_2 are introduced in a phenomenological way, it must be understood in this theory that the predicted natural resonance line shapes will always be described by a damped oscillator resonance function (Lorentz) in the steady state. This corresponds to the observed exponential decay of free induction signal amplitudes in the transient case. This concept does not necessarily apply in general, especially as regards magnetic resonance line shapes in solids in the steady state. It remains to be shown that the properties of free nuclear induction signals in solids are explained by a transient analysis which gives results equivalent to the general steady state treatment formulated by Van Vleck[24] and others.[3,25]

With further refinements in technique for obtaining a sufficiently fast response of the r-f circuits to very short r-f pulses of large intensity (t_w of the order of microseconds at $H_1 \sim 20$ to 100 gauss), it may be possible to obtain informative data from free nuclear induction signals in solids which have a T_2 of the order of 10^{-5} to 10^{-6} seconds. It will be of profit to investigate the induction decay which follows single pulses (already found for protons in powdered crystals of NH_4Cl, $(NH_4)_2SO_4$, $MgSO_4 \cdot 7H_2O$ and for F^{19} in CaF_2) without the attempt to observe echoes.

IV. EXPERIMENTAL TECHNIQUE

The block diagram in Fig. 12 indicates the necessary components for obtaining the echo effect. All features of the sample, inductive coil, and methods of coupling from the oscillator and to the receiver are typical of nuclear induction techniques and have been discussed in detail elsewhere. A great simplification is introduced here because only a single LC tuned circuit is necessary. However, in the case where a narrow band receiver is used ($\sim$8 kc) in place of a very broad band i.f. strip ($\sim$5 Mc) and where r-f pulses are particularly intense, it is convenient to use a bridge balance in order to minimize overloading of the receiver. Two methods have been used to provide r-f power in the form of pulses. A method best suited for r-f amplitude and frequency stability employs a 7.5 Mc crystal-controlled

[24] J. H. Van Vleck, Phys. Rev. 74, 1168 (1948).
[25] G. E. Pake, J. Chem. Phys. 16, 327 (1948).

oscillator whose frequency is quadrupled to 30 Mc and amplified r-f power is then gated through stages whose grids are biased by square wave pulses from a one shot multivibrator controlled in turn by timer pulses. This method is essential for studies of the phases of various echo signals and other effects. The crystal maintains a source of coherent r-f oscillations which can be used to heterodyne weakly with the nuclear signal that has a phase determined initially by these oscillations in the form of intense pulses. The phases of the resulting audio beat frequency oscillations seen superimposed on the echoes then yield certain interesting proofs.

(a) The phase of the audio modulation on all echoes is invariant to any time variation in the spacing between r-f pulses. With respect to a fixed reference in a rotating coordinate system, all echoes form a resultant which is constant in direction due to the fact that the accumulated phase differences before the r-f pulse (at $t=\tau$) are exactly neutralized after it (referring to discussion in III-G).

(b) The negative sign of the "image echo" term (Eq. 22-B) signifies that the resultant of this signal is 180° out of phase relative to the resultants of all other echoes formed before or after it. This is borne out by the fact that the phase of the observed audio modulation on the image echo is exactly 180° out of phase with respect to the modulation pattern on all other echoes. Otherwise, the modulation patterns appear to be identical.

(c) Echoes at $t=2\tau$ are observed not to fluctuate in amplitude when $\tau<T_2^*$ due to the fact that the r-f is coherent for successive pulses, and the phase of the moment configuration prior to the r-f pulse at τ has a definite relationship with respect to the phase of the echo which follows.

With the above method, precautions must be taken to prevent r-f power leakage to the sample during the absence of pulses. The necessity for this precaution is eliminated by turning on and off a high power oscillator (by gating the oscillator grid bias, Fig. 13) which drives the LC circuit directly. This method, although not as stable, makes available higher r-f power, and the ability to vary the driving frequency ω is convenient. Although r-f pulses are produced now in random phase, experimental results are the same as long as $\tau\gg T_2^*$. Both

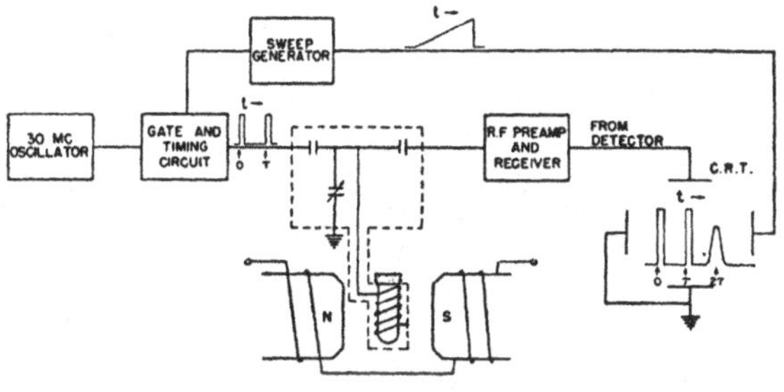

FIG. 12. Arrangement for obtaining spin echoes.

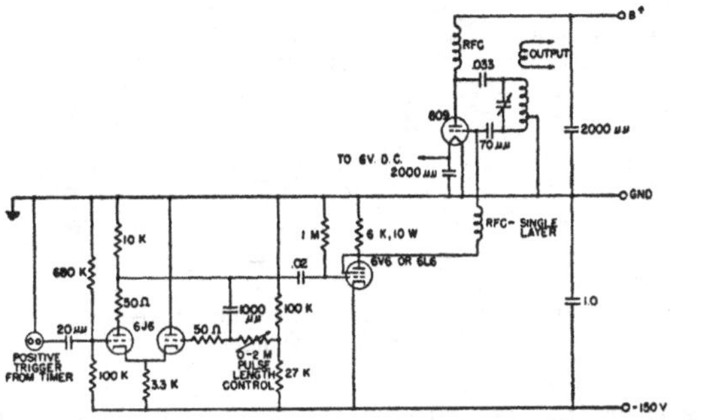

FIG. 13. Gated oscillator.

pulse methods combined provide pulses of $t_w\sim 20$ μsec. to a few milliseconds and $H_1\sim 0.01$ to 50 gauss.

In order to obtain accurate and reproducible data with echoes it is necessary that the dc magnetic field be held constant to at least one part in 10^5 over the length of time during which a set of echo data is being photographed. One might say that some field drift is tolerable within the limits set by the condition $\omega_1\gg(\Delta\omega)_{\frac{1}{2}}$. However, it is advisable even to guard against field drifts less than H_1 gauss because the Fourier amplitudes of all r-f frequency components which resonate with the given Larmor spin frequencies will vary to some degree as the dc field varies. In cases where the decay of echoes is plotted for nuclei in liquids having a long T_1 (several seconds for protons in most organic liquids) and where maximum available signal is desired, the spins must be allowed to return to complete thermal equilibrium between applications of paired pulses. Therefore, if one waits at least five half-lives to obtain a plot such as is shown in Fig. 10, a total time of 17 minutes is required during which the magnetic field must not drift appreciably. In order to minimize the effect of slow field drifts it is convenient to apply paired pulses at a repetition rate whose period is some constant fraction of T_1. During this period the operator has sufficient time to adjust timer switches (reset switches on a conventional scalar unit[17]) in order to provide increasing integers of time between the two pulses. The sample is therefore partially saturated at a level which is practically constant when the pulses are applied, although a small but negligible variation in the level of saturation is introduced as the time τ is systematically increased. The over-all signal to noise ratio of the pattern is reduced but data can be recorded at a convenient speed.

For these experiments the magnetic field is stabilized by means of a separate proton resonance regulator[26] which monitors practically the same magnetic field which is present at the echo sample. The regulator resonance sample is located in the same magnet gap and is subjected to 30 Mc r-f power which is well shielded from the experiment sample. The regulator sample is placed in the inductance of the tuned circuit

[26] To be discussed in a later paper.

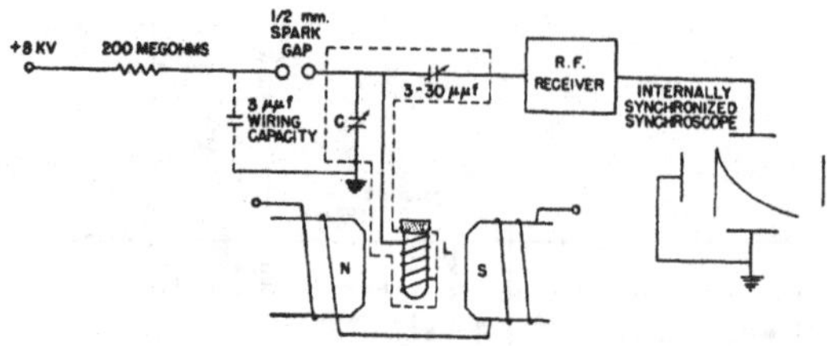

FIG. 14. Simple spark method for obtaining free nuclear induction decay signals.

of a transitron oscillator. A sinusoidally vibrating reed capacitively modulates the frequency of this oscillator within the line width of the regulator proton resonance. When the magnetic field is brought into the resonance value it is locked in and controlled by the regulator. The transitron oscillator r-f voltage level decreases due to resonance absorption and is modulated at the frequency of the vibrating reed. A discriminator circuit utilizes this signal to control a correction current to the magnet in a conventional manner.[27]

In Fig. 14 a sparking technique is noted purely for its novel features of simplicity in demonstrating, qualitatively, free nuclear induction decay signals directly following single random r-f pulses. The spark generated across the gap contains essentially all frequencies and excites for a very short time the tuned circuit in which the sample is located. After the spark extinguishes, the sample in the inductive coil transmits a decaying r-f induction signal to the receiver at the Larmor frequency determined by H_0. Capacitor C needs adjustment such that the tank circuit will resonate approximately in the region of the Larmor frequency (with no spark). Signals can be obtained over a broad range of H_0 without requiring a retuning of C. The observed signal, of course, has random amplitudes since the r-f energy transferred by the sparks is random within a certain range.

[27] M. Packard, Rev. Sci. Inst. **19**, 435 (1948).

V. CONCLUDING REMARKS

Simple principles of the free nuclear induction technique have been described and tested, principally with proton and fluorine (F^{19}) signals in liquids. Data which is made available by this technique is to be presented later in more systematic detail. The echo technique appears to be highly suitable as a fast and stable method in searching for unknown resonances. Intense pulses of H_1 provide a broad spectrum of frequencies. This makes possible the observation of free induction signals far from exact resonance. Echo signals have proved useful for the measurement of relaxation times under conditions where interference effects (microphonics, thermal drifts, oscillator noise) encountered in conventional resonance methods are avoided. The self-diffusion effect in liquids of low viscosity offers a means of measuring relative values of the self-diffusion coefficient D, a quantity which is very difficult to measure by ordinary methods. It is of technical interest to consider the possibility of applying echo patterns as a type of memory device.

The formal analysis of the signal-to-noise ratio of the echo method is nearly identical to the treatment already given by Torrey with regard to transient nutations.[9] However, a great practical improvement in eliminating noise is available with the echo technique which cannot be assessed from formal analysis; namely, that H_1 is absent during the observation of nuclear signals, and noise or hum that may be introduced by the oscillator and associated bridge components is avoided.

The author wishes to thank Professor J. H. Bartlett for his counsel in carrying out this research, and is grateful to Dr. C. P. Slichter for the benefit derived from many clarifying discussions with him regarding this work. The author is indebted to H. W. Knoebel for his excellent design and construction work on the electronic apparatus.

FREE NUCLEAR INDUCTION

By *E. L. Hahn*

Unusually clear measurements of the time precession of nuclear magnetic moments and other properties of nuclei have become possible with the development of the free induction "spin echo" technique in the still new science of nuclear magnetic resonance.

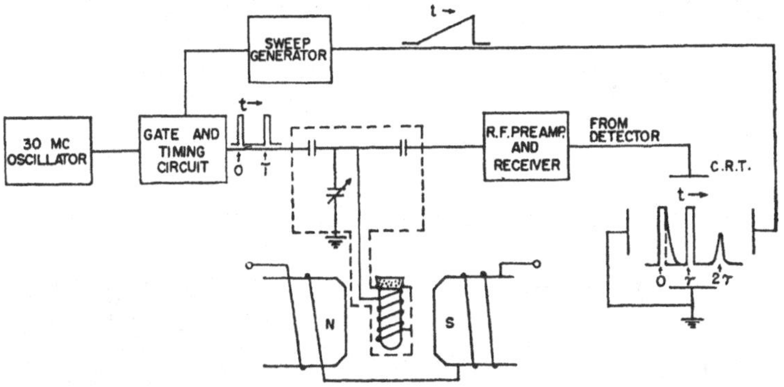

Fig. 1. Arrangement for obtaining spin echoes.

THE STUDY of nuclear magnetic resonance or nuclear induction, a recent field of research for which F. Bloch [1] and E. M. Purcell [2] have been awarded the Nobel Prize, has been carried out by a variety of techniques. The usual approach has been to observe the nuclear resonance of an ensemble of nuclear moments in a large static magnetic field as a function of a slow change in this field. Meanwhile, a small radio-frequency field is applied continuously to the nuclear sample in a direction perpendicular to the large field. An alternative method to this steady state or "slow passage" technique is one by which the radio-frequency energy is applied to the sample in the form of short, intense pulses, and nuclear signals are observed after the pulses are removed. The effects which result can be compared to the free vibration or "ringing" of a bell, a term often applied to the free harmonic oscillations of a shocked inductive-capacitive (LC) circuit. The circuit is first supplied with electrical energy from some source, and the supply of energy is suddenly removed. The LC circuit then remains for a time in the "excited state", and the energy is gradually dissipated into heat, mostly in the circuit resistance. Similarly the atom or nucleus in the excited state can store energy for a time before it is completely dissipated, and in the case discussed here, the free oscillation or precession of an ensemble of nuclear spins in a large static field provides the ringing process.

It appears that the topic of free nuclear precession or "spin echoes", as it will be called here, can be classified under one of the following two approaches to the study of excited states: (a) A study can be made of the absorption or emission of radiation by a system. The system gains or loses radiation (or particles), and the experiment involves a measurement of the energy and intensity of radiation. (b) The behavior of some systems can be studied, not by observing directly whether they gain or lose radiation, but by detecting their mode of motion under the influence of applied *static* electric and magnetic fields. While observing such motion it is possible to infer whether or not the system has gained or lost radiation in the past.

It is well known that the latter approach is involved in the Rabi molecular beam technique [3] in which molecules and atoms are deflected in space. Also, this approach applies to the free nuclear precession or induction effect. The viewpoint of the experiment has been particularly emphasized by Bloch,[1] and is very much like the scheme for detecting the ringing of a tuned LC circuit. A pickup loop can be coupled to the magnetic flux about the inductance and a voltage of induction is measured. In the actual experiment the magnetic induction is provided by the precession of an ensemble of nuclear moments in a static magnetic field after the moments are "shocked" into a coherent state of precession by one or more pulses of radio-frequency magnetic field. One might classify this measurement under (a) above, and deduce that here the ringing of the system is measured by detecting the stored magnetic energy which the system dissipates. Certainly some radio-frequency (rf) energy is consumed by the loop which couples to the ringing LC circuit, but this can be made

E. L. Hahn, a physicist at the Watson Laboratory and a member of the physics staff at Columbia University, received his PhD at the University of Illinois in 1949. He was a National Research Council Fellow and instructor at Stanford University in 1951–52.

Reprinted from Physics Today 6, 4–9 (1953); ©American Institute of Physics.

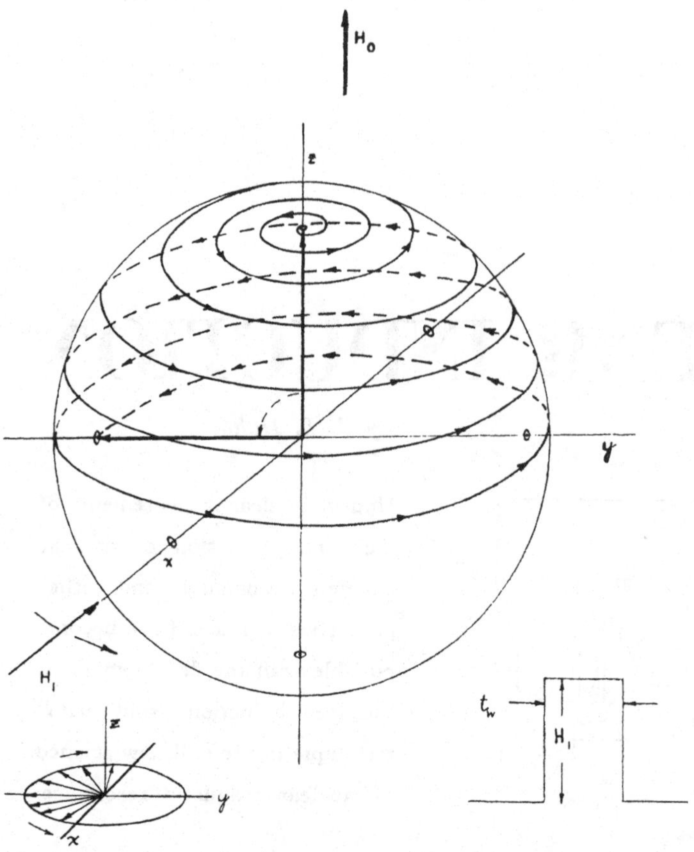

Fig. 2a. As seen in the laboratory frame of reference the classical magnetic moment precesses toward the *xy* plane in a spiral motion due the torque effect of the rf field H_1 at nuclear resonance. In the frame of reference rotating with H_1 the magnetic moment appears to precess in a plane perpendicular to H_1.

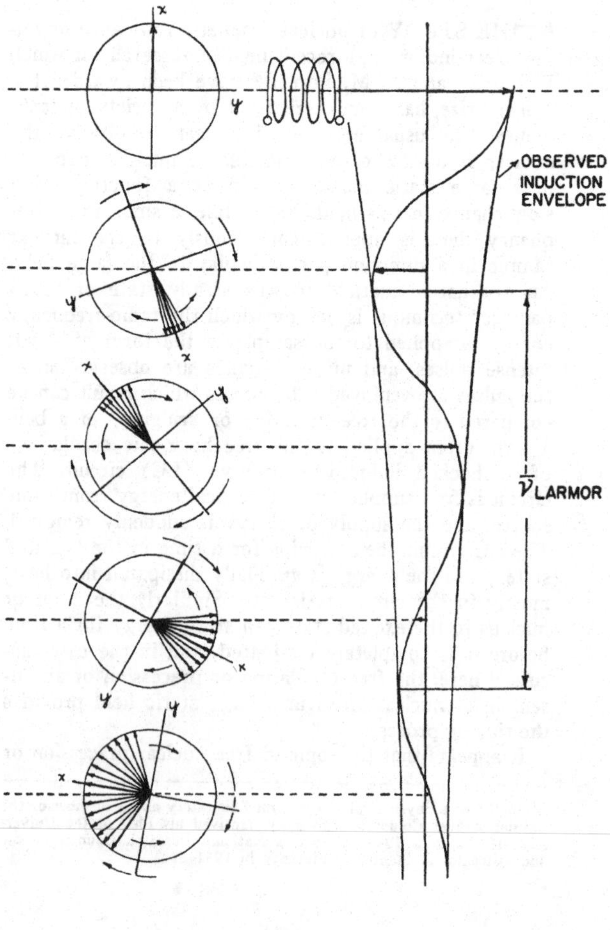

Fig. 2b. Following the orientation of the total magnetic moment into the *xy* plane by a 90° pulse the spectrum of moment vectors provides a free nuclear induction signal. The precessional motion relative to the moving and fixed frames of reference is indicated.

negligibly small by using a loop of high-circuit impedance. In the free precession experiment, a sample of nuclei is placed in a pickup loop comprising the inductance of an LC circuit tuned to the Larmor precession frequency. A low voltage of induction (about a millivolt at most due to protons) is produced across the inductive coil by the nuclear spins, and the current which flows in the coil produces a weak rf magnetic field throughout the volume of the sample. This field does react upon the nuclear spins, but causes only a negligibly small rate of induced emission. The ringing oscillator, in this case the nuclear moments, is again very weakly loaded by the pickup coil.

If any clear distinction can be made between the terms "nuclear magnetic resonance absorption" and "nuclear induction", it can be made in the free precession experiment. Nuclear resonance absorption belongs under heading (a) above. Radio-frequency energy in the form of pulses, absorbed by the nuclear moments. prepares the moments for the ringing process which is observed only by nuclear induction under heading (b).

A qualitative description of the experiment involves many of the basic features of nuclear resonance techniques. A large constant magnetic field H_0 must be available in which polarization and precession of the nuclear spins take place. In practice this field is never perfectly homogeneous for all of the nuclei throughout the volume of the sample. Instead the field varies throughout space in a manner which is determined by the inhomogeneity of the magnet or also by local magnetic fields in the lattice of the sample. In liquids and gases, however, except for certain special but small molecular effects, any local magnetic field in the lattice

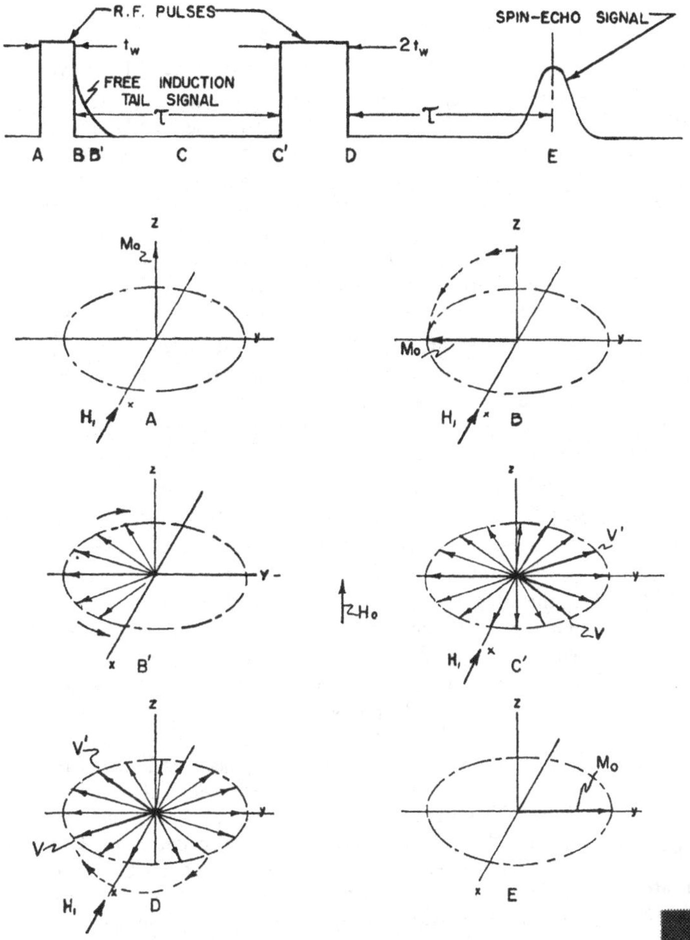

Fig. 3. Vector schematic of the spin echo formation.

Fig. 4. Oscillographic display of the spin echo due to protons in glycerine. The rf pulse widths are narrow compared to the duration of the signals.

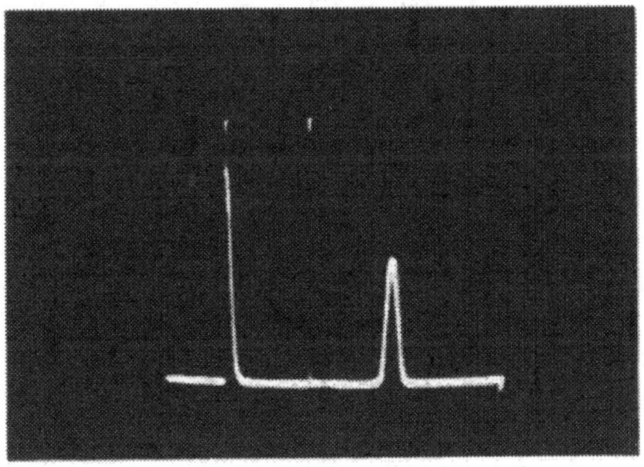

which a given nuclear moment sees as a result of its neighbors will average completely to zero over a Larmor period. This happens because the tumbling and translational frequencies common to liquid molecules at normal temperatures, for example, are extremely large compared to the Larmor frequency of precession (greater by a factor of 10^4 to 10^5). During one Larmor period of precession the field caused by a neighboring dipole does not remain at a given value long enough to influence the rate of precession determined by the externally applied field H_0. It becomes possible, therefore, to ascribe to each volume element of the liquid an isochromatic or classical magnetic moment, which is due to the preponderance of nuclear moments pointing in the direction of the external field. This field can be assigned as homogeneous over the volume element. The entire liquid sample provides a distribution of magnetic moments according to how much volume of spin sample is assigned to each value of H_0 as it varies in space. Such a spectrum may be described by some symmetric distribution, where the maximum number of nuclear moments may be subject to a field of 7000 gauss, for example, and fewer moments see values of H_0 smaller or greater than this average value.

At thermal equilibrium the net magnetic moment M_0 which is aligned with the field can be compared to the "sleeping" mechanical top that spins with its axis along the direction of the gravitational field. If the top is perturbed by applying torque perpendicular to its spin axis, it will precess in a certain direction at a given frequency for any angle θ which exists between the spin axis and the direction of the gravitational field. The nuclear magnetic top has a torque exerted upon it by the magnetic field H_0 in place of gravity, and when it is tipped away from the H_0 or z axis direction by rotating rf magnetic fields in the xy plane, it consequently precesses about the z axis at the Larmor frequency given by $\omega_0 = \gamma H_0$ after the xy perturbing field is removed. The constant γ is the gyromagnetic ratio defined by $\gamma = \mu / I\hbar$, where μ is the magnetic moment and $I\hbar$ is the nuclear spin angular momentum. The time that the induction signal due to a classically precessing $\overline{M}_0$ vector can persist is also the time for which constituent nuclear spins precess in phase before damping effects due to the lattice become appreciable. This coherence time is given by T_2, often referred to as the "transverse relaxation time". Another relaxation time of importance, which determines in part the value of T_2, is the longitudinal or thermal relaxation time T_1, the time in which a precessing spin remains in the magnetically excited state regardless of its phase. In liquids both of these relaxation times may vary from fractions of milliseconds to several seconds.

After a few preliminary definitions, we shall impose special experimental conditions for purposes of clearly explaining the echo nuclear induction effect. An inductive driving coil which surrounds the nuclear sample is tuned to the Larmor frequency $\omega = \omega_0'$ where ω_0' is the average angular Larmor frequency of a sample of nuclear moments. rf pulses applied to the tuned LC circuit each last for t_w seconds during which time approximately $\omega_0' t_w$ Larmor oscillations take place. The effective rf magnetic field is referred to as the $\bar{H}_1$ Gauss vector which precesses in the direction that $\bar{M}_0$ precesses, and is the vector which remains essentially perpendicular to the plane defined by $\bar{M}_0$ and $\bar{H}_0$ (see Fig. 2). In a coordinate system which rotates with frequency $\omega_0'/2\pi$ about the z axis, the $\bar{M}_0$ vector will then appear to precess about $\bar{H}_1$ through angle $\theta = \gamma H_1 t_w$ from the time $\bar{H}_1$ is turned on. This $\bar{H}_1$ field is one of two circularly polarized field components which sum to provide the alternating magnetic field $2H_1$ along the axis of the inductive coil. The other $\bar{H}_1$ field component, rotating in the opposite direction, can be ignored because its torque acting upon $\bar{M}_0$ is alternately positive and negative, and averages to zero for all practical purposes. We shall assume that $\bar{H}_1$ is turned on infinitely fast, and that it shall be removed after t_w seconds infinitely fast. During the time t_w, all isochromatic moments, initially aligned along $\bar{H}_0$, will be turned away from the z axis toward the xy plane. The driving radiofrequency pulses may be transmitted by the same coil used for receiving signals, or two separate coils perpendicular to each other may be used, one for transmitting and one for receiving. If $\bar{H}_1$ rotates in the xy plane at a frequency ω, all those isochromatic moments which are tuned to Larmor frequency ω_0, different from ω, will, strictly speaking, not precess exactly in a plane perpendicular to the direction of $\bar{H}_1$. However, $\bar{H}_1$ is chosen to be sufficiently intense so that most of the isochromatic moments are rotated toward the xy plane in a time short compared to the time in which they would get out of phase with $\bar{H}_1$ (i.e. they would scarcely deviate from the plane defined by $\bar{M}_0$ and $\bar{H}_0$, perpendicular to $\bar{H}_1$). Therefore all the isochromatic moments are substantially in phase at the time t_w when they have reached the xy plane. At this time t_w the field $\bar{H}_1$ is suddenly removed. A nuclear induction signal in the receiver or pickup coil will persist after the field $\bar{H}_1$ is removed, but will finally die out because each isochromatic moment is now free to precess at its natural frequency. Since these frequencies differ for each isochromatic moment, the moments will, after a time, get out of phase and their inductive effects will interfere or cancel among themselves.[4] The time required for this loss of a net observed induced output signal is usually determined by the inhomogeneity of the magnet. After such a time, the isochromatic moment vectors are uniformly distributed about the z axis, but the original magnitudes of these isochromatic moments are preserved if it is assumed that there are no relaxation effects. Although

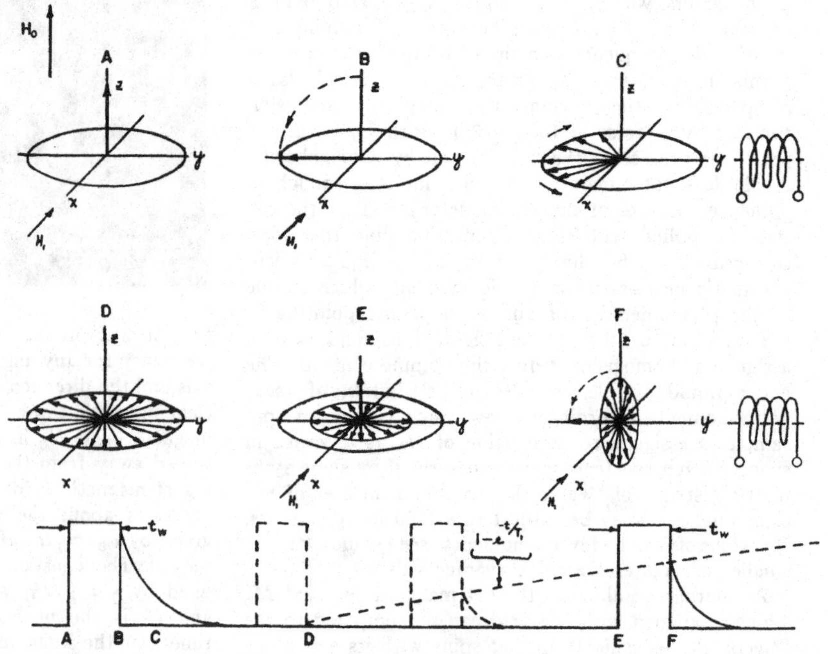

Fig. 5. Scheme for measurement of T_1. The sequence from A through D is the same as in Fig. 3. At the time a pulse is applied at time E a fraction of the total nuclear magnetism has returned to thermal equilibrium along the z axis. Following this pulse at F a free induction signal results which is proportional to this magnetism.

these isochromatic moment vectors do not provide a resultant moment, they do have definite phase relations among themselves, and each vector occupies a position which has been determined by its past history. If at a time τ after cessation of the first pulse, a second pulse is transmitted to the driving coil, this past history is manifested in what is called the "spin echo".

The echo effect [5] can be explained from a very simple analogy. Let a team of runners with different but constant running speeds start off at a time $t = 0$ as they would do at a track meet (see the cover of this issue). At some time T these runners will be distributed around the race track in apparently random positions. The referee fires his gun at a time $t = \tau > T$, and by previous arrangement the racers quickly turn about-face and run in the opposite direction with their original speeds. Obviously, at a time $t = 2\tau$, the runners will return together precisely at the starting line. This will happen once and only once, just as it will be shown in the case of two rf pulses and the echo. From this analogy, one can see that if even more than two pulses are applied to the ensemble, a pattern of echoes or constructive interference events will occur which will uniquely be related to the pulses which were applied in the past. For example, if the referee again fired his gun for a third time after the racers came together at the starting point and fanned out again around the track, and the runners again repeated the about-face procedure, they would again come back to the starting line.

For the purpose of simplifying the description of the actual echo formation, as shown in Fig. 3, the second pulse is made either of the same intensity and twice as long as the first pulse or twice as intense with the same duration as the first pulse. From the time $\bar{H}_1$ appears again in the xy plane, each isochromatic moment vector precesses in a cone whose axis is the direction of $\bar{H}_1$. At the instant the second pulse is removed, all vectors will have been rotated from whatever xy plane quadrants they happened to have been in (at the onset of the second pulse) on one side of $\bar{H}_1$ to a mirror image position on the opposite side of $\bar{H}_1$. The second pulse, so to speak, has flipped the "pancake" of isotropically distributed moments by 180°. When this flip has occurred it can be seen that if we refer to the central average isochromatic moment, each isochromatic moment which lay ahead of this average moment by a given angle before the second pulse now lies behind it by the same angle. Furthermore, each isochromatic moment which lay behind the average isochromatic moment by a given angle will lie ahead by the same angle. Now if these isochromatic moments continue to precess as before, those behind the reference vector or average isochromatic moment will be catching up and those ahead will be falling back. Hence, at time τ beyond the second pulse, all the moment vectors will be back in phase and the echo of the first pulse will occur at time 2τ where τ is the time between the first pulse and the second or reversing pulse. This can be seen by tracing the history of a pair of vectors from the figure. At the onset of the first pulse, the moments M_0 lie on the z axis at thermal equilibrium at A. Following a rotation of 90°, completed at B by the first pulse, the isochromatic moments spread out as shown in B'. During this time, an induction signal forms as a "tail" following the first pulse. At C and C' the tail is absent because the isochromatic moments are evenly distributed in the xy plane. Follow, for example, the precessional motion of isochromatic moment vectors V and V' shown at C'. Here they happen to be oriented in positions indicated at the onset of the second pulse, which now rotates them and the whole array in the xy plane by 180°, as shown at D. After the pulse is removed at D, the vectors V and V' will proceed to precess through angles in the xy plane again in a time τ as they did after the first pulse. They obviously must coincide at the time of the echo at E. This argument holds for every pair of vectors in the ensemble for the special case given here. The spin echo has a shape which grows and dies out symmetrically in the time it takes for the isochromatic moments to get in phase and then out of phase.

It should be noted at this point that although it has been assumed for the sake of simplicity that the first pulse rotates all the vectors by 90°, and the second pulse rotates them by 180°, these rotations happen to be the ones which give the maximum available echo. Useful results may also be obtained by use of other arbitrary combinations of t_w and H_1 giving different angles of rotation. For example, the second pulse may be of equal length and equal intensity as the first pulse, in which case the array is rotated 90° rather than 180° as described above.

Useful information about the local magnetic fields due to chemical environment about the precessing nucleus can be obtained in certain cases from measurements of shapes and amplitudes of free induction echo signals. Steady state resonance techniques of course can provide the same information, which in some cases is more direct, particularly when a number of closely spaced transitions between stationary states of nuclear Zeeman levels are indicated directly by resonance lines.[6] The equivalent of such small differences in energies of closely spaced Zeeman levels is manifested in the echo method by interference beats [7] between the various components of precessing magnetic moments which precess at different Larmor frequencies in the same sample of spins subjected to a given external H_0. If the maximum echo signal amplitude is measured for increasing values of τ, where for each setting of τ the ensemble is initially at thermal equilibrium, a plot of the echo signal amplitude as a function of τ displays predominantly a monotonic decay. In many cases the decay is exponential and serves as a direct measure of the nuclear spin relaxation parameter T_2, but the decay may also be determined in part by other factors. There are other special ways of studying echo signal plots which do not require that the ensemble be at equilibrium for just a pair of pulses, where the nature of the information obtained is essentially the same.

The decay of the echo may be understood in terms of the race track analogy if it is assumed now that the runners become fatigued after the start of the race. For this reason they may change their speeds erratically or even drop out of the race completely. Consequently, following the second gun shot (the second pulse) some of the racers may return together at the starting line but not all of them. In the nuclear spin system a similar situation prevails: either (1) the nuclear spins return to thermal equilibrium or (2) they lose phase memory of Larmor precession. The effect (1) occurs when the magnetic energy of precession contained by the moment is transferred to a molecule completely in the form of kinetic energy. The time in which this effect occurs is called T_1, the spin-lattice thermal relaxation time. The effect (2) arises when magnetic energy is transferred from spin to spin. One must also add to this the effect due to fluctuating local magnetic fields caused by neighboring moments and paramagnetic substances. The over-all time which relates to processes in the lattice which shorten the phase memory of precession has been denoted by T_2, which includes the effect of T_1 as well.

Another influence which destroys the phase memory of precession is that due to the self diffusion of molecules which contain resonant nuclei. Since there is an established gradient of the magnetic field over the volume of the sample, a molecule whose nuclear moment has been flipped initially in a field H_0, may, in the course of time 2τ, drift by Brownian motion into a randomly differing field H_0. Therefore, as τ is increased, a lesser number of moments participate in the generation of in-phase nuclear radio-frequency signals. The theory of the diffusion effect can be incorporated into the nuclear equations, and a useful expression is obtained by which the self-diffusion coefficient of molecules can be measured from the plotted envelope curve and known parameters.

It is possible to measure the thermal relaxation time T_1, independently of all other effects, by measuring signals proportional to the excess population of moments in the z direction at any time, as shown in Fig. 5. This can be done by measuring the amplitude at some arbitrary point on the free induction tail following the second radio-frequency pulse. This is compared to the amplitude at the corresponding point on the free induction tail on the induced signal following the first pulse, which is the amplitude proportional to the maximum available moment M_0. It can be shown that if $\gamma H_1 t_w = \pi/2$ for both pulses, then the tail signal following the second pulse is proportional to the number of gyromagnetic moments which have been thermally relaxed during the time τ. There are alternate methods of measuring T_1, for instance, from the observation of echoes obtained from an application of more than two radio-frequency pulses.

Recently a very interesting phenomenon of a ringing system in the case of an ensemble of molecular electric dipole moments has been demonstrated by R. H. Dicke at Princeton. The principle is very much like the case for pulsed nuclear induction except that a coherent electric field due to molecular rotation induces a signal in a microwave cavity following a strong microwave pulse at resonance. One aim of Dicke's method is to obtain higher resolution in spite of Doppler broadening. Similarly the pulsed echo method permits high resolution where long relaxation times (the equivalent of narrow natural line widths) can be measured in spite of artificial line broadening due to an inhomogeneous magnetic field.

Bibliography

1. F. Bloch, *Physical Review* **70**, 460 (1946).
2. Bloembergen, Purcell, and Pound, *Physical Review* **73**, 679 (1948).
3. Rabi, Millman, Kusch, and Zacharias, *Physical Review* **55**, 526 (1939).
4. E. L. Hahn, *Physical Review* **77**, 297 (1950).
5. E. L. Hahn, *Physical Review* **80**, 580 (1950).
6. H. S. Gutowsky, D. W. McCall, and C. P. Slichter, *Journal of Physical Chemistry* **21**, 279 (1953).
7. E. L. Hahn and D. E. Maxwell, *Physical Review* **88**, 1070 (1952).

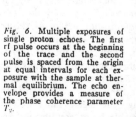

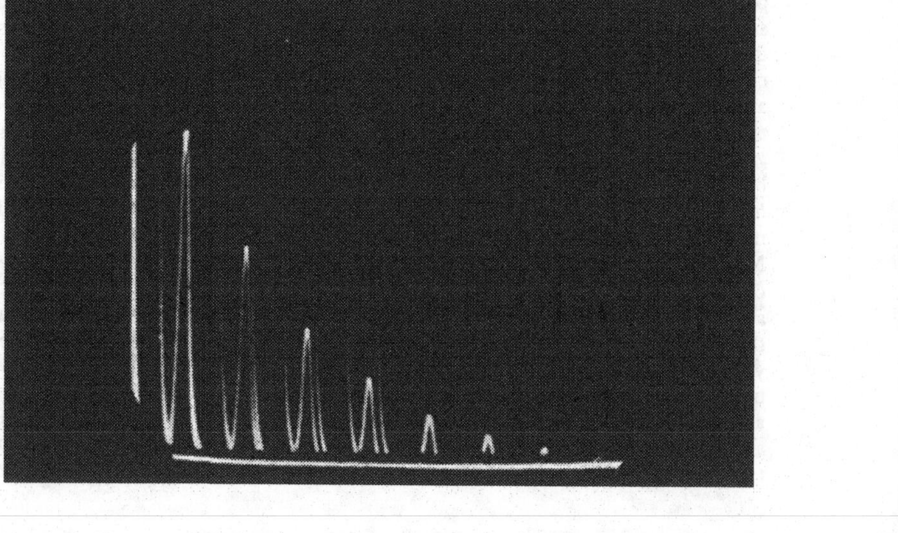

Fig. 6. Multiple exposures of single proton echoes. The first rf pulse occurs at the beginning of the trace and the second pulse is spaced from the origin at equal intervals for each exposure with the sample at thermal equilibrium. The echo envelope provides a measure of the phase coherence parameter T_2.

Effects of Diffusion on Free Precession in Nuclear Magnetic Resonance Experiments*†

H. Y. Carr, *Department of Physics, Rutgers University, New Brunswick, New Jersey*

AND

E. M. Purcell, *Lyman Laboratory of Physics, Harvard University, Cambridge, Massachusetts*

(Received January 19, 1954)

Nuclear resonance techniques involving free precession are examined, and, in particular, a convenient variation of Hahn's spin-echo method is described. This variation employs a combination of pulses of different intensity or duration ("90-degree" and "180-degree" pulses). Measurements of the transverse relaxation time T_2 in fluids are often severely compromised by molecular diffusion. Hahn's analysis of the effect of diffusion is reformulated and extended, and a new scheme for measuring T_2 is described which, as predicted by the extended theory, largely circumvents the diffusion effect. On the other hand, the free precession technique, applied in a different way, permits a direct measurement of the molecular self-diffusion constant in suitable fluids. A measurement of the self-diffusion constant of water at 25°C is described which yields $D=2.5(\pm 0.3)\times 10^{-5}$ cm^2/sec, in good agreement with previous determinations. An analysis of the effect of convection on free precession is also given. A null method for measuring the longitudinal relaxation time T_1, based on the unequal-pulse technique, is described.

I. INTRODUCTION

In order to explain the spin-echo envelopes obtained from certain samples in his nuclear magnetic resonance free precession experiments, Hahn[1] found it necessary to consider the diffusion of the molecules through the inhomogeneous external field. Recent experiments using a somewhat modified free precession technique have allowed further observation and analysis of the diffusion effects. These studies are interesting for two reasons. First, they provide a firmer basis for the use of the free precession method in measuring nuclear magnetic relaxation times. Secondly, for suitable samples a relatively direct measurement of the diffusion coefficient is possible.

In order to introduce terminology and notation, the principles of the free precession method will be briefly reviewed. Hahn[1] has described typical apparatus. The only additional equipment used in our experiments consists of two independent pulse width controls and more flexible triggering circuits.

We will denote the total magnetic moment of the nuclear sample by the vector $\mathbf{M}$. At thermal equilibrium, $\mathbf{M}$ is parallel to the strong applied field, $\mathbf{H}=H_z\mathbf{k}$, and its magnitude is given by the product of H_z and the static nuclear susceptibility. We denote this equilibrium value by M_0.

In describing the precessional motions of the nuclei of the sample it is necessary to take account of the inhomogeneity of the applied field. The fraction of the nuclei associated with a value of H_z in a small interval about a particular H_z will be denoted by $f(H_z)dH_z$. The normalized distribution function $f(H_z)$ is considered centered at $H_z=H_{z0}$. The field deviations are $\Delta H_z=H_z-H_{z0}$. The root-mean-square deviation σ may be used as a measure of the width of the distribution. With each value of H_z and with an increment dH_z, we associate an incremental magnetic moment $d\mathbf{M}(H_z)$, which is the vector sum of the moments of all nuclei within regions where the field strength is between H_z and H_z+dH_z. The motions of these incremental net magnetic moment vectors are described by the equation

$$\frac{d}{dt}d\mathbf{M}(H_z)=-\gamma\mathbf{H}\times d\mathbf{M}(H_z), \qquad (1)$$

where γ the gyromagnetic ratio is the ratio of the magnetic moment to the nuclear angular momentum. Thus an incremental magnetic moment vector making a finite angle with $\mathbf{H}$ will precess in a negative sense for $\gamma>0$ with angular frequency of magnitude $\gamma|\mathbf{H}|$ about the direction of $\mathbf{H}$. In nuclear magnetic resonance studies it is customary to describe this precession in a rotating coordinate system in which all or part of the magnetic field is effectively eliminated. This procedure has been described elsewhere.[2] In the transformations used in our discussion only H_z or portions of H_z will be eliminated. For example, if $d\mathbf{M}(H_{z0})$ makes an angle θ with $\mathbf{H}=H_{z0}\mathbf{k}$, it will precess in the laboratory coordinate system about the z axis with angular frequency having magnitude γH_{z0}. By transforming to a rotating coordinate system having the angular frequency $\mathbf{\Omega}=-\gamma H_{z0}\mathbf{k}$, $d\mathbf{M}(H_{z0})$ will appear stationary, as though H_{z0} had been eliminated. This may also be seen from the transformation equation

$$\frac{\delta}{\delta t}d\mathbf{M}(H_{z0})=-\gamma\left(\mathbf{H}+\frac{\mathbf{\Omega}}{\gamma}\right)\times d\mathbf{M}(H_{z0}). \qquad (2)$$

* This work was partially supported by the joint program of the U. S. Office of Naval Research and the U. S. Atomic Energy Commission, by The Radio Corporation of America, by the Rutgers University Research Council, and by the United States Air Force under a contract monitored by the Office of Scientific Research, Air Research and Development Command.

† Portions of this work have been previously described by H. Y. Carr, thesis, Harvard University, 1952 (unpublished).

[1] E. L. Hahn, Phys. Rev. 80, 580 (1950).

[2] R. K. Wangsness, Am. J. Phys. 21, 274 (1953).

However, for the more general incremental vector $dM(H_z)$ not all of the H_z is eliminated in the above rotating coordinate system. As a result, $d\mathbf{M}(H_z)$ precesses with the small angular frequency

$$\omega_\Delta = -\gamma(H_z - H_{z0})\mathbf{k}'.$$

This is illustrated in Fig. 1B.

In free precession experiments the radio-frequency energy is applied to the sample in short intense pulses. The rf magnetic field, as in other nuclear resonance methods, is applied to the sample at right angles to the strong static field. Its magnitude will be denoted by $2H_1$. The frequency of the rf field satisfies the resonant condition $\omega_{rf} = \gamma H_{z0}$. As is well known,[3] only one of the two circularly polarized components composing the linearly polarized rf field is effective in nutating the net magnetic moments. This component, with magnitude H_1, is the one which rotates in phase with the precessing moments. The phase of the rotating system will be so chosen that this H_1 is along the x' axis. With the rf pulse applied, the net field $\mathbf{H} = \Delta H_z \mathbf{k}' + H_1 \mathbf{i}'$ will determine the precession of the incremental moment vectors. This is illustrated in Fig. 1C. However, a special approximation is possible in the present work. The magnitude of the rf field is made so large ($H_1 \gg \sigma$), and the duration of the rf pulse τ_w so short ($\tau_w \ll 1/\gamma\sigma$), that the precession about the net field is effectively equivalent to precession about $H_1\mathbf{i}'$. This precession during the time the rf pulse is applied might be referred to as pure nutation. Furthermore, in the above approximation all incremental moments will rotate together as a single total magnetic moment vector. Figure 2A illustrates the combined precession about $\mathbf{k}$ and nutation about $\mathbf{i}$ as seen from the laboratory frame of reference while Fig. 2B illustrates the pure nutation about $\mathbf{i}'$ as seen from the rotating frame of reference. In a time τ_w the angle of nutation is $\theta = \gamma H_1 \tau_w$. In the present experiments only two angles of nutation are used. These are 90° and 180°. When the rf oscillator gating circuit is so adjusted that

$$\gamma H_1 \tau_w = \tfrac{1}{2}\pi, \tag{3}$$

the net magnetic moment vector will rotate through an angle $\theta = \frac{1}{2}\pi$, for example, from the polar z' axis down to the equatorial plane of the reference sphere. This is referred to as a 90° pulse. Similarly, if τ_w is such that the vector rotates, for example, from the positive z' axis to the negative z' axis, one has a 180° pulse.

Consider a situation in which free precession is initiated with a 90° pulse. Following the removal of the pulse the total net magnetic moment precesses in the equatorial plane and induces a rf signal in the coil surrounding the sample. For liquid and gaseous samples this signal decays, in a time long compared to the τ_w used in our experiments, because of the differences in the precessional frequencies of the incremental moment vectors. Viewed from the rotating frame of reference, the incremental moment vectors appear to fan out. This is illustrated in *C* and *D* of Fig. 3. The magnitude of the total magnetic moment vector, and hence the induced signal, is thereby reduced. The decaying nuclear signal following the 90° pulse is referred to as a "tail." For a symmetrical distribution function no x' component of the total moment vector exists. The y' component is given by

$$M_{y'} = M_0 \int f(H_z) \cos\gamma(H_z - H_{z0})t \, dH_z. \tag{4}$$

This expression describes the time dependence of the tail. For example, if the distribution is an error function, the tail has the form

$$M_{y'} = M_0 \exp(-\sigma^2\gamma^2 t^2/2), \tag{5}$$

while if $f(H_z)$ is a rectangular function of width $2h$,

$$M_{y'} = M_0[(\sin\gamma ht)/\gamma ht]. \tag{6}$$

Wide distributions correspond to fast decays and narrow, or more homogeneous distributions, to slow decays. This is illustrated in Fig. 4. If the distribution function consists of two or more distinct spin groups, such as those caused by different diamagnetic effects

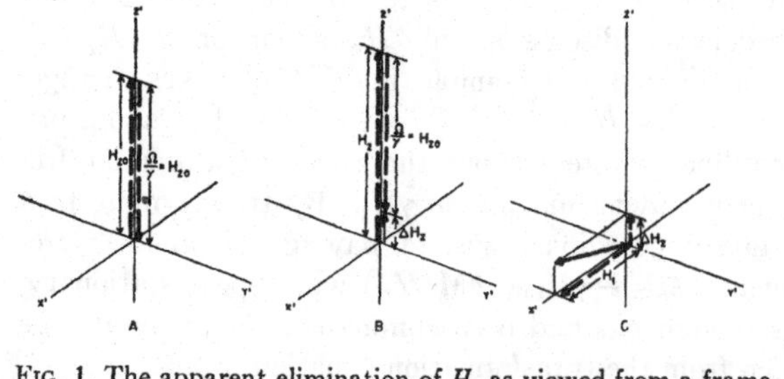

FIG. 1. The apparent elimination of H_z as viewed from a frame of reference rotating with the angular frequency $\mathbf{\Omega} = -\gamma H_{z0}\mathbf{k}$. For $H_z = H_{z0}$, there is no resultant field (*A*); but in a general case (*B*), there is a resultant field $\Delta H_z = H_z - H_{z0}$. When the rf field H_1 is applied, ΔH_z combines with H_1 to determine the direction (*C*) about which rotation occurs.

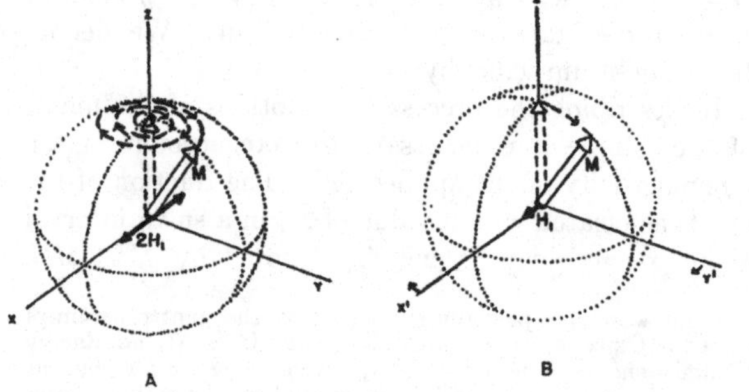

FIG. 2. (*A*) The combination of precession and nutation of the net magnetic moment as viewed from the laboratory frame of reference. (*B*) Pure nutation of the net magnetic moment as viewed from the frame of reference rotating at the precessional frequency.

[3] G. E. Pake, Am. J. Phys. 18, 438 (1950).

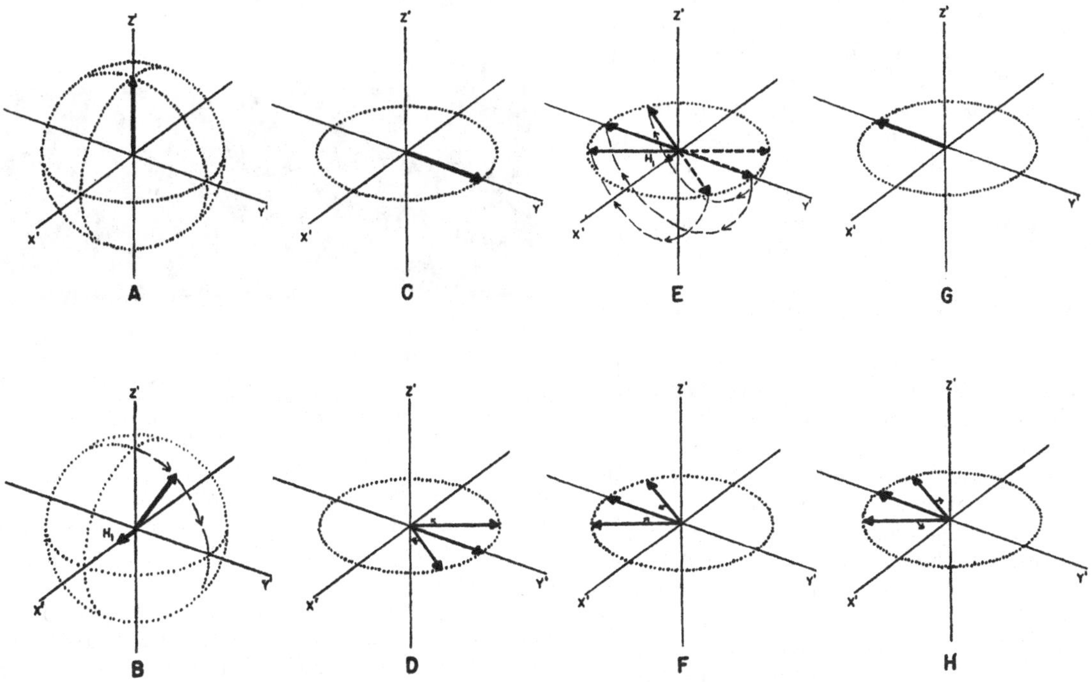

FIG. 3. The formation of an echo. Initially the net magnetic moment vector is in its equilibrium position (A) parallel to the direction of the strong external field. The rf field H_1 is then applied. As viewed from the rotating frame of reference the net magnetic moment appears (B) to rotate quickly about H_1. At the end of a 90° pulse the net magnetic moment is in the equatorial plane (C). During the relatively long period of time following the removal of H_1, the incremental moment vectors begin to fan out slowly (D). This is caused by the variations in H_z over the sample. At time $t=\tau$, the rf field H_1 is again applied. Again the moments (E) begin to rotate quickly about the direction of H_1. This time H_1 is applied just long enough to satisfy the 180° pulse condition. This implies that at the end of the pulse all the incremental vectors are again in the equatorial plane. In the relatively long period of time following the removal of the rf field, the incremental vectors begin to recluster slowly (F). Because of the inverted relative positions following the 180° pulse and because each incremental vector continues to precess with its former frequency, the incremental vectors will be perfectly reclustered (G) at $t=2\tau$. Thus maximum signal is induced in the pickup coil at $t=2\tau$. This maximum signal, or echo, then begins to decay as the incremental vectors again fan out (H).

in different parts of a molecule, the tail will be modulated by their beat pattern. The tail in Fig. 5 illustrates such a beat pattern observed with a proton sample of acetic acid CH_3COOH.

II. METHODS FOR MEASURING TRANSVERSE RELAXATION TIME

In liquids and gases the natural lifetimes of the nuclear signals ordinarily cannot be inferred from the decay of the tail. This decay is usually determined, as described above, by the inhomogeneous external field. However, by means of the spin-echo effect discovered by Hahn, this artificial decay can be effectively eliminated and the natural decay of the nuclear signal observed. The spin-echo effect is associated with the successive application of two or more rf pulses. Hahn has described the effect for equal pulse widths. In the present work combinations of 90° and 180° pulses have been used. This greatly simplifies the explanation of the effect and the interpretation of the observed data.

To obtain an echo using this technique, a 180° pulse is applied at a time τ after the 90° pulse. τ is chosen larger than the artificial decay time of the tail but smaller than the natural lifetime of the nuclear signal. As illustrated in E of Fig. 3, the incremental vectors rotate 180° about the x' axis. When the rf field is removed at the end of the pulse, all incremental vectors are again in the equatorial plane. Since it is assumed that each nucleus remains in the same H_z, each will continue to precess in the same sense and with exactly the same angular frequency as it did before the 180° pulse. Because of this memory and because of their new relative positions after the 180° pulse, the incremental moment vectors will all recluster together in exactly τ units of time after the 180° pulse. This is at time 2τ. At this time there will be a maximum total magnetic moment vector and hence maximum induced signal or an echo. Following $t=2\tau$ the incremental vectors again fan out, and the echo decays in the same manner that it formed. An echo simply appears as two tails back to back. This is vividly illustrated by the echo in Fig. 5.

In order to utilize the above effect in observing the natural lifetimes of nuclear signals, Hahn introduced a method of measurement which will be referred to as

Method A. In this method the amplitude of the echo following a single two-pulse sequence is plotted as a function of time. The data for the different points are obtained by using different values of τ.

The plot may be made photographically by taking a multiple exposure picture of the oscilloscope screen displaying the detected nuclear signals. Each exposure corresponds to a different two-pulse sequence and echo at 2τ. The envelope of the photographed echoes is the desired curve. From considerations mentioned thus far, one would expect that this plot would indicate the natural decay or lifetime of the transverse polarization $M_{y'}$.

This natural decay is expected to proceed exponentially. It is caused, if one restricts the discussion to nuclei without electric quadrupole moments, by the local magnetic field which originates in moments carried by neighboring atoms and molecules. Owing to the Brownian motion of these other dipoles as well as that of the nucleus in question, this local magnetic field varies randomly both in magnitude and in direction.[4] In liquids and gases the variation is usually so much more rapid than the Larmor precession that the local field has practically the same character whether viewed from the stationary or the rotating frame of reference. In this case it is easy to understand what happens. A moment directed originally along the y' axis, exposed to this random field, executes on the reference sphere a random walk made up of very short steps. The whole cluster of moments thus spreads by diffusion over the sphere. The exponential decay follows as a direct and general result. To illustrate with a simple if somewhat artificial model, suppose that the local field **h** at a given nucleus changes exactly once every τ_c seconds to a new magnitude and direction randomly selected. The other nuclei experience similar, but uncorrelated, variations in their local fields. It is then easy to show that

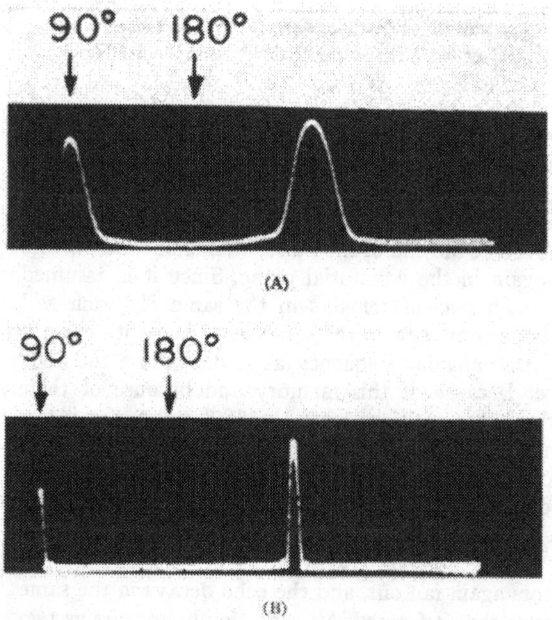

FIG. 4. Effect of the homogeneity of the field on the widths of tails and echoes. The tails and echoes associated with a sample (A) in a very homogeneous field are wide compared to those associated with a sample (B) in a less homogeneous field.

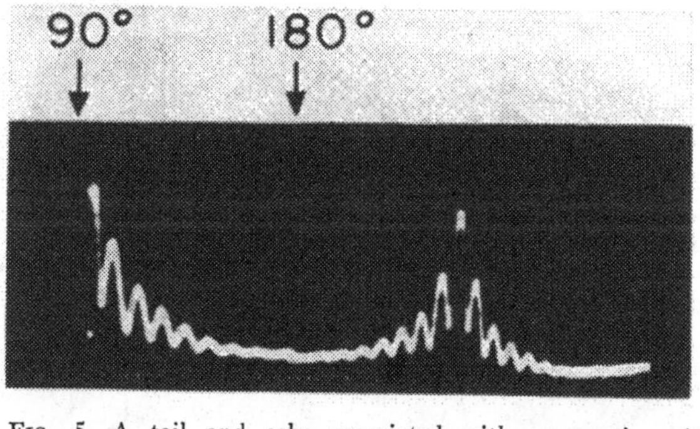

FIG. 5. A tail and echo associated with an acetic acid (CH_3COOH) sample in a very homogeneous field. The nuclear signal displays modulation typical of the beating of two signals of nearly equal frequencies, the magnitude of one being three times that of the other. The beat frequency shown, corresponding to room temperature and 7000 gauss, is approximately 220 cps.

$$M_{y'}(t)=M_0\exp(-\gamma^2\langle h^2\rangle_{Av}\tau_c t/3), \qquad (7)$$

where

$$\langle h^2\rangle_{Av}=\langle h_x^2\rangle_{Av}+\langle h_y^2\rangle_{Av}+\langle h_z^2\rangle_{Av}. \qquad (8)$$

Many other approaches lead to the same result.[5,6]

In the cases of certain samples the echo envelopes obtained by Hahn's method, Method A, have essentially the above exponential form. However, for certain other nonviscous samples such as pure water the envelope is more nearly of the form $\exp(-kt^3)$. See Fig. 7A. Hahn was successful in showing that this is the form of the decay to be expected if the decay is caused by diffusion of the molecules through the inhomogeneous external field H_z. Therefore if diffusion is important, a decay or relaxation time measured by Method A is "artificially" determined by the external magnet.

In the present work a new method for obtaining an echo envelope which reveals the "natural" decay even in the presence of diffusion has been introduced. This method of measurement will be referred to as Method B. In Method A a number of 90° pulses must be applied to the sample in order to obtain sufficient data to plot the envelope. Each 90° pulse is followed by an 180° pulse and the accompanying echo. Between each 90°–180° sequence sufficient time must elapse to allow the sample to return to its thermal equilibrium condition. In Method B a single 90° pulse is used to obtain an entire echo envelope without requiring the sample to return to thermal equilibrium. This is accomplished by initiating the measurement with the usual 90° pulse at $t=0$. A 180° pulse is then applied at time $t=\tau$. The

[4] Bloembergen, Purcell, and Pound, Phys. Rev. **73**, 679 (1948).

[5] R. K. Wangsness and F. Bloch, Phys. Rev. **89**, 728 (1953).

[6] A. Abragam and R. V. Pound, Phys. Rev. **92**, 943 (1953).

usual echo appears at time $t=2\tau$. Next, to illustrate with a particularly simple case, an additional 180° pulse is applied at time $t=3\tau$. By reasoning identical to that previously used in connection with the first echo, it can be seen that the incremental vectors will recluster at time $t=4\tau$. Thus, a second echo is formed. If this process is continued throughout the natural lifetime of the nuclear signal, additional echoes of decreasing amplitude are formed. The envelope of these echoes

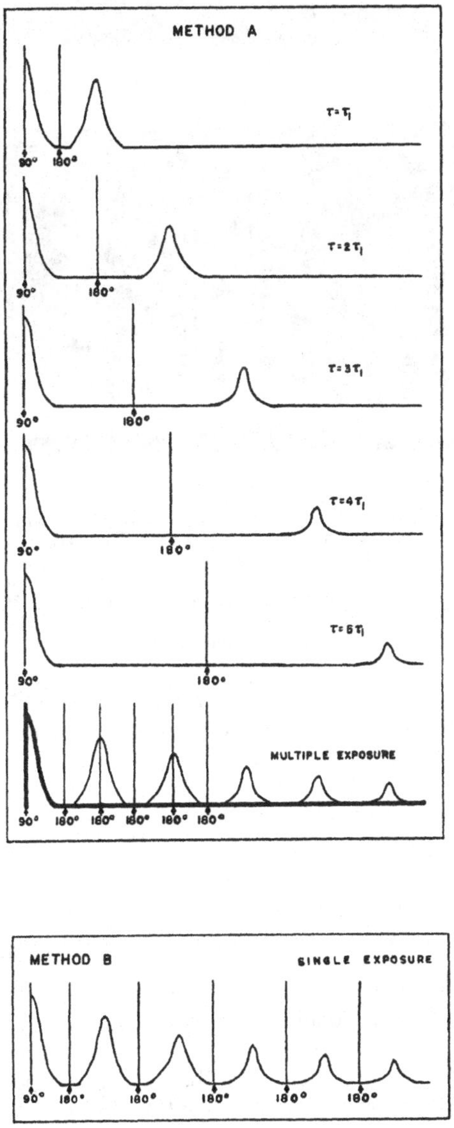

FIG. 6. Comparison of the two free precession methods for observing the decay time of the horizontal component of nuclear magnetic polarization. In "Method A" the sample must return to its equilibrium condition each time an additional echo is to be observed. In "Method B" the sample need only start from equilibrium once. The separate oscilloscope traces of "Method A" are usually displayed superimposed in a multiple exposure picture. Only a single trace and hence a single exposure picture is required in "Method B." However this latter method requires more than one 180° pulse.

indicates the decay of the polarization in the equatorial plane. In some experiments as many as 30 echoes have been formed following a single 90° pulse. Methods A and B are compared in Fig. 6.

In the case of viscous samples with negligible diffusion, the echo envelope obtained by Method A is identical to that obtained by Method B. This is not so if certain nonviscous samples are used. Figure 7 vividly illustrates the difference in envelopes obtained by the two methods if a water sample is used. The top photograph illustrating Method A is a multiple exposure picture of a series of thirteen 90°–180° pulse sequences, each with an accompanying echo at a different time 2τ. The decay is of the form $\exp(-kt^3)$. The time constant is approximately 0.2 sec. The lower photograph is a single exposure picture of a single sequence of pulses consisting of one 90° pulse followed by fifteen 180° pulses. The decay is of the form $\exp(-Kt)$. The time constant is approximately 2.0 sec!

III. THE EFFECT OF DIFFUSION

The above difference between Method A and Method B may be explained by analyzing the effect of the molecular diffusion on the distribution-in-phase of the precessing moments. One may replace the diffusion by a random walk of discrete steps. This makes it easier to formulate certain parts of the problem. As we shall pass finally to the limit of infinitesimal steps, introducing at that point the ordinary molecular self-diffusion constant D, the details of the random walk model are not critical. For simplicity we assume a random walk of the following sort: a molecule remains at a given z position for exactly τ seconds, when it abruptly jumps to a new position whose z coordinate differs from the previous one by ζa_i, where ζ is a fixed distance and a_i a random variable whose value is 1 or -1. The constant linear gradient is in the z direction. ζ might more properly be defined by $(\langle z_i^2\rangle_{\text{Av}})^{\frac{1}{2}}$, the rms z component of a three-dimensional jump of magnitude $\delta=(3\langle z_i^2\rangle_{\text{Av}})^{\frac{1}{2}}$.

The effect we are interested in is the change in the rate of nuclear precession, and the resulting discrepancy in phase, caused by the transport of the nucleus into a region where the applied field is slightly different. Let the gradient of H_z in the z direction be constant, and, in magnitude, G gauss/cm, and let $H_z(0)$ be the field in which a given nucleus finds itself at $t=0$. At some later time $t=j\tau$ the nucleus will find itself in a field $H_z(j\tau)$ given by

$$H_z(j\tau)=H_z(0)+G\zeta\sum_{i=1}^{j}a_i.$$

After N steps, that is after a time $t=N\tau$, the phase ϕ of the precessing moment of this nucleus will differ from the value ϕ_0 it would have had at this same instant if the nucleus had remained in the same place by an angle

$$\phi_D=\phi-\phi_0=\sum_{j=1}^{N}\gamma\tau[H_z(j\tau)-H_z(0)]$$

$$=G\zeta\gamma\tau\sum_{j=1}^{N}\sum_{i=1}^{j}a_i=G\zeta\gamma\tau\sum_{j=1}^{N}(N+1-j)a_j. \quad (10)$$

As the a_j's are random variables anyway, the last sum can just as well be written

$$G\zeta\gamma\tau\sum_{j=1}^{N}ja_j. \quad (11)$$

Now it can be shown that the central limit theorem holds for this sum. Therefore, if at time $t=N\tau$ we consider the distribution in phase of an ensemble of moments, we can assert that the distribution will be Gaussian in the limit of large N. It only remains to calculate $\langle\phi_D^2\rangle_{Av}$. This is easily done:

$$\langle\phi_D^2\rangle_{Av}=G^2\zeta^2\gamma^2\tau^2\sum_{j=1}^{N}j^2$$

$$=G^2\zeta^2\gamma^2\tau^2N(N+1)(2N+1)/6. \quad (12)$$

If N is large—and in this problem it is in fact enormous—we may not only drop the lower powers of N, but we may at the same time pass over to a description in terms of a continuous diffusion rather than discrete random steps. By comparing the solution to the diffusion equation $D\nabla^2f=\partial f/\partial t$, where $f(x,y,z)$ is the probability density, with the solution to the random walk problem of the assumed type, it is easy to show that the appropriate diffusion constant is

$$D=\delta^2/6\tau=\zeta^2/2\tau. \quad (13)$$

We thus have

$$\langle\phi_D^2\rangle_{Av}=2G^2\gamma^2Dt^3/3, \quad (14)$$

and for the distribution-in-phase after time t we have

$$P(\phi_D)=(4\pi\gamma^2G^2Dt^3/3)^{-\frac{1}{2}}\exp(-3\phi_D^2/4G^2\gamma^2Dt^3). \quad (15)$$

This is identical with the result given by Hahn,[1] and it applies directly to the behavior of the signal in the tail after a 90° pulse.

We now extend the analysis to the case in which a number of 180° pulses are applied after the initial 90° pulse. Let the 90° pulse occur at $t=0$ as before. If we return to the discrete model, the effect of a 180° pulse applied at $t=t_1$, for example, is simply to *reverse the signs* of all terms in

$$\phi_D=\sum_{j=1}^{N}\gamma\tau[H_z(j\tau)-H_z(0)] \quad (16)$$

beyond $j=t_1/\tau$. Let us specialize at once to the interesting case in which, within the whole interval $t=N\tau$, n 180° pulses are applied; the first at $t=N\tau/2n$, the second at $t=3N\tau/2n$, and so on. Echoes will then occur midway between the 180° pulses and, in particular, an echo will occur at $t=N\tau$. We are interested in finding the effect of diffusion on this final echo. For this case we must now rewrite ϕ_D as

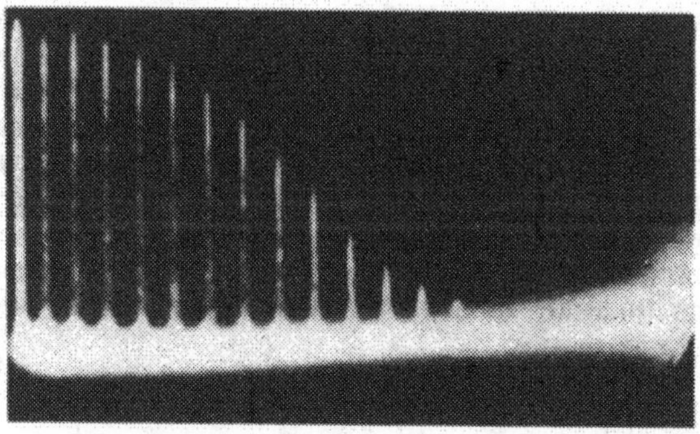

(A)

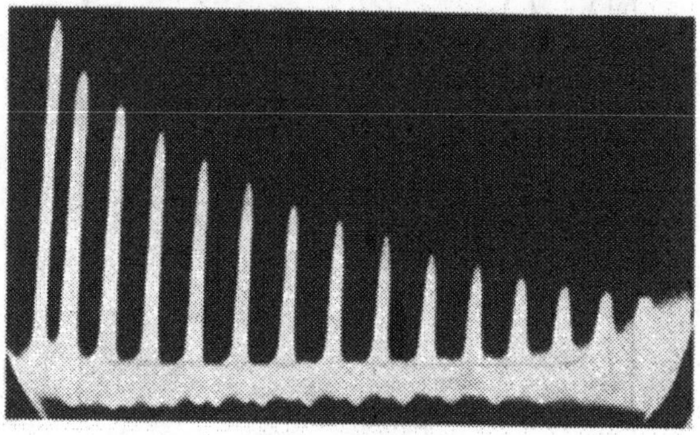

(B)

FIG. 7. (A) A "Method A" decay associated with water at 25°C. The time constant is approximately 0.2 sec. The decay is largely determined by the molecular diffusion through the 0.28 gauss/cm gradient. The decay is dominated by the factor $\exp(-kt^3)$. (B) A "Method B" decay associated with water at 25°C. The time constant of the decay is approximately 2.0 sec. The effect of diffusion has been largely circumvented.

$$\phi_D=\phi-\phi_0=G\zeta\gamma\tau\sum_{j=1}^{N}b_j\sum_{i=1}^{j}a_i, \quad (17)$$

where

$b_j=1$ for $0<j\leqslant N/2n$, $3N/2n<j\leqslant 5N/2n$, etc.,

$b_j=-1$ for $N/2n<j\leqslant 3N/2n$, $5N/2n<j\leqslant 7N/2n$, etc.

By writing out the terms in an array, it is easy to see that the above double sum involving the random variables can be replaced by the sum of n independent sums of the type $\sum_{k=1}^{(N/2n)-1}ka_k$, and n independent sums of the type $\sum_{k=1}^{N/2n}ka_k$. We thus find at once that

$$\langle\phi_D^2\rangle_{Av}=G^2\zeta^2\tau^2\gamma^2[(N^3/12n^2)-N/6], \quad (18)$$

or, in the limit of large N, by

$$\langle\phi_D^2\rangle_{Av}=G^2\gamma^2Dt^3/6n^2, \quad (n\neq 0). \quad (19)$$

The effect of interpolating a larger number n of 180° pulses in the interval of duration t is to reduce the mean square phase dispersion by a factor determined by $1/n^2$. As far as diffusion is concerned, the intensity of the echo at time t is determined by

$$M_{y'}(t)=M_0\int_{-\infty}^{+\infty}\cos\phi_D P(\phi_D)d\phi_D = M_0\exp(-\gamma^2G^2Dt^3/12n^2). \quad (20)$$

Thus decay caused by diffusion can theoretically be eliminated by simply making n large enough. If the effect of the natural decay is included we have

$$M_{y'}(t)=M_0\exp[(-t/T_2)+(-\gamma^2G^2Dt^3/12n^2)]. \quad (21)$$

This quantitative theory successfully explains the echo amplitudes obtained in both Method A and Method B. The form of the envelope obtained in Method A is given when $n=1$. There is a difference in numerical factors in the exponent of the above result and the result given by Hahn.[1] Hahn did not take into account the effectiveness of the 180° pulse in partially eliminating the artificial decay caused by the diffusion.

IV. MEASUREMENT OF A DIFFUSION CONSTANT

As an application of the above considerations, the diffusion coefficient of water at 25°C has been measured. To do this Method A was used to observe the decay of the transverse polarization. Water is a suitable sample for such an experiment. This is determined from the values of T_2 and D. For the effect to be pronounced one must require that

$$12/\gamma^2G^2D \leqslant T_2^3. \quad (22)$$

A gradient G may be obtained at the sample by placing symmetrically on either side of the sample two long current carrying wires or two circular turns of wire. The current directions should be such that the fields oppose. It is not sufficient to satisfy the above condition by simply increasing G. One must keep G small enough to insure that the condition $H_1 \gg \sigma$ of Sec. I is also satisfied. This implies that our analysis in terms of simple 90° and 180° pulses is applicable. Furthermore, if one uses the value of G calculated from the current in the wires, it is necessary that this G be large compared to the average gradient of the field due to the magnet. To satisfy this condition, it is wise to search for a very homogeneous spot in the magnet.

In our measurements H_1 was the order of 10 gauss. This can be determined by measuring τ_w for a 90° pulse. The average gradient of the field due to the magnet was approximately 0.03 gauss/cm. This value can be determined either from the width of a tail following a 90° pulse, or from a Method A envelope with no current flowing in the parallel wires or the circular turns. Since a typical sample dimension is $\frac{1}{2}$ cm, it can be seen that a gradient the order of $\frac{1}{2}$ gauss/cm would satisfy both of the above conditions relating directly to G. T_2 for water at 25°C has been measured by Method B. Diffusion effects were thus eliminated. T_2 has a value of 2.4 sec. D for water at this temperature is the order of 2×10^{-5} cm²/sec. With this information available it can be seen that condition (22) is also satisfied.

The value of the diffusion constant for water at 25°C as determined by this nuclear resonance method is $2.5(\pm0.3)\times10^{-5}$ cm²/sec. The error which is indicated is the estimated maximum limit of error. This result is in good agreement with previous determinations.[7] The slope of a plot of $\ln(M_{y'}/M_0)+t/T_2$ *vs* t^3 is used to obtain D. Errors in determining this slope introduce the major portion of the uncertainty in the final value of D. For increasing values of the gradient, the experimental curve follows the straight line given by the assumed theory less well. Errors associated with the determination of the slope become greater. A gradient of 0.28 gauss/cm was used in the above determination.

The second important source of error occurs in the determination of the gradient. The gradient may be calculated from the current in the wires and the geometrical dimensions of the circuits. Corrections for the image currents in the pole faces of the magnet must be included. The estimated error in the calculation of the gradient is 2 percent. If one uses two circular turns or rings of current creating opposing fields, the gradient may be made very uniform over the volume of the sample by making the square of the diameter of the rings equal to 4/3 times the square of the ring separation. Such a uniform gradient may be measured quite directly by a second method. This involves the observation of the modulation of a tail following a 90° pulse. For a rectangular sample the tail has a form given by Eq. (6) of Sec. I. For a cylindrical sample the tail has the form

$$M_{y'}=M_0 2J_1(\tfrac{1}{2}\gamma Gdt)/\tfrac{1}{2}\gamma Gdt, \quad (23)$$

where $J_1(\frac{1}{2}\gamma Gdt)$ is the first-order Bessel function and d is the diameter of the sample. The value of G obtained by this method has an estimated maximum limit of error of 5 percent. The values obtained by the two methods agree within experimental error.

It should be noted that this method for measuring the diffusion constant is apt to fail if the echo envelope exhibits modulation.[8] Such modulation is likely to occur if the resonant nucleus interacts with another nucleus of the same species which is in a chemically nonequivalent position in the molecule. It also sometimes occurs when the resonant nucleus interacts with another nucleus of a different species.

All methods for measurement of diffusion depend on somehow "labelling" molecules. To the extent that the "label" has a negligible effect on the diffusion process itself, a method may be said to measure truly the "self-diffusion" constant. In the method here described a molecule is in effect labelled by the direction of the nuclear magnetic moment it carries; a more innocuous label would be difficult to imagine.

[7] W. J. C. Orr and J. A. V. Butler, J. Chem. Soc. **1935**, 1273.
[8] E. L. Hahn and D. E. Maxwell, Phys. Rev. **88**, 1070 (1952).

V. THE EFFECT OF CONVECTION

With any method for measuring the diffusion coefficient, it is desirable to be able to check that convection effects are not present. Fortunately we have such a check readily available here. If convection currents exist, the second echo of Method B may be larger than the first echo! This effect is most conspicuous when τ_2, the time between the first and second 180° pulses, is exactly equal to $2\tau_1$, where τ_1 is the time between the 90° pulse and the first 180° pulse. This is illustrated in Fig. 8.

A mathematical explanation of the effect is easily given. Consider the changing phase angle of a group of nuclei which are moving uniformly into larger values of the external field of H_z. For the short distances which the nuclei move in the time between rf pulses, one may assume that the field gradient is linear. Assume that the field increases by h units per unit time. At time τ_1 the net magnetic moment vector of these nuclei will have in the rotating coordinate system a phase angle given by

$$\phi=\int_0^{\tau_1}\gamma(\Delta H_z+ht)dt=\gamma\Delta H_z\tau_1+\tfrac{1}{2}\gamma h\tau_1^2=\phi_0+\phi_C. \quad (24)$$

ϕ_0 is the phase angle which the moment vector would have had if it had remained stationary. However, the vector is out of phase by $\frac{1}{2}\gamma h\tau_1^2$ radians more than ϕ_0. Since this portion of the phase angle is due to convection, it will be denoted by ϕ_C. After a 180° pulse is applied at $t=\tau_1$, the phase angle begins to decrease in magnitude. At time $t=2\tau_1$ the stationary component has been exactly compensated for, and ϕ_0 vanishes. However ϕ_C, being proportional to t^2, has more than compensated for the value it obtained between 0 and τ_1. It now has the value

$$\phi_C=\int_{\tau_1}^{2\tau_1}\gamma htdt-\tfrac{1}{2}\gamma h\tau_1^2=\gamma h\tau_1^2. \quad (25)$$

As time continues, both $|\phi_0|$ and $|\phi_C|$ increase. At time $\tau_1+\tau_2$ just before the second 180° pulse is applied, the value of ϕ_C is given by

$$\phi_C=\int_{\tau_1}^{\tau_1+\tau_2}\gamma htdt-\tfrac{1}{2}\gamma h\tau_1^2=\tfrac{1}{2}\gamma h(\tau_1+\tau_2)^2-\gamma h\tau_1^2. \quad (26)$$

In addition to this, the moment is out of phase by the stationary amount ϕ_0. The second 180° pulse is then applied, and again the phase angles begin to decrease in magnitude. At the time $t=2\tau_2$, ϕ_0 will have been reduced to zero, but the convection will have caused ϕ_C to change by an amount

$$\int_{\tau_1+\tau_2}^{2\tau_2}\gamma htdt=2\gamma h\tau_2^2-\tfrac{1}{2}\gamma h(\tau_1+\tau_2)^2 \quad (27)$$

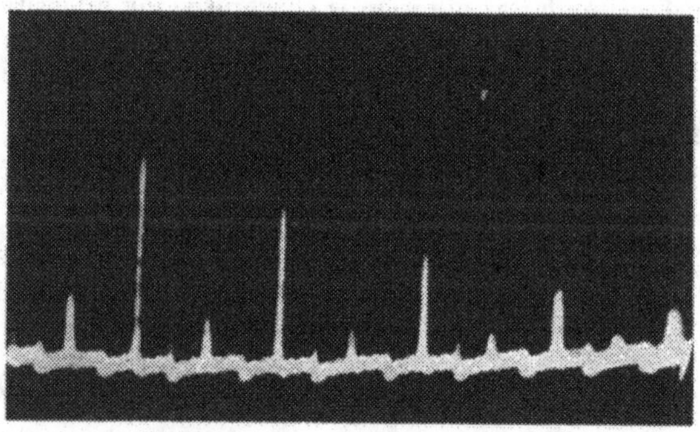

FIG. 8. The effect of molecular convection on a "Method B" decay. The odd-numbered echoes are much smaller than the even-numbered echoes.

radians. Thus, at time $t=2\tau_2$, when the second echo normally occurs, the phase of the net magnetic moment will be given by $\varphi=\varphi_0+\varphi_C$, where

$$\varphi_0=0, \quad \varphi_C=\gamma h\tau_2^2-2\gamma h\tau_1\tau_2. \quad (28)$$

It is clear that it is possible to have all moments perfectly reclustered or in phase, at the time of the second echo, if $\tau_2=2\tau_1$. Such perfect reclustering is not possible at the time of the first echo because of the variations in h over the sample. This implies that with $\tau_2=2\tau_1$, the second echo may have a larger amplitude than the first, in agreement with the observed effect. Thus this effect can be used as an indicator of the presence of convection.

VI. A METHOD FOR MEASURING THE LONGITUDINAL RELAXATION TIME

All discussion thus far has dealt with the equatorial or horizontal component of polarization and the associated relaxation time T_2. In conclusion, it will be mentioned that the use of 90° and 180° pulses makes possible a null method for the measurement of the longitudinal or spin-lattice relaxation time of the sample. This type of measurement is initiated by a 180° pulse. The total magnetic moment vector is inverted from the north pole to the south pole of the reference sphere. Because of the spin-lattice relaxation processes, the z component begins to return to its original value. A 90° pulse is then applied at various times τ. For very short values of τ the 90° pulse nutates the total magnetic moment vector of nearly maximum amplitude from the south pole to the equatorial plane. A tail with nearly maximum amplitude follows the 90° pulse. For very large values of τ, compared to the spin-lattice relaxation time, the 90° pulse nutates the reformed total magnetic moment vector of nearly maximum amplitude from the north pole down to the equatorial plane. Again a tail of nearly maximum value follows the 90°pulse. For intermediate values of τ the tails will have smaller amplitudes. For one value in particular, which will be designated τ_{null}, there will be no tail. Since the system relaxes or returns to equilibrium exponentially with the time constant T_1, it is easily shown that T_1 may be calculated directly from the measured value of τ_{null} by using the relation $\tau_{\text{null}} = T_1 \ln 2$. A 180° pulse at $t=\tau+\tau_2$ will cause an echo to be formed at $t=\tau+2\tau_2$. If $2\tau_2 \ll T_2$ this echo may be used as an indicator of the growth of M_z from $-M_0$ to M_0. The inhomogeneity of H_z plays no role in the decay and growth of the longitudinal component of the polarization; diffusion and convection may here be ignored. One precaution must be noted, however. It has been assumed that each sequence of pulses is initiated only after the system has returned to thermal equilibrium. Thus if the above relation is used to compute T_1, the time interval between sequences must be large compared to T_1.

One of the authors (HYC) wishes to acknowledge many beneficial discussions with H. C. Torrey and P. R. Weiss during the latter period of this research.

Modified Spin-Echo Method for Measuring Nuclear Relaxation Times*

S. Meiboom and D. Gill†
Department of Applied Mathematics, The Weizmann Institute of Science, Rehovot, Israel
(Received May 9, 1958; and in final form, June 6, 1958)

A spin echo method adapted to the measurement of long nuclear relaxation times (T_2) in liquids is described. The pulse sequence is identical to the one proposed by Carr and Purcell, but the rf of the successive pulses is coherent, and a phase shift of 90° is introduced in the first pulse. Very long T_2 values can be measured without appreciable effect of diffusion.

INTRODUCTION

In liquids the relaxation of the magnetic polarization of a system of equivalent nuclei can be described by two relaxation times T_1 and T_2 introduced by Bloch.[1] The longitudinal relaxation time T_1 characterizes the approach to equilibrium of the polarization component in the direction of the external magnetic field. The transverse relaxation time T_2 characterizes the rate at which the transverse component decays to zero, and determines the line width of the nuclear magnetic resonance. In liquids T_2 can amount to many seconds, corresponding to a line width of a fraction of a cycle. Such long relaxation times cannot be determined by direct line-width measurement, as even in the best magnets the inhomogeneity of the magnetic field over the sample will give a very appreciable contribution to the observed line width.

A number of different methods for measuring the "natural" T_2 have been described.[2–6] Of these, the most widely used is the spin echo method due to Hahn.[7] The precision of this method, when measuring very long T_2's, is limited by the self-diffusion in the liquid. A modification described by Carr and Purcell[8] reduces the error caused by diffusion. In the Carr and Purcell scheme a series of rf pulses are applied, the first pulse flipping the polarization through 90° ("90° pulse"), and the following pulses flipping through 180° ("180° pulses"). The time interval between successive 180° pulses is twice that between the 90° pulse and the first 180° pulse. Echos are observed midway between the 180° pulses, the amplitude of successive echos decaying exponentially with a time constant equal to T_2.

In the actual application of the Carr and Purcell method to the measurement of long relaxation times, it was found that the amplitude adjustment of the 180° pulses was very critical. This is because a small deviation from the exact 180° value gives a cumulative error in the result, and the number of 180° pulses should be large in order to eliminate the effects of diffusion. The reproducibility of the measurements was accordingly low.

DESCRIPTION OF THE METHOD

In the modification of the Carr-Purcell method described here, the necessity of a very accurate adjustment of the 180° pulses has been eliminated and the reproducibility improved. The pulse sequence used is identical with that of the original Carr and Purcell scheme: a 90° pulse at time $t=0$ and a series of 180° pulses at times $t=(2n+1)\tau$, $(n=0, 1, 2 \cdots)$. The present method differs in two respects:

1. The successive pulses are coherent.

* This research has been sponsored in part by the Air Force Office of Scientific Research of the Air Research and Development Command, U. S. A. F., through its European Office under Contract No. AF61 (052)-03.

† Taken in part from the dissertation submitted by D. Gill to the Hebrew University, Jerusalem, in partial fulfillment of the requirements for the Ph.D. degree.

[1] F. Bloch, Phys. Rev. **70**, 460 (1946).
[2] H. C. Torrey, Phys. Rev. **76**, 1059 (1949).
[3] R. Gabillard, Compt. rend. **233**, 39 (1951).
[4] J. S. Gooden, Nature **165**, 1014 (1950).
[5] E. E. Salpeter, Proc. Phys. Soc. (London) **A63**, 337 (1950).
[6] Strick, Bradford, Clay, and Craft, Phys. Rev. **84**, 363 (1951).
[7] E. L. Hahn, Phys. Rev. **80**, 580 (1950).
[8] H. Y. Carr and E. M. Purcell, Phys. Rev. **94**, 630 (1954).

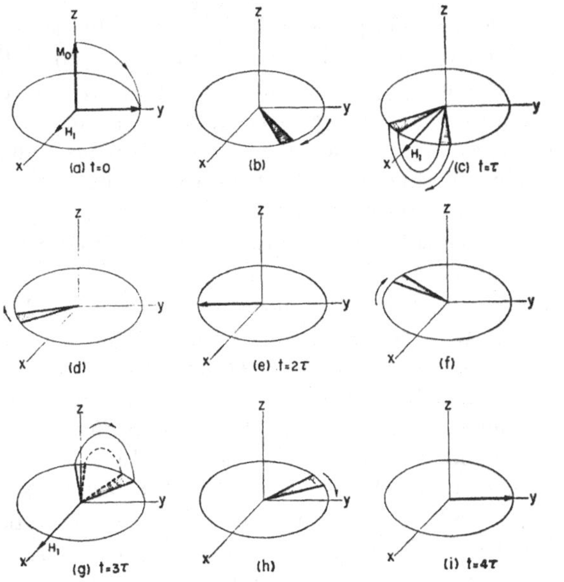

FIG. 1. Behavior of the nuclear polarization in a spin-echo scheme with coherent pulses. A 90° pulse is applied at time $t=0$ and 180° pulses at times $t=(2n+1)\tau$, $(n=0, 1, 2 \cdots)$.

2. The phase of the rf of the 90° pulse is shifted by 90° relative to the phase of the 180° pulses.

The principle of the method is indicated in Figs. 1 and 2. These figures show the behavior of the nuclear polarization relative to a coordinate system rotating with the frequency of the applied rf.[9] Figure 1 is drawn for the case that the pulses are coherent, but without the 90° phase shift of the 90° pulse mentioned under 2. As the pulses are coherent, the direction of the rf field (H_1) relative to the rotating coordinate system is the same for all pulses. This direction is chosen as the x axis of the rotating coordinate system. Figure 1(a) shows the 90° flip of the polarization on application of the 90° pulse. During the interval τ between the 90° pulse and the first 180° pulse, the polarization vector will precess through an angle $\theta=\tau(\gamma H_0-\omega)$, where H_0 is the constant magnetic field and ω the rf frequency of the pulses (and accordingly the angular velocity of the rotating coordinate system). As H_0 is not quite homogeneous, the polarization vectors belonging to the different volume elements of the sample will precess at different rates. The resulting fanning out of the polarization vectors is indicated in Fig. 1(b) by the shaded sector. (In actual cases the fanning out may be much more pronounced and amount to many turns. The extent of the fanning out is however irrelevant for the following discussion.)

Figure 1(c) shows the effect of the first 180° pulse. During the next interval the polarization vectors of the different volume elements in the sample will continue to precess, [Fig. 1(d)], each at its individual rate, and at the end of this interval all will reach the negative y axis simultaneously, so producing the echo [Fig. 1(e)]. Figure 1(f) shows the situation after the echo, Fig. 1(g) the effect of the second 180° pulse, Fig. 1(h) the polarization after the pulse, and Fig. 1(i) the second echo. The process will then repeat itself with a period 4τ.

During this process the magnitude of the polarization, and thus the amplitude of the echos, will decrease with the natural relaxation time T_2.

Two conclusions can be drawn from the above picture. (1) The polarization vector at the time of the echos is alternatingly in the $+y$ and $-y$ directions. That this is actually the case is shown in Fig. 2(a) which shows a record made with a synchromous detector, whose reference voltage is derived from the same rf oscillator as the pulses. The oscillator frequency was adjusted to be slightly different from the Larmor frequency of the nuclei, so that beats with the difference frequency are observed. Note the alternating phase of these beats at successive echos. (2) If the "180°

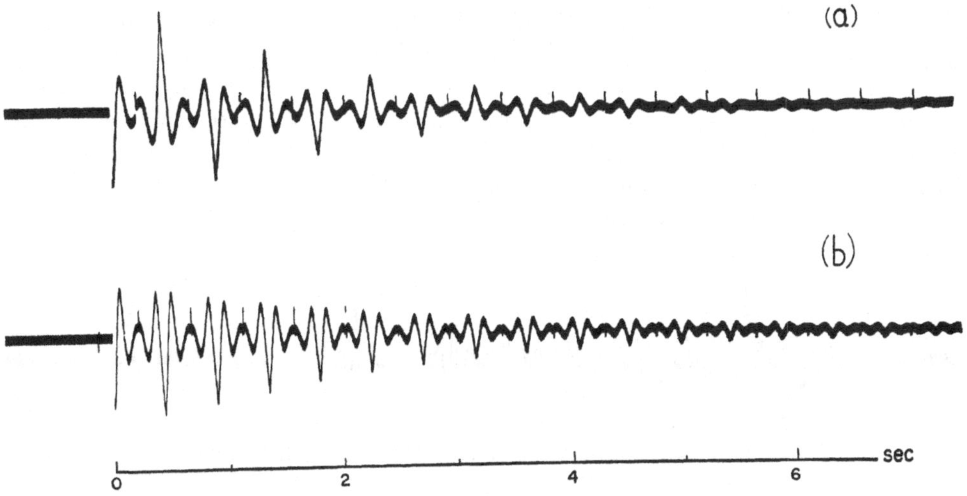

FIG. 2. Spin-echo patterns in water with coherent pulses and a relatively homogeneous magnetic field. Pulse rate about 2 per second. The records were made using a synchronous detector. The two traces were made under identical conditions, except for the introduction of a 90° phase shift in the first (90°) pulse for the bottom record.

[9] Rabi, Ramsey, and Schwinger, Revs. Modern Phys. **26**, 167 (1954).

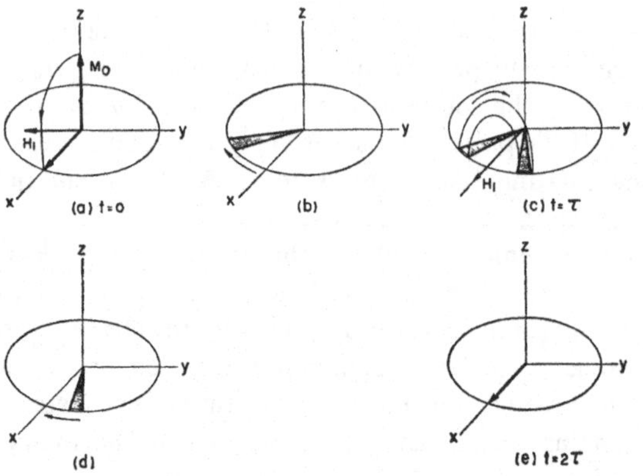

FIG. 3. Behavior of the nuclear polarization if a 90° phase shift is introduced in the rf of the first pulse.

pulses" deviate from the true 180° amplitude, the corresponding error is cumulative, and successive pulses will rotate the polarization vectors more and more out of the xy plane. This is because the sense of rotation is the same in Figs. 1(c) and 1(g). The decay of the echo amplitude will then not correspond to the true T_2.

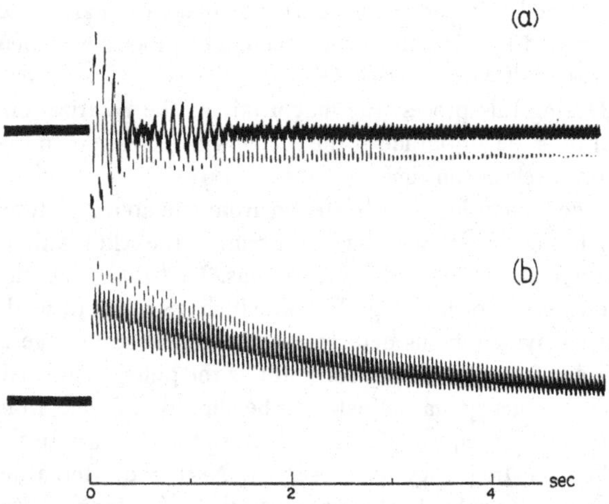

FIG. 4. As Fig. 2, but with pulse rate of 22 per second. In these records the "180° pulses" were intentionally made too weak, and gave actually about 171° flips. The error is cumulative in record a, but not in record b, which was made with 90° phase shift of the first pulse.

In Fig. 3 the behavior of the polarization vector is sketched when the second modification is also introduced. The modification consists of a 90° rf phase shift in the 90° pulse, relative to the phase of the 180° pulses. In the rotating coordinate system this means that the H_1 vector of the 90° pulse is in the y direction, while the H_1 vector of the 180° pulses is in the x direction. Figure 3(a) shows the 90° flip on application of the 90° pulse, Fig. 3(b) the fanning out a time τ later, Fig. 3(c) the effect of the first 180° pulse, Fig. 3(d) the polarization vector after the pulse and Fig. 2(c) the echo. The process will repeat itself with a period of 2τ.

In contrast to the first scheme, described in Fig. 1, all the echos have the same phase. This fact is illustrated in Fig. 2(b), which was made under conditions identical to those of Fig. 2(a), except for the introduction of the 90° phase shift of the 90° pulse. With this modification an amplitude deviation of the 180° pulses will not be cumulative in its effect: if the pulses are for instance somewhat less than 180°, the first pulse will leave the polarization vector above the xy plane [Fig. 3(c)], but the polarization will be returned to this plane on the next pulse. That the resulting improvement is very real is shown in Figs. 4(a) and 4(b). The records in these figures were made with the "180° pulses" intentionally shortened so as to give an actual flip of about 171°. Figure 4(a) was recorded without the 90° phase shift, while in the record of Fig. 4(b) the phase shift was applied. All other conditions were identical. Figure 5 shows a record of benzene, made at a pulse rate of about 2 per second.

APPARATUS

The main item of interest is the transmitter gate. A schematic diagram is given in Fig. 6. The 15-Mc input for this gate is derived from a quartz controlled Meacham oscillator at 5 Mc, followed by a tripler stage. The first stage of the gate is a straight push-pull amplifier, while the second stage is a doubler stage. The plate voltage of both stages is pulsed according to the desired pulse scheme. Rf leakage in the interval between pulses is practically zero, without the need for careful screening. This is because a 15-Mc leakage is harmless, and doubling to 30 Mc can take place only when a pulse is applied.

The 90° phase shift during the first (90°) pulse is obtained by slightly detuning the input circuit of the gate during this pulse. This is done by pulsing a 1N34 diode,

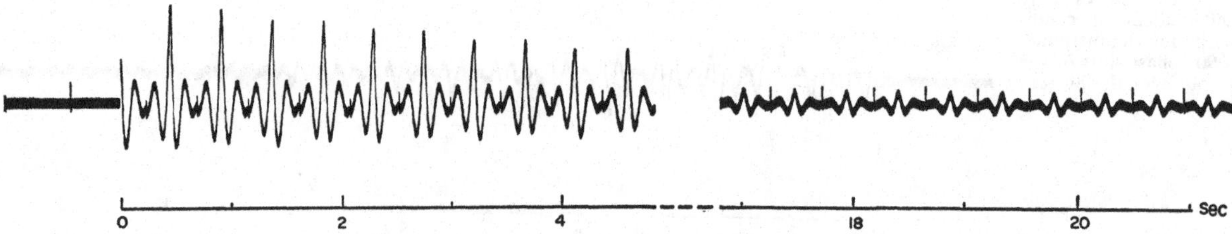

FIG. 5. A record of degassed benzene.

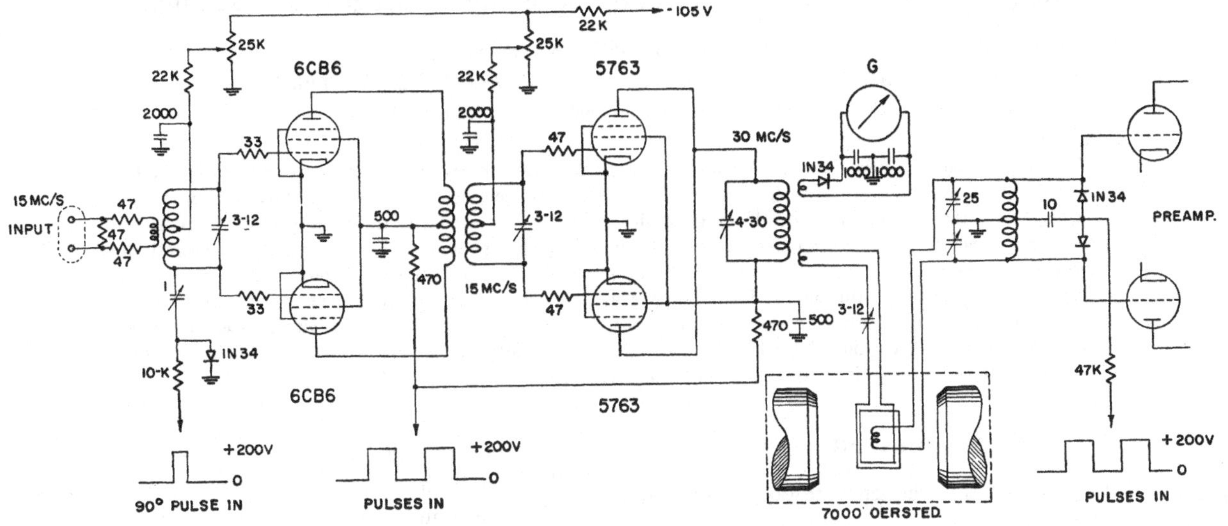

FIG. 6. Schematic diagram of transmitter gate.

which effectively switches a 1-$\mu\mu$f trimmer in and out the input circuit (see Fig. 6). Adjustment to the desired 90° phase shift can easily be obtained by comparing the phase relation of successive echos (compare Fig. 2).

The head used is of the crossed coils type. This construction, rather than a single coil, was chosen because, by making the transmitter coil larger than the receiver coil, a homogeneous rf field (H_1) over the active part of the the cylindrical sample can be obtained. The transmitter coil is made up of two sections, one on each side of the sample tube. Each section consists of two turns of copper wire and is of rectangular shape, 4×2 cm, the long dimension being parallel to the sample tube. The distance between the two sections is 2 cm. The receiver coil is cylindrical, diameter 1 cm and length 1 cm. It consists of 6 turns of No. 34 copper wire. The small wire diameter was chosen in order to avoid screening of the rf field. With the arrangement described an rf field of about 0.3 oersted is obtained, corresponding to a pulse length of 250 μsec for the 180° pulses. Good decoupling between transmitter and receiver coils is necessary, as even a relatively small coupling causes large induced currents in the tuned receiver coil, resulting in a nonhomogeneous H_1. This adjustment could however be made much less critical by the introduction of the two 1N34 diodes in the receiver input (see Fig. 6). The same pulses which operate the transmitter gate were applied to these diodes, resulting in a reduction of the Q of the receiver circuit during the pulses, and a corresponding reduction of the induced currents.

Application of Fourier Transform Spectroscopy to Magnetic Resonance

R. R. Ernst and W. A. Anderson
Analytical Instrument Division, Varian Associates, Palo Alto, California 94303
(Received 9 July 1965; and in final form, 16 September 1965)

The application of a new Fourier transform technique to magnetic resonance spectroscopy is explored. The method consists of applying a sequence of short rf pulses to the sample to be investigated and Fourier-transforming the response of the system. The main advantages of this technique compared with the usual spectral sweep method are the much shorter time required to record a spectrum and the higher inherent sensitivity. It is shown theoretically and experimentally that it is possible to enhance the sensitivity of high resolution proton magnetic resonance spectroscopy in a restricted time up to a factor of ten or more. The time necessary to achieve the same sensitivity is a factor of 100 shorter than with conventional methods. The enhancement of the sensitivity is essentially given by the square root of the ratio of line width to total width of the spectrum. The method is of particular advantage for complicated high resolution spectra with much fine structure.

I. INTRODUCTION

It is well-known that the frequency response function and the unit impulse response of a linear system form a Fourier transform pair. Both functions characterize the system entirely and thus contain exactly the same information. In magnetic resonance, the frequency response function is usually called the spectrum and the unit impulse response is represented by the free induction decay. Although a spin system is not a linear system, Lowe and Norberg[1] have proved that under some very loose restrictions the spectrum and the free induction decay after a 90° pulse are Fourier transforms of each other. The proof can be generalized for arbitrary flip angles.[2]

For complicated spin systems in solution, the spectrum contains the information in a more explicit form than does the free induction decay. Hence it is generally assumed that recording the impulse response does not give any advantages compared to direct spectral techniques. The present investigations show that the impulse response method can have significant advantages, especially if the method is generalized to a series of equidistant identical pulses instead of a single pulse. In order to interpret the result, it is usually necessary to go to a spectral representation by means of a Fourier transformation. The numerical transformation can conveniently be handled by a digital computer or by an analog Fourier analyzer.

Here are some of the features of the pulse technique: (1) It is possible to obtain spectra in a much shorter time than with the conventional spectral sweep technique. The required time T for one trace is roughly given by the resolution Δ (in cps) to be achieved, $T \sim 1/\Delta$ sec. This allows the study of time dependent systems such as chemical reactions and transitory saturation experiments. (2) The achievable sensitivity of the pulse experiment is higher, providing the investigated spectrum possesses much fine structure (e.g., high resolution NMR spectroscopy). All spins with resonance frequencies within a certain region are simultaneously excited, increasing the information content of the experiment appreciably compared with the spectral sweep technique where only one resonance is observed at a time. (3) The method allows the investigation of internuclear Overhauser effects. (4) The total integral and the higher moments of the spectrum can easily be determined from the initial values of the impulse response and of its derivatives, respectively. (5) The accurate frequency calibration of the spectrum is simplified and consists of an accurate time measurement.

The investigations described here employ a series of equal rf pulses equidistant in time. This is a special case of a wide band frequency source with an arbitrary, discrete or continuous power spectrum.[3] The optimum choice of the power spectrum depends on the purpose. The present choice has the advantages of instrumental simplicity and ease of interpretation.

Section II describes the response of a simple spin system to a series of short rf pulses. One aspect, improvement of the sensitivity, is treated here in detail. The possible enhancement is calculated in Sec. III. An experimental setup is described in Sec. IV and the results are presented in Sec. V.

II. BEHAVIOR OF AN ISOLATED SPIN SUBJECTED TO A REPETITIVE SEQUENCE OF rf PULSES

This section describes some of the elementary features of the response of a spin system to a sequence of equally spaced, identical rf pulses. The calculations are done within the classical frame, based on the Bloch equations. Intercorrelation effects between different transitions such as Overhauser effects necessarily need a quantum mechanical theory for a satisfactory explanation.[2]

Suppose a system of isolated spins in a magnetic field H_0

[1] I. J. Lowe and R. E. Norberg, Phys. Rev. **107**, 46 (1957); compare also A. Abragam, *The Principles of Nuclear Magnetism* (Oxford University Press, New York, 1961), p. 114.

[2] R. R. Ernst, "Density Operator Theory of Fourier Transform Spectroscopy" (to be published).

[3] The case of a "white" power spectrum is treated in R. R. Ernst and H. Primas, Helv. Phys. Acta **36**, 583 (1963).

Reprinted from Review of Scientific Instruments 37, 93–102 (1966); ©American Institute of Physics.

along the z axis is subjected to a sequence of rectangular rf pulses along the x axis with frequency ω_1, magnetic field amplitude $2H_1$, duration τ, and period T. The system eventually reaches a stationary state, independent of the initial conditions, with a periodic motion of the magnetization.

The motion of the magnetization vector $\mathbf{M}(t)$ is best described in a frame rotating with the frequency ω_1 around the z axis. The motion consists of two phases: (1) During the presence of the rf pulse, there is a precession around the effective field $\mathbf{H}_{1\text{eff}}$,

$$|H_{1\text{eff}}| = [(\Omega_i-\omega_1)^2/\gamma^2+H_1^2]^{\frac{1}{2}}, \qquad (2.1)$$

where Ω_i is the resonance frequency of spin i and γ the common gyromagnetic ratio; and (2) in the absence of the rf pulse, there is a free induction decay. For a stationary state, the initial and final positions must be identical.

If γH_1 is made much larger than the total width of the spectrum, it is possible to choose ω_1 such that the relation

$$(\Omega_i-\omega_1) \ll \gamma H_1 \qquad (2.2)$$

is fulfilled for any line of the spectrum, so that the effective magnetic field in the rotating frame lies along the x axis and its absolute value becomes

$$|H_{1\text{eff}}| \sim H_1 \qquad (2.3)$$

independent of the line position. If the pulse length τ is short compared to the relaxation times T_1 and T_2, relaxation is negligible during the pulse and the angle of rotation α, caused by the pulse, is equal for all the nuclei,

$$\alpha = \gamma H_1 \tau. \qquad (2.4)$$

The assumption that the period of the precessing magnetization is the same as the period of the pulse sequence is not necessarily true, in principle. It could happen that the magnetization returns to the original position only after n successive rf pulses or that no stationary state is reached at all. A simple algebraic analysis of the following Eqs. (2.5) and (2.6) shows that a stationary state is reached with the same periodicity as the pulse sequence if one of the side bands of the pulsed rf carrier lies in the center of the resonance line. The same is probably also true for the off-resonance case. This is in accord with the experimental investigations.

The motion of the magnetization vector $\mathbf{M}(t)$, in a frame rotating with frequency ω_1, can be represented by the following transformations as given by the Bloch equations

$$\mathbf{M}_2 = R_x(\alpha)\mathbf{M}_1 \qquad (2.5)$$

and

$$\mathbf{M}_3 = R_z(\vartheta)S(T,T_1,T_2)\mathbf{M}_2 + (1-e^{-T/T_1})\mathbf{M}_0 = \mathbf{M}_1. \qquad (2.6)$$

$\mathbf{M}_1$, $\mathbf{M}_2$, $\mathbf{M}_3$ are the positions at the beginning of the rf pulse, at the end of the rf pulse, and at the end of the free induction decay, respectively. $R_x(\alpha)$ represents the rotation around the x axis by the angle α caused by the rf pulse. $R_z(\vartheta)$ is the operator of rotation around the z axis, ϑ is the precession angle of the magnetization during the time T around the effective field $(\Omega_i-\omega_1)/\gamma$ in the rotating frame,

$$\vartheta = (\Omega_i-\omega_1)\cdot T. \qquad (2.7)$$

The time τ is here neglected compared with T. The operator $S(T,T_1,T_2)$ represents relaxation during the time T and has, in the (x,y,z) basis, the following representation,

$$S(T,T_1,T_2) = \begin{bmatrix} e^{-T/T_2} & 0 & 0 \\ 0 & e^{-T/T_2} & 0 \\ 0 & 0 & e^{-T/T_1} \end{bmatrix}. \qquad (2.8)$$

$\mathbf{M}_0$ is the equilibrium magnetization which lies along the z axis. These equations allow us to calculate the y component of the magnetization, $M_y(t)$, which is observed in a cross coil method. Assuming that an rf pulse occurs at $t=0$, the y magnetization after the pulse is given by

$$M_y(t) = M_y(+0)\cos[(\Omega_i-\omega_1)t]e^{-t/T_2} + M_x(+0)\sin[(\Omega_i-\omega_1)t]e^{-t/T_2}, \quad +0 \le t \le T. \qquad (2.9)$$

$M_y(t)$ is the y magnetization in the rotating frame. It can be measured in the laboratory frame by phase sensitive detection of the y magnetization with a reference signal in phase with the transmitter signal. The x and y components of the magnetization at the end of the rf pulse, $M_x(+0)$ and $M_y(+0)$, are found to be

$$M_x(+0) = M_0\frac{-(1-E_1)E_2\sin\vartheta\sin\alpha}{(1-E_1\cos\alpha)(1-E_2\cos\vartheta)-(E_1-\cos\alpha)(E_2-\cos\vartheta)E_2}, \qquad (2.10a)$$

$$M_y(+0) = M_0\frac{(1-E_1)(1-E_2\cos\vartheta)\sin\alpha}{(1-E_1\cos\alpha)(1-E_2\cos\vartheta)-(E_1-\cos\alpha)(E_2-\cos\vartheta)E_2}. \qquad (2.10b)$$

E_1 and E_2 are abbreviations for the exponentials e^{-T/T_1} and e^{-T/T_2}. Equation (2.9) shows that the phases of the oscillations after each pulse depend on the frequency deviation $\Omega_i-\omega_1$. The phase differs thus from line to line within the same spectrum. The deviation depends on the pulse spacing T. If T is much longer than the relaxation time T_1, the response of each pulse is independent of the preceding pulses and the initial magnetization reduces to

$$M_x(+0)=0, \quad M_y(+0)=M_0 \sin\alpha. \tag{2.11}$$

In applying the pulse technique to enhance the sensitivity (Sec. III), it is important to determine the optimum performance conditions for maximum signal. The magnetization depends on two free parameters, the pulse spacing T and the flip angle α. The pulse spacing is usually fixed by the necessary frequency range and by resolution requirements (as discussed in Sec. III). Both components, $M_x(+0)$ and $M_y(+0)$, depend in the same manner on the flip angle α. The optimum angle for producing the maximum magnetization in the x-y plane is easily found to be

$$\cos\alpha_{\text{opt}}=\frac{E_1+E_2(\cos\vartheta-E_2)/(1-E_2\cos\vartheta)}{1+E_1E_2(\cos\vartheta-E_2)/(1-E_2\cos\vartheta)}. \tag{2.12}$$

For $T \gg T_1$ the optimum flip angle approaches 90°, which is the optimum flip angle for a single pulse experiment. For sufficiently short pulse spacings, $T \ll T_2$, the average rf amplitude $(\gamma H_1)_{\text{opt}}(\tau/T)=\alpha_{\text{opt}}/T$ can be shown to fulfill the well-known saturation condition

$$[(\gamma H_1)_{\text{opt}}(\tau/T)]^2 T_1 T_2=1, \tag{2.13}$$

providing one of the pulse side bands coincides with the center of the resonance line. The optimum flip angle depends on the precession angle ϑ which varies from line to line within the same spectrum. An average optimum flip angle $\alpha_{\text{opt av}}$, independent of ϑ, is given by

$$\cos\alpha_{\text{opt av}}=E_1. \tag{2.14}$$

The Fourier transform of the response to a sequence of rf pulses, given by Eq. (2.9), now is determined. As long as the pulse spacing T is much longer than T_1, the Fourier transform is identical with the spectrum of the spin system, according to Ref. (1). The Fourier cosine transform yields the absorption mode spectrum, and the Fourier sine transform yields the dispersion mode spectrum. If the pulse spacing is comparable with T_1, distortions occur in the spectrum. The Fourier transform of a periodic signal with period T is, of course, a discrete spectrum with the frequencies $f=n/T$ $(n=0,1,2,\cdots)$. The corresponding coefficients C_n of the Fourier cosine transform of the free induction decay [Eq. (2.9)] are given by[4]

$$C_n=M_0\frac{T_2}{(\Delta\omega T_2)^2+1}\cdot\frac{1}{(2T)^{1/2}}\cdot\frac{(1-E_1)(1-2E_2\cos\vartheta+E_2^2)\sin\alpha}{(1-E_1\cos\alpha)(1-E_2\cos\vartheta)-(E_1-\cos\alpha)(E_2-\cos\vartheta)E_2}, \tag{2.15}$$

with $\Delta\omega=\Omega_i-\omega_1-2\pi(n/T)$, and where the term containing $\Delta\omega'=\Omega_i-\omega_1+2\pi(n/T)$ has been neglected.

This expression shows the interesting fact that the resulting line shape is identical to the line shape observed in a slow pass, low power spectral experiment. The line shape is independent of T, α, and ϑ. Since it is also independent of the applied rf power no line broadening occurs in contrast to the conventional slow passage saturation experiments. However, the line intensities depend in a complicated manner on all three parameters.

The manner in which the line depends on these three parameters as well as on the relaxation times T_1 and T_2 is discussed for optimum flip angle $\alpha_{\text{opt av}}$ [Eq. (2.14)]. The Fourier coefficients in this case are given by

$$C_n=M_0\frac{T_2}{(\Delta\omega T_2)^2+1}\frac{1}{(2T)^{1/2}}\left(\frac{1-E_1}{1+E_1}\right)^{1/2} \times\left(1+E_2\frac{E_2-\cos\vartheta}{1-E_2\cos\vartheta}\right). \tag{2.16}$$

The dependence of the signal intensity on ϑ can cause inaccurate measurements of relative line intensity. The maximum variation of the intensities is given by the ratio of maximum to minimum Fourier coefficient C_n as a function of ϑ,

$$[C_n(\vartheta)_{\max}/C_n(\vartheta)_{\min}]_{\alpha_{\text{opt av}}}=(1+E_2)/(1-E_2). \tag{2.17}$$

This variation is negligible for a sufficiently long pulse spacing ($T/T_2=3$ causes a maximum variation of ±5%). The dependence of the intensity on T_2 is also negligible if $e^{-T/T_2} \ll 1$. The dependence on T_1 can be important if T_1 is much longer than T_2 and if T_1 varies strongly from line to line within the same spectrum. This problem occurs in a similar manner in slow passage experiments if the rf power is optimized for maximum intensities. The influence of T_1 on the intensities is small if $e^{-T/T_1} \ll 1$ (for $T/T_1 \geq 3$, the influence is less than ±2.5%).

The true free induction decay after a single rf pulse extends, in principle, to infinity and the corresponding Fourier transform is a continuous spectrum. Since the

[4] The Fourier coefficients are calculated by means of the formula

$$C_n=(2/T)^{1/2}\int_0^T \cos(2\pi nt/T)M_y(t)dt]$$
$$=(2T)^{1/2}(1/c)\sum_{k=0}^{c-1}\cos(2\pi kn/c)M_y(Tk/c) \quad (n=1,2,\cdots).$$

The last equality holds if $M_y(t)$ does not contain frequencies higher than $c/2T$, where c is the number of samples within the time T. This is a consequence of the sampling theorem. The factor $2/T$ is chosen such that

$$\int_0^T M_y^2(t)dt=\Sigma\, C_n^2+S_n^2,$$

where S_n are the corresponding Fourier sine coefficients.

impulse response, given by Eq. (2.9), is defined only in the time interval 0–T, the Fourier spectrum consists of discrete lines with a spacing of $1/T$ cps. The values approximately represent discrete points of the true spectrum.

To obtain a continuous spectrum it is necessary to make assumptions about the impulse response for $t>T$. Under the usual experimental conditions, the free induction decay signal is small for $t>T$. It is reasonable to assume that the impulse response can be set equal to zero for $t>T$,

$$M_y(t)=0, \quad t>T. \tag{2.9a}$$

The Fourier transformation of $M_y(t)$, given by Eqs. (2.9) and (2.9a), produces a continuous spectrum. The accuracy of the points of the continuous spectrum which lie between the values given by Eq. (2.15) is somewhat less than the accuracy of the harmonics of $1/T$. The accuracy again depends on the ratio T/T_2 and is improved by making T large.

The Fourier cosine transform, corresponding to the absorption spectrum, now has the form

$$C(\omega)=M_0\frac{1}{(2T)^{\frac{1}{2}}}\frac{(1-E_1)\sin\alpha}{(1-E_1\cos\alpha)(1-E_2\cos\vartheta)-(E_1-\cos\alpha)(E_2-\cos\vartheta)E_2}$$
$$\times(\{T_2/[(\Delta\omega T_2)^2+1]\}\{1-E_2[\cos\vartheta+\cos(\Delta\omega T)]+E_2^2\cos(\omega T)\}$$
$$+\{\Delta\omega T_2^2/[(\Delta\omega T_2)^2+1]\}[\sin(\Delta\omega T)-\sin\vartheta+E_2\sin(\omega T)]E_2) \tag{2.18}$$

with $\Delta\omega=\Omega_i-\omega_1-\omega$. This formula reduces to Eq. (2.15) for $\omega=2\pi n/T$. The spectrum consists of absorptive and dispersive parts whose relative coefficients are of the order 1 and E_2, respectively. The dispersive part affects chiefly the wings and can be neglected if $e^{-T/T_2}\ll 1$. The maximum variation of the absorptive part as a function of ϑ is again given by Eq. (2.17), for the flip angle $\alpha_{\text{opt av}}$ determined by Eq. (2.14). Here not only the intensities but also the line shape is affected.

It is possible to improve the resolution, given so far by the spacing $1/T$ of the Fourier coefficients [Eq. (2.15)], by calculating some additional points of the continuous spectrum [Eq. (2.18)]. However when the inherent error, expressed by Eq. (2.18), becomes larger than the intensity difference between two adjacent points there is no advantage in calculating further intermediate points.

III. APPLICATION FOR SENSITIVITY IMPROVEMENT

The main stimulus to investigate this pulse method was the expected increase in sensitivity, which makes the method of interest to chemists and biologists.

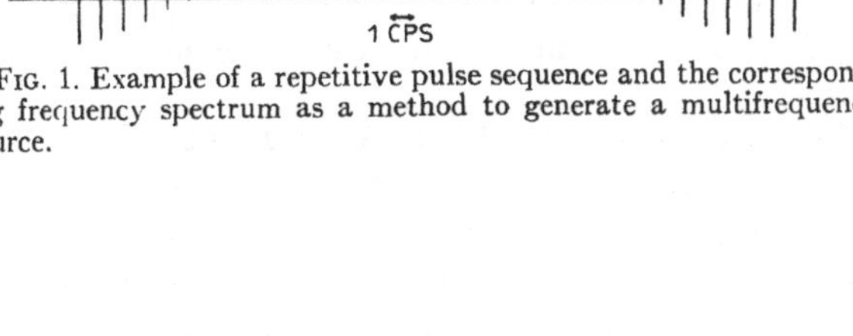

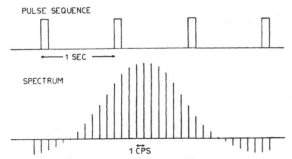

FIG. 1. Example of a repetitive pulse sequence and the corresponding frequency spectrum as a method to generate a multifrequency source.

To make the possible enhancement of the sensitivity intuitively clear, it is helpful to consider the problem from another viewpoint. The conventional spectral methods in magnetic resonance use a field or frequency sweep to record a spectrum. At any particular time, the transmitter frequency corresponds to one single point in the spectrum and the rest of the spectrum is for the most part disregarded. This corresponds in optical spectroscopy to a very narrow slit through which the spectrum is observed. It is obvious that the information rate of such an experiment could be greatly increased by investigating the entire spectrum during the total available time. This could be done, at least in principle, by using several transmitters and receivers with different frequencies which correspond to different points in the spectrum. This kind of spectrometer could be called a multichannel spectrometer. The sensitivity improvement which is achievable depends on the total number of channels.

Applying a sequence of equally spaced short rf pulses corresponds to a multichannel experiment. This becomes clear by considering the frequency spectrum of an rf pulse sequence with the pulse length τ and the pulse spacing T. It consists of numerous side bands of the carrier frequency ω_1, $1/T$ cps apart. The Fourier coefficients of the frequencies $f_n=\omega_1/2\pi+n/T$ are given by

$$A_n=[\sin(n\pi\tau/T)]/(n\pi\tau/T) \tag{3.1}$$

(see Fig. 1). To obtain a side band spectrum of constant amplitude throughout F, the width of the spectrum being investigated, it is necessary that

$$\pi F\tau\ll 1. \tag{3.2}$$

The resolution and the faithful representation of the spectrum are determined by the density of these side band frequencies; e.g., by the spacing T of the pulses.

Each side band produces a response which is determined by the value of the spectrum at its frequency. The amplitudes of these responses can be determined by a Fourier analysis and allow the spectrum to be determined. This treatment can be used only for very crude qualitative arguments because of the nonlinearity of the spin system. The influence of the different frequency components of the pulse sequence cannot be considered separately unless the inequality

$$\gamma H_1 \tau \ll 1 \tag{3.3}$$

is satisfied. If this inequality does not hold, the treatment of Sec. II must be used.

To obtain an improvement in sensitivity it is essential that the responses after each pulse are added together coherently. This can conveniently be done by using a time averaging computer. The final step is then the Fourier transformation of the time averaged free induction decay. The Fourier analysis is equivalent to using a multichannel receiver.

In order to show the possible improvement of the signal-to-noise ratio obtainable with the pulse method, it is useful to compare the pulse method with a single scan, slow passage experiment done in the same time T_t. It is assumed that the noise has a white power spectrum with the power density P in $W/$(cps). The total sweep width to be covered is denoted by F (cps). The signal-to-noise ratio S/N is defined here as the ratio between the peak signal amplitude and the rms noise amplitude.

The S/N ratio of the single scan experiment is calculated first. It is of advantage to use a linear filter at the output of the spectrometer to limit the noise bandwidth. The theoretically optimum filter delivering the maximum available S/N ratio is the matched filter.[5] Its shape depends on the line shape to be filtered. One can show that the maximum available S/N ratio is given by

$$\left(\frac{S}{N}\right)^2 = \frac{1}{RP}\int_{-\infty}^{\infty} v(t)^2 dt, \tag{3.4}$$

where R is the resistance over which the voltages are measured and $v(t)$ is the absorption mode signal in question. It is known[6] that in a true slow passage experiment the maximum S/N ratio is achieved for a saturation parameter $S=(\gamma H_1)^2 T_1 T_2 = 2$. The absorption mode signal $v(t)$ is in this case given by

$$v(t) = M_0\left(\frac{T_2}{T_1}\right)^{\frac{1}{2}} \frac{\sqrt{2}}{3+(2\pi a t)^2 T_2^2}.$$

The sweep rate a, in cps/sec, can be expressed by the total available time T_t and the sweep range F, $a=F/T_t$. If one inserts this expression into Eq. (3.4), one obtains for the S/N ratio of a single scan

$$(S/N)_s^2 = (1/6\sqrt{3})(M_0^2/RP)(T_t/T_1 \cdot F). \tag{3.5}$$

In most spectrometers, a modulation method is used for baseline stabilization. It delivers a somewhat smaller S/N ratio.[7] Assuming a centerband method, the above calculated S/N ratio has to be corrected by the factor $1.14/\sqrt{2}$,

$$(S/N)_s^2 = (1.30/12\sqrt{3})(M_0^2/RP)(T_t/T_1 \cdot F). \tag{3.6}$$

The S/N ratio of the corresponding pulse experiment, performed in the same total time T_t by recording and summing all the responses to n rf pulses, is given by

$$(S/N)_p^2 = (n \cdot s_{\max})^2/(RPBT_t/2). \tag{3.7}$$

Here, $s_{\max}$ is the maximum Fourier component of a single impulse response, $n \cdot s_{\max}$ is the total accumulated signal height and corresponds to $v_{\max}$ in the single scan experiment. $PBT_t/2$ is the total noise energy accumulated during the time T_t with the effective bandwidth B of a single Fourier component. The equidistant Fourier components with the frequencies $f=m/T$, where $T=T_t/n$ is the pulse spacing, have the effective bandwidth B,

$$B = 2/T. \tag{3.8}$$

The factor 2 in Eq. (3.7) occurs because the Fourier cosine transformation [Eq. (2.15)] halves the noise power by selecting only one phase. The factor 2 in Eq. (3.8) occurs because both of the sidebands, $\omega_1/2\pi + f$ and $\omega_1/2\pi - f$, contribute to the noise at the frequency f after demodulation. Equation (3.8) assumes the existence of a band limiting low pass filter at the frequency F (as explained in Sec. IV) which prevents the down conversion of high frequency noise in the digitizing process.

The noise in the Fourier spectrum is the Fourier transform of the original noise at the output of the spectrometer. In general, it can change its character entirely by the Fourier transformation. Stationary nonwhite noise becomes nonstationary, whereas stationary white noise remains stationary and white after the Fourier transformation.

To achieve the maximum possible S/N ratio with the pulse technique, it is again necessary to use a matched filter, determined by the line shape. However, in the present case the filter is readily achieved. The filtering process corresponds to a convolution integral between the spectrum and the filter impulse response. In the Fourier time domain, this corresponds to multiplication of the free induction decay with the frequency response function which is a trivial process for the computer to perform. In the case of Lorentzian lines with a common relaxation

[5] S. Goldman, *Information Theory* (Prentice-Hall, Inc., Englewood Cliffs, New Jersey, 1953), p. 230; and L. S. Schwartz, *Principles of Coding, Filtering and Information Theory* (Spartan Books, Inc., Baltimore, 1963), p. 136.

[6] R. R. Ernst and W. A. Anderson, Rev. Sci. Instr. **36**, 1696 (1965).

[7] W. A. Anderson, Rev. Sci. Instr. **33**, 1160 (1962).

time T_2, the envelope of the decay is given by e^{-t/T_2}. Since the S/N ratio is proportional to the envelope height, it is obvious that it deteriorates towards the end of the trace. It is intuitively clear that the best S/N ratio after the Fourier transformation is obtained by weighting each point of the free induction decay with its own S/N ratio. For a Lorentzian line this results in multiplication of the trace by e^{-t/T_2}. It can be proved rigorously that this is exactly equivalent to the use of a matched filter.

To determine the maximum Fourier component s_{max} of Eq. (3.7), the y component of the free induction decay [Eq. (2.9)] must be multiplied by the matched filter function e^{-t/T_2} and then Fourier transformed,

$$s_{max}=\left(\frac{2}{T}\right)^{\frac{1}{2}}\int_0^T \cos\left(2\pi n\frac{t}{T}\right)M_y(t)e^{-t/T_2}dt$$
$$=M_y(+0)\frac{1}{(2T)^{\frac{1}{2}}}\frac{T_2}{2}(1-E_2^2). \quad (3.9)$$

It is assumed here that one of the pulse modulation side bands lies exactly on resonance $\Omega_i-\omega_1=2\pi n/T$. This gives $\cos\vartheta=1$. In this case $M_x(+0)$ disappears and the magnetization $M_y(t)$ is determined alone by $M_y(+0)$, Eq. (2.10). Deviations from this assumption can easily be included by using the results of Sec. II. The initial magnetization $M_y(+0)$ is calculated for the optimum flip angle α_{opt} which is given in Eq. (2.12).

The action of the matched filter on the noise is essentially a uniform attenuation of all noise frequencies. The power spectrum remains unchanged providing it is a sufficiently smooth function of frequency. The power density P of the white noise is changed to P_f according to

$$P_f=P\cdot\frac{1}{T}\int_0^T e^{-2t/T_2}dt=P\cdot\frac{T_2}{2T}(1-E_2^2). \quad (3.10)$$

Inserting Eqs. (3.8)–(3.10) into Eq. (3.7), the maximum S/N ratio of the pulse experiment becomes

$$\left(\frac{S}{N}\right)_p^2=\frac{nT_2}{4RP}M_0^2\frac{(1-E_1)^2}{1-E_1^2}. \quad (3.11)$$

The ratio of the sensitivities of the pulse experiment and of the steady state experiment performed in the same total time can be found after slight rearrangements to be

$$\frac{(S/N)_p}{(S/N)_s}=0.799\left(\frac{F}{\Delta}\right)^{\frac{1}{2}}G\left(\frac{T}{T_1}\right),$$

with

$$G(x)=\left[2-\frac{1}{x}\frac{(1-e^{-x})^2}{(1-e^{-2x})}\right]^{\frac{1}{2}}. \quad (3.12)$$

The function $G(T/T_1)$ is displayed in Fig. 2. It varies slowly and can often be approximated by one. The possible

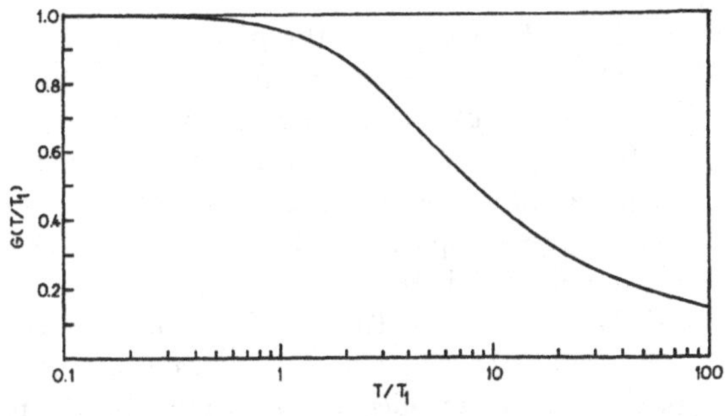

FIG. 2. S/N improvement by means of the pulse method, dependence on the ratio of pulse spacing T to relaxation time T_1, represented by the function $G(T/T_1)$ which is given by Eq. (3.12).

gain in sensitivity is thus essentially given by the ratios of the total sweep width F to a characteristic line width $\Delta=1/\pi T_2$. This is a plausible result which shows that the gain is equal to the square root of the total available time to the time actually spent within a single line of the width Δ using the ordinary spectral sweep technique. The time actually saved in achieving a certain S/N ratio is roughly the ratio of total sweep width to a characteristic linewidth. A factor of 100 in time can be gained easily.

Some important facts in connection with Eq. (3.12) have to be remembered:

(1) *Sweep width F.* In favorable cases, the sweep rate F is identical with the width of the investigated spectrum. Often, the width of the spectrum is not known in advance and can not be determined easily. Here, F is more likely the maximum range of possible chemical shift values.

(2) *Choice of the pulse spacing T.* According to Eq. (3.12), the gain in S/N ratio increases slightly by using a shorter pulse spacing. However, there are factors which limit the minimum useful pulse spacing. In Sec. II, it was shown that the longer T is, the better the calculated points represent actual points of the spectrum. The optimum choice depends on the relative emphasis with respect to resolution and to sensitivity. In addition, in many experiments the pulse spacing is determined by the sweep range F necessary to cover the entire spectrum and by the number of channels c of the digital storage used to add the impulse responses together. (With this method it is impossible only to analyze a fraction of a spectrum.) The total spectrum has to lie on one side of the carrier frequency ω_1 and within $\omega_1\pm f_{max}$, where $f_{max}=c/2T$ is the maximum frequency component which can be analyzed with c samples within the time T. This requires that the pulse spacing be equal to or shorter than $T_{max}=c/2F$.

(3) *Deviations from the slow passage conditions.* This comparison was based on a true slow passage experiment. It is well-known that intermediate passage

conditions allow considerable improvement in the S/N ratio.[6] This is more pronounced for larger ratios of T_1/T_2. By comparing the pulse experiment with an intermediate passage experiment the actual gain is smaller. Thus, the pulse method may not be as useful for improving the sensitivity in systems with very high T_1/T_2 ratios. But it has the advantage that it gives no appreciable line deformations as they occur in intermediate or rapid passage experiments.

(4) *Line broadening.* The use of an optimum saturation parameter $S=2$ in slow passage experiments produces a line broadening by a factor $\sqrt{3}$. The pulse experiment described here does not show any line broadening, irrespective of the applied rf field strength. The inherent resolution of this experiment is therefore somewhat higher.

IV. EXPERIMENTAL

This section describes an experimental arrangement which was successfully used to enhance the sensitivity of proton magnetic resonance spectra using a Varian Associates double purpose 60 Mc spectrometer DP60. A block diagram of the modified spectrometer is shown in Fig. 3. The pulsed 60 Mc transmitter frequency is generated by pulsing a 30 Mc carrier and afterwards doubling the frequency to eliminate any 60 Mc leakage.[8] The 30 Mc signal is taken from the transmitter unit V4311 and fed into a pulse gate consisting of two gate stages, frequency doubler, and power amplifiers, which deliver a peak power of 200 W into a load of 100 Ω. A typical pulse length is $\tau=100$ μsec. The pulse spacing T is typically 0.5–2 sec and the peak-to-peak rf voltage 50–250 V.

The rf pulses are fed into the transmitter coil of the cross coil arrangement of the high resolution NMR probe. The induced signal in the receiver coil is fed to the receiver, where it is amplified, converted to 5 Mc, and demodulated in a phase sensitive detector. The existing phase sensitive detector of the V4311 unit is changed to have a passband of 5 kc. The free induction decay, containing frequencies up to 500 cps, is fed into the time averaging computer C1024 (Varian Associates) where the signal is digitized in 1024 sample points, equidistant in time, and added to the values already stored.

The highest frequency which can be retrieved from the computer is $c/2T$ cps, c is the number of channels ($c=1024$) and T is the pulse spacing. If higher frequencies are fed in, they are down converted into the frequency range $0\rightarrow c/2T$ and appear in the output as lower frequencies $f'=|f-nc/T|$ ($n=0, 1, 2, \cdots$). To avoid possible false resonances originating from this, the impulse response should not contain frequencies higher than $F=c/2T$. In

[8] S. Meiboom and D. Gill, Rev. Sci. Instr. 29, 688 (1958).

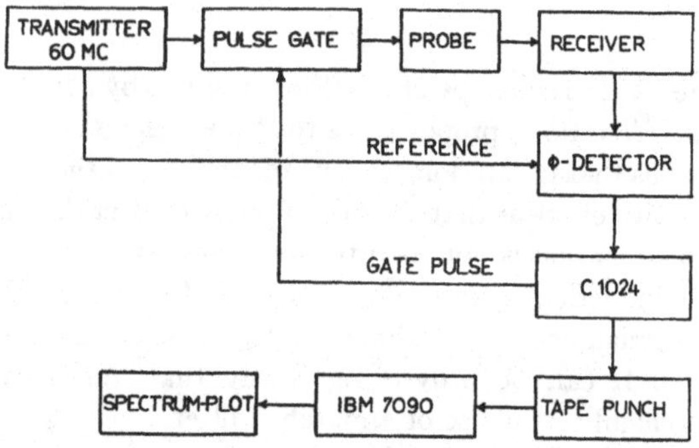

FIG. 3. Instrumental arrangement to record and Fourier transform the impulse response, using a modified Varian DP60.

addition, noise at frequencies above F is down converted in the frequency range below F and increases the power density of the noise in the critical range. Thus, it is important that the bandpass is limited before sampling in the time averaging computer.

The necessary low pass filter with a 3 dB point in the region of F introduces an undesirable frequency dependent phase shift into the passband. This complicates the selection of the true absorption mode signal during the Fourier transformation. To minimize the phase shift, a simple RC filter is used in front of the sampling device with a time constant of RC $\sim 1/2\pi F$.

To accumulate the impulse response, the time averaging computer is driven in the repetitive mode using the internal trigger. The same trigger pulse is used to open the transmitter gate for 100 μsec at the beginning of each trace to generate the rf pulse. After a sufficient number of scans, the accumulated signal in the memory of the C1024 is read out through the data output unit model 220 C[9] which converts the binary code into bcd numbers. They are punched on paper tape using a perforator model 420.[10] The punched tape is used as the input for an IBM 7090 digital computer which calculates the Fourier transform of the impulse response and plots the spectrum automatically by means of an incremental curve plotter.

Computer Program for the Fourier Transformation

A simplified block diagram of the computer program is shown in Fig. 4. Only the 512 ($=c/2$) Fourier components corresponding to harmonics of the fundamental frequency $1/T$ were calculated.

The adjustment of the phase to get a pure absorption mode signal can cause severe difficulties. The first correction to be made concerns the phase shift introduced by the band limiting filter mentioned above. The total phase shift and attenuation of this filter and the receiver were measured as a function of the frequency. The computer separately calculates the sine and cosine transform of the

[9] Technical Measurement Corporation (North Haven, Connecticut).
[10] Tally Corporation (Seattle, Washington).

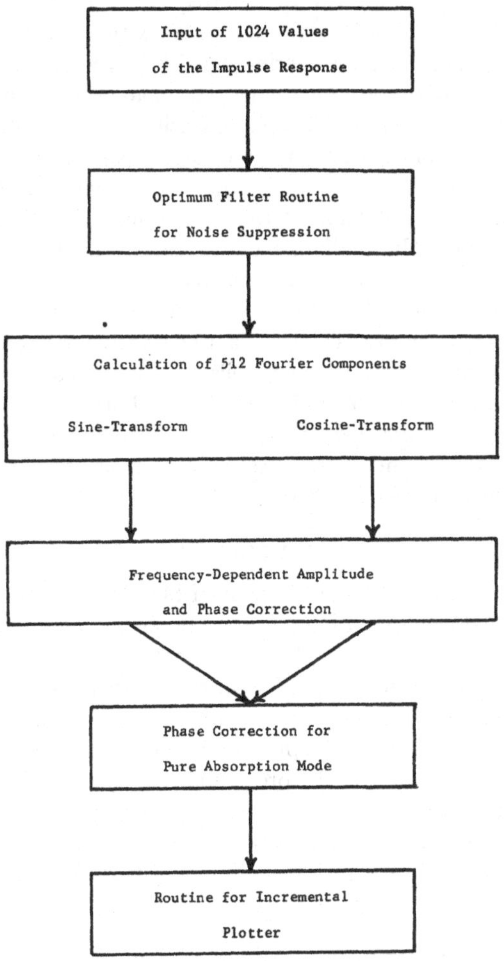

FIG. 4. Simplified block diagram of the computer program for the Fourier transformation.

impulse and then combines them to obtain a spectrum with frequency independent phase.

It is not practical to adjust manually the phase of the 5 Mc phase sensitive detector to obtain a pure absorption mode signal. Consequently, one usually obtains a mixture of absorption and dispersion modes. However, it is easy to include a routine in the computer program to correct the phase. For this purpose the computer calculates the Fourier cosine transform $s_c(\omega)$ and the Fourier sine transform $s_s(\omega)$ of the impulse response. The true absorption mode signal $s_a(\omega)$ is then given by

$$s_a(\omega)=s_c(\omega)\cos\varphi+s_s(\omega)\sin\varphi.$$

The correct angle φ of the phase adjustment is determined by the computer by maximizing the ratio $R(\varphi)=A_u(\varphi)/A_l(\varphi)$ as a function of the phase φ. $A_u(\varphi)$ and $A_l(\varphi)$ are the areas above and below the corrected signal trace $s_a(\omega)$ limited by the maximum and the mini-

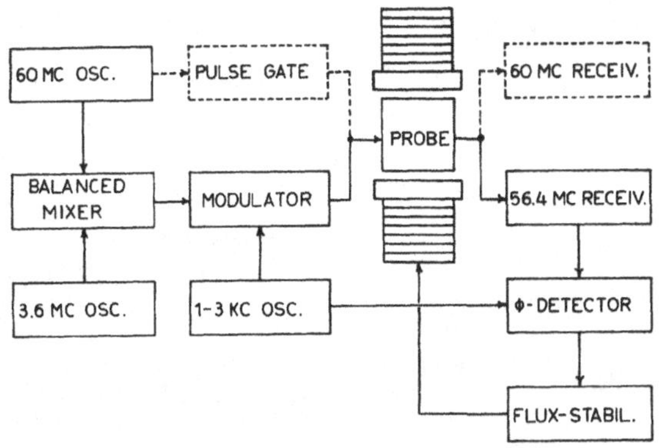

FIG. 5. Instrumental arrangement for the stabilization of the magnetic field in proton resonance. The magnetic field is locked on an internal fluorine reference signal.

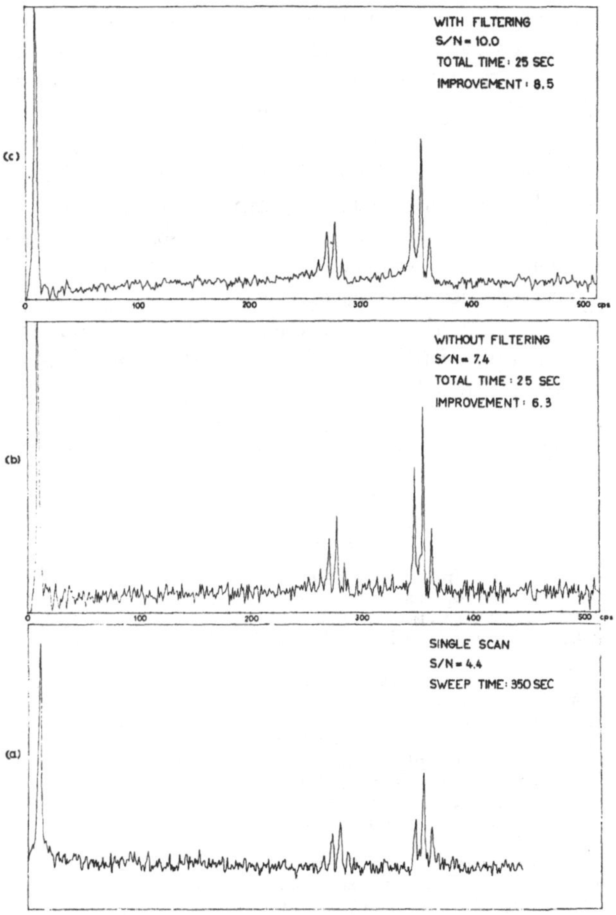

FIG. 6. Three spectra of 1% ethylbenzene dissolved in hexafluorobenzene. [The S/N ratio is indicated for the strongest line of the quartet, using the customary practical definition: $S/N=2.5\times$peak value of the signal/peak-to-peak value of the noise.] (a) Single scan, the absorption mode signal is directly recorded in 350 sec. (b) Pulse method, applying 25 pulses, 1 sec apart and adding the 25 responses together. The spectrum is obtained by means of a Fourier transformation. 512 points with 1 cps distance are calculated and plotted. The zero point marks the position of the 60 Mc carrier frequency. (c) Same spectrum as in (b), using a filtering procedure in the computer calculation to suppress the noise.

mum, respectively, of $s_a(\omega)$ and are given by

$$A_u(\varphi)=\sum_k [s_{a\,\max}-s_a(2\pi k/T)],$$
$$A_l(\varphi)=\sum_k [s_a(2\pi k/T)-s_{a\,\min}].$$

This method is independent of the actual line shape. It gives the correct phase angle for pure absorption mode if the S/N ratio is sufficiently high. The sensitivity of this method is higher for a wide sweep width.

Field-Frequency Stability

For coherent summation of different traces of the same impulse response, it is necessary to maintain high field-frequency stability for the total time T_t. The variations must be small compared to the line width of the observed spectrum. In the case of high resolution proton magnetic resonance, stability of ±0.1 cps is necessary. This is usually achieved by means of an internal field-frequency lock,[11] which locks the field on the dispersion mode signal of a reference line in the analytical sample. Because the spectrum to be investigated is covered with side bands due to the pulsing, it is of advantage to stabilize on the resonance of a different nuclear species. In the present experiments, investigating proton resonance, the field is locked on a strong fluorine resonance by using a fluorine-containing, proton-free compound as solvent. A convenient solvent was found to be hexafluorobenzene.[12] The instrumentation is sketched in Fig. 5. The transmitter frequency for fluorine resonance is generated by mixing the proton transmitter frequency of 60 Mc with a stable, quartz controlled frequency of 3.6 Mc in a balanced mixer, suppressing the carrier frequency. To be able to position the 60 Mc carrier anywhere with respect to the proton spectrum, the fluorine signal is locked on an audio side band of the 56.4 Mc carrier. The modulation frequency of 1.5–3 kc is detected in a phase sensitive detector and the dispersion mode signal is used to correct the magnetic field. The audiofrequency is adjusted so that the 60 Mc carrier is just outside of the total range of the proton spectrum. (All the lines have to lie on one side of the carrier because after phase sensitive demodulation of the signal by means of the transmitter frequency it is not possible to distinguish negative from positive frequencies.)

Transmitter and receiver coils are used in common for both frequencies, 60 and 56.4 Mc, and both are tuned to 60 Mc to give a good S/N ratio for proton spectra. The S/N ratio of the fluorine resonance is poor and a strong reference line is necessary.

V. RESULTS

To test the theoretical result of Sec. III, spectra were recorded using both the pulse technique and conventional spectral recording. All the experiments were done on the same spectrometer with the same rf units to make sure that the inherent instrumental sensitivity was constant. (No attempts were made to optimize the common parts of both experiments such as the preamplifier, etc., since these do not affect the ratios of the sensitivities.) The single section RC filter used in the conventional spectrometer and the rf power in both experiments were adjusted empirically to give the best S/N ratios.

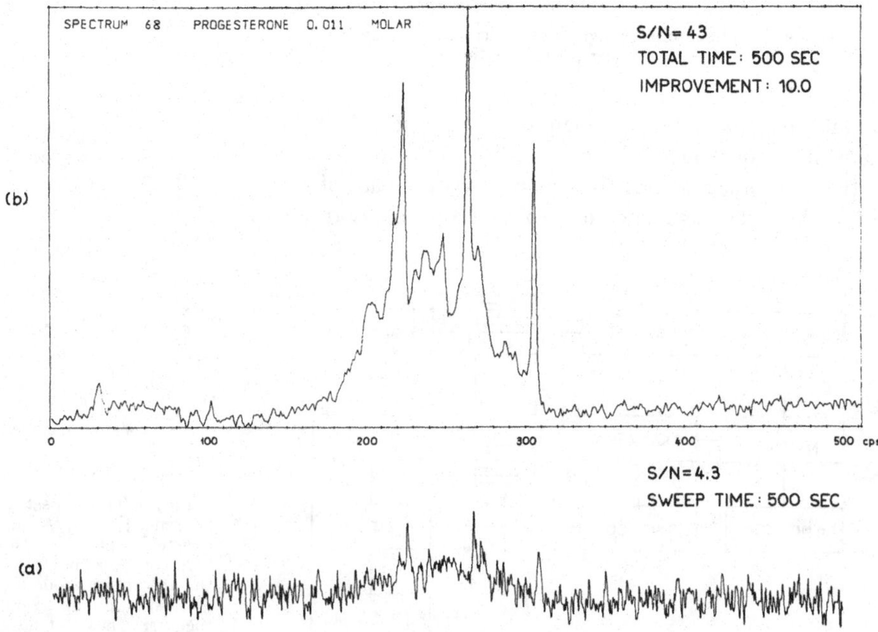

FIG. 7. Two spectra of a 0.011 M solution of progesterone in hexafluorobenzene, both spectra performed in 500 sec. [The S/N ratio is indicated for the strongest line of the spectrum, using the customary practical definition: $S/N=2.5\times$peak value of the signal/peak-to-peak value of the noise.] (a) Single scan, the absorption mode signal is recorded directly. (b) Pulse method, applying 500 pulses, 1 sec apart and adding the 500 responses together. The spectrum is obtained by means of a Fourier transformation. 512 points with 1 cps distance are calculated and plotted. The zero point marks the position of the 60 Mc carrier frequency.

[11] W. A. Anderson (unpublished work); and H. Primas, 5th European Congr. Mol. Spectry. (June 1961).
[12] City Chemical Corporation (New York).

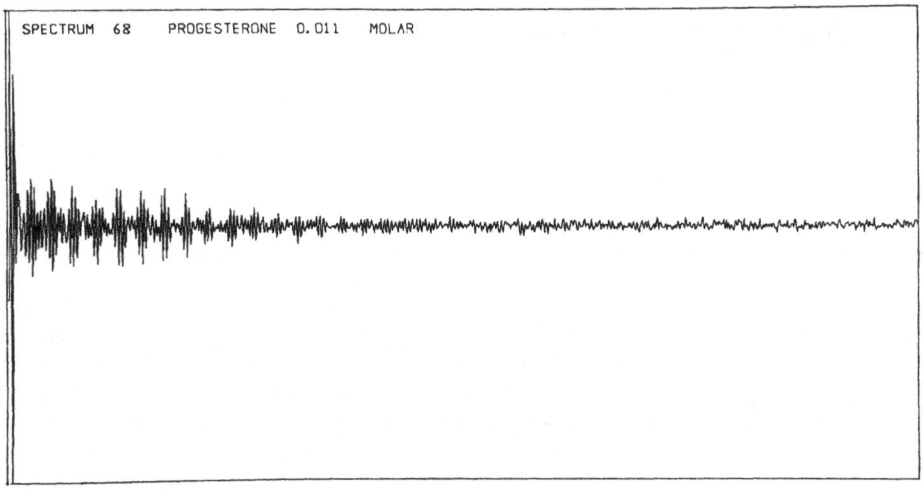

FIG. 8. Impulse response of a 0.011 M solution of progesterone in hexafluorobenzene. It consists of the sum of 500 traces, each trace of 1 sec length. The Fourier transform of this impulse response is shown in Fig. 7 (b).

The sweep range was chosen to be 512 cps. This corresponds closely to the sweep range used in routine measurements of proton resonance spectra at 60 Mc. It is usually necessary to use a sweep range considerably wider than the width of the investigated spectrum because of the uncertainty of the accurate line positions.

Figure 6 shows three spectra of a solution of 1% ethylbenzene in hexafluorobenzene. The lowest trace corresponds to a single scan in 350 sec for a sweep range of 512 cps. The middle trace was taken with the pulse technique and results from 25 pulses spaced 1 sec apart. Filtering was not included in the computer routine. The improvement of the S/N ratio, considering the shorter performance time, is 6.3. The S/N ratio can further be improved, as shown in the top trace, by using a filter procedure in the computer program. The total observed improvement in S/N is a factor 8.5. The theoretical improvement, assuming a line width of 2 cps for the quartet should be 0.799 $\times(512/2)^{\frac{1}{2}}=12.8$. The slight discrepancy is probably due to the deviation from the true slow passage conditions in the single scan experiment.

Figure 7 compares two spectra of a 0.011 M solution of progesterone in hexafluorobenzene taken with the two methods, single sweep and pulse method, using the same total time of 500 sec. The observed improvement is a factor 10.0 and corresponds closely to the theoretically expected value. The peak on the left side of the spectrum in Fig. 7 is caused by the single proton at the double bound of progesterone. The corresponding impulse response used to calculate the spectrum in Fig. 7 is displayed in Fig. 8.

VI. DISCUSSION

The theoretical results of Sec. III and the experimental results of Sec. V show that Fourier transform spectroscopy is able to improve the sensitivity of magnetic resonance experiments effectively if certain conditions are fulfilled which can be summarized as follows:

(1) The longitudinal relaxation time T_1 must not be too long compared with T_2. Otherwise, it is possible to gain about the same improvement by an intermediate passage experiment, combined with time averaging. Intermediate passage causes line shifts and deformations[6] which could be prevented with the pulse technique.

(2) The spectrum must have sufficient fine structure to give a high ratio of total sweep width to line width F/Δ since this determines the achievable improvement [compare Eq. (3.12)].

(3) The technical requirements are mainly concerned with the high field-frequency stability and with the availability of a suitable time averaging computer which must have a number of channels at least equal to $2F/\Delta$ in order to give sufficient resolution and to cover the total frequency range of the spectrum.

These conditions are easily fulfilled for proton magnetic resonance in liquids. Here it is possible to gain up to a factor of 10 in sensitivity in a restricted time and up to a factor of 100 in time for a given sensitivity. There are some other systems where the same method could be used also, for example, boron resonance or deuterium resonance.

Fourier transform spectroscopy used in the described manner shows several side effects which have not been analyzed here in detail. The intensities of the resonance lines are often affected by Overhauser effects within the investigated spin system. Studies of these effects, both experimentally and theoretically, are in progress and will be published later. Using strong samples, maser effects due to radiation damping[13] can distort the spectrum. It should be emphasized that despite the presence of several transmitter frequencies no double or multiple resonance effects can occur. Furthermore, multiple quantum transitions do not occur either, irrespective of the rf field strength.[2]

[13] S. Bloom, J. Appl. Phys. **28**, 800 (1957); and A. Szöke and S. Meiboom, Phys. Rev. **113**, 585 (1959).

A fast recovery probe and receiver for pulsed nuclear magnetic resonance spectroscopy

I J Lowe and C E Tarr
Department of Physics, University of Pittsburgh, Pennsylvania, USA

MS *received 1 September 1967, in revised form 13 November 1967*

Abstract A single coil probe and receiver for use in pulsed nuclear magnetic resonance experiments are described. The impedance transformation properties of a one-quarter wavelength line are used to protect the receiver during the radio-frequency pulse. The probe is very efficient in its use of r.f. power to create an oscillating magnetic field, and the overall system has a fast recovery time (less than 4 μs at 30 MHz).

1 Introduction

Various techniques for designing sample probes for use in pulsed n.m.r. measurements on solids are described in the literature. A review of these methods may be found in papers by Clark (1964), Lowe and Barnaal (1963), Mansfield and Powles (1963), McKay and Woessner (1966) and Trappaniers *et al.* (1964) and in the references cited in these works. Each of these methods, however, is deficient in one or more of the following properties: recovery time, efficient impedance matching to the appropriate r.f. transmitter and/or receiving amplifier, ease of use and suitability for use over a wide temperature range.

This paper describes a simple single coil probe and receiver system that is useful over a wide temperature range. This system provides short recovery times (4 μs at 30 MHz), ease of tuning and good impedance matching.

The parameters in the design described here are adjusted for operation at a frequency of 30 MHz, but they may be easily changed for operation at other frequencies.

2 Circuit description

Figure 1 shows a block diagram of the probe and associated components.

The output impedance of the transmitter is matched to the transmission line and is equal to the input impedance of the receiver. The impedance at resonance of the sample tank circuit is also equal to that of the transmission line.

The operation of the system depends upon the nonlinear properties of semiconductor diodes. Semiconductor diodes have a sharp corner in their current–voltage characteristic curve. Thus, when a small voltage ($\ll 0 \cdot 5$ v) is applied to a diode, little current flows and the diode appears as a high impedance. When a large voltage ($\gg 0 \cdot 5$ v) is applied, a large current flows, and the diode appears as a low impedance.

During the application of a high-powered r.f. pulse from the transmitter, both sets of crossed diodes in figure 1 are conducting heavily and have low impedances. Thus, the r.f. pulse passes through the diodes at A without appreciable attenuation, while the diodes at B prevent the r.f. level at the receiver input from rising above about 1 v. Since point A is a quarter wavelength from B, the effective short circuit at B appears as an open circuit at A and thus does not affect the voltage at A.

Following the r.f. pulse and after the transients have died down, the diodes stop conducting, and the series crossed diodes present a large impedance at point A, while the impedance at point B becomes the characteristic line impedance. The small signal voltages (here typically about 1 mv) are not large enough to lower the impedance of the diodes. Thus, in the absence of an r.f. pulse, the transmitter is isolated from the system, while the receiver is matched to the sample tank circuit.

It should be observed that the length of the $\lambda/4$ line between points A and B is not especially critical, as the purpose of this line is to allow the use of the diode short at B without presenting a low impedance shunt at the sample coil. Thus,

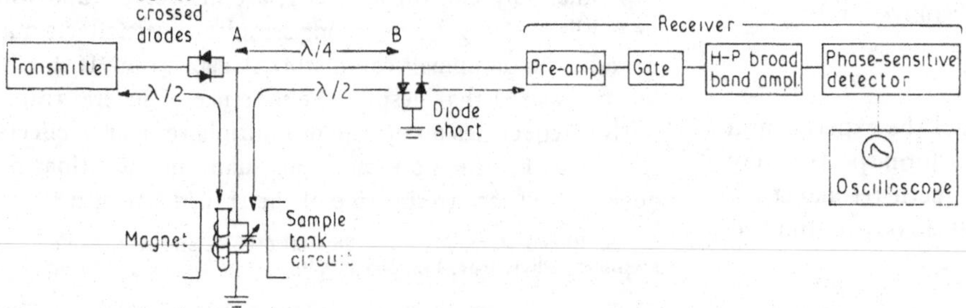

Figure 1 Block diagram of probe and receiver

since the transformed impedance of the diode short need only be large compared with 185 Ω, within the bandwidth of the tuned sample coil, the cable need only be of the order of $\lambda/4$, and the use of the quarter wave cable does not entail any significant narrow-banding of the system.

The electrical lengths of the transmission lines from the transmitter to the sample coil and from the sample coil to the receiver are each one half wavelength. This minimizes reflections due to slight mis-matches in the connections, etc.

Multiple pairs of crossed diodes were used at A and B to reduce their impedances in the conducting state. The series crossed diodes contained 5 pairs of 1N3604's, and the diode short contained 8 pairs.

3 Probe design

Suppose the intrinsic fall time of the transmitter pulse of peak amplitude V_0 into a purely resistive load is shorter than the ringing time constant of the sample tank circuit. Then, if the recovery time of the receiver is shorter than the time required for the r.f. pulse to decay to a value of the order of the induced nuclear signal voltage V_n, the recovery time T_R of the system is given by:

$$T_R = (Q/\pi f) \ln (V_0/V_n) \qquad (1)$$

where $Q = R/\omega L$.

For the recovery figures quoted in this work we have $f = 30$ MHz, $R = Z_0 = 185\ \Omega$, $V_0/V_n \sim 10^7$ and $T_R \sim 3\mu$s, which yields $Q \simeq 17$.

The probe consisted of a variable capacitor and a sample coil that were suspended from a Dewar top plate by a rigid coaxial line. The complete mechanical details of the probe construction will not be given here, as the dimensions must be tailored to individual experimental situations.

The capacitor consisted of several copper shim stock plates separated by annealed glass microscope slide covers 0·006 in thick. The plates were rigidly clamped at one end and their separation varied by means of a screw adjustment at the other end. This screw was operated from the top plate by a stainless-steel tube passed through an O-ring seal so that the probe could be operated in a vacuum if necessary.

In order to provide good homogeneity of the oscillating r.f. magnetic field and a good filling factor, the sample coil was close wound over the desired sample dimensions. In order to keep the impedance and Q at the desired values, however, the inductance was kept much lower than is practical for a wire wound coil. For this reason the coil was wound of copper shim stock. In the present work the coil was three turns, wound in a single layer 1 cm in diameter and 1·2 cm in length from 0·007 in copper shim stock, as shown in figure 2. The Q of the coil was adjusted by means of a short length of manganin wire placed in series with the winding so that Q was approximately constant from 1°K to 300°K. A ¼ in length of no. 20 (0·032 in) manganin wire was found to be satisfactory in producing the required Q of about 17.

To ensure mechanical rigidity, the coil was potted in sulphur. This was accomplished by winding the coil on a Teflon mandrel. The mandrel was placed in a Teflon sleeve with an inner diameter about twice that of the coil, and the resulting annulus was filled with molten sulphur. Sulphur was chosen as a potting agent as it has no nuclei with a magnetic moment.

The coil leads were flat copper strips that were brought out of the coil parallel and close together to reduce lead inductance. The insulation necessary to separate the leads was glass filter paper, as shown in figure 2.

Finally, the coil was enclosed in a brass shield about 2·5 cm in diameter and about 7 cm in length for electrical shielding and mechanical rigidity.

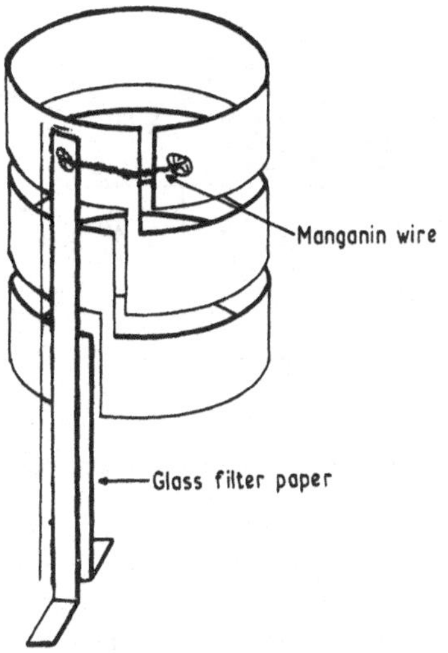

Figure 2 Coil construction

4 Receiver

4.1 *Preamplifier*

A two stage preamplifier with a recovery time of about 1 μs from an overload of 1 v and with an input impedance of 185 Ω was constructed for use in the system. The preamplifier was stagger tuned to provide a flat response over a bandwidth of 4 MHz (3 dB down), centred at 30 MHz. The resulting voltage gain was 135, and the noise figure less than 3·5 dB.

A standard 30 MHz preamplifier (Linear Electronics Laboratories, with a voltage gain of 1000) was also tested in the system. However, it was not entirely satisfactory, because of small amplitude ringing following partial recovery from an r.f. pulse. This ringing was due to decoupling chokes used in the B^+ lines between stages.

4.2 *Diode gate*

The broadband amplifier (Hewlett-Packard 461–A) recovers in less than 1 μs from the overload presented by the r.f. pulse feedthrough, but the gain is rather unstable for about 10 μs following the pulse. To prevent distortion of the free induction decay resulting from this gain instability, a balanced diode gate was used to decouple the preamplifier from the broadband amplifier for the duration of the r.f. pulse and for a short time thereafter. The gate was placed between the preamplifier and the amplifier so that the noise generated in the diodes by the current passed through them in their conducting state would be injected at a relatively high signal level.

Since the diode gate is completely balanced, it does not produce any ringing or blocking effects due to turn-on transients. Also, since the gate prevents overloads from the r.f. pulses from reaching the detector, the system is especially well suited to use with signal averaging equipment, as there are no large transients to interfere with the analogue-to-digital converter.

The gate circuit is shown in figure 3. When the gate is in its normal (closed) stage, side C of the diode bridge is negative with respect to side D. Thus, current flows around the diode bridge, and a signal can pass from the input branch to the output branch. When a positive pulse (+ 4 v minimum) is applied to the phase splitting network, the transistors conduct, and the bias on the bridge is reversed, and no signal can pass the gate.

The 1 kΩ potentiometer across the gate provides d.c.

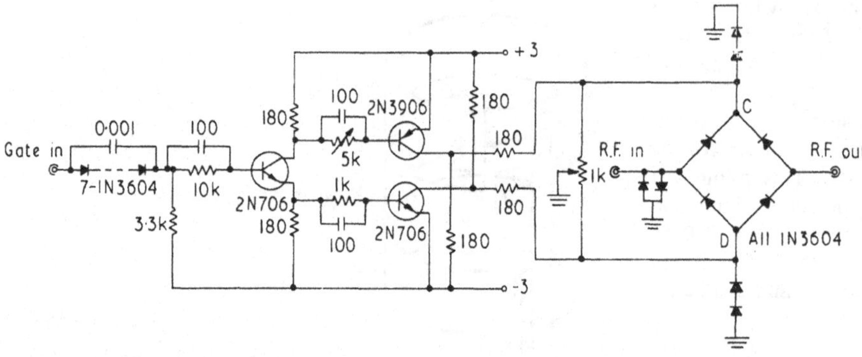

Figure 3 Schematic diagram of diode gate used to protect broadband amplifier

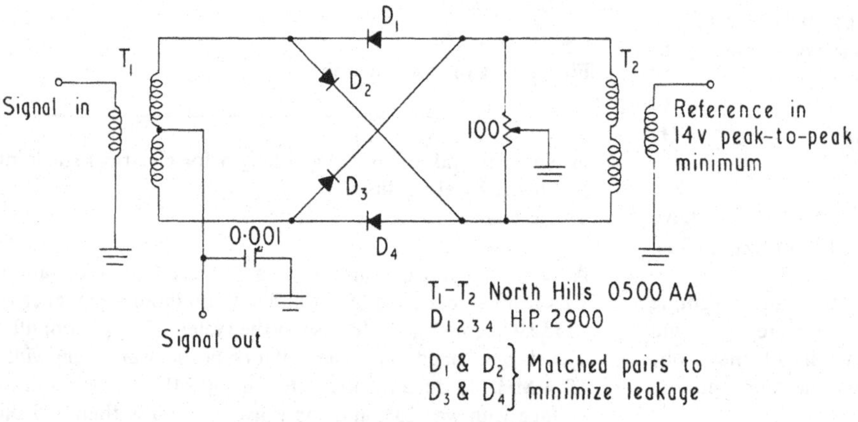

Figure 4 Schematic diagram of phase sensitive detector

balance for the gate. The 5 kΩ potentiometer in the base of the 2N3906 sets the drive to that transistor and is set to balance out the turn-on transient of the bridge. The feed-through of the r.f. pulse can be seen at the beginning of figure 5 and is less than one tenth the amplitude of the detected signal. This feedthrough of the r.f. pulse to the input of the broadband amplifier is less than 0·010 v peak-to-peak over a broad range (500–1200 v peak-to-peak) of r.f. pulse amplitudes. The turn-on transients of the gate may be balanced to less than ±0·001 v.

4.3 *Phase sensitive detector*

The circuit of the phase sensitive detector is shown in figure 4. The circuit has a flat frequency response (0·1 dB) from 1 to 100 MHz. This broadband response is provided by the two North Hills 0500 AA transformers (North Hills Electronics Inc., Glen Cove, L.I., New York), which have a flat frequency response (0·1 dB) from under 1 MHz to over 100 MHz (down 1 dB at approx. 0·1 MHz and 200 MHz) and have an input impedance of 50 Ω (unbalanced) and an output impedance of 200 Ω.

The detector requires a minimum r.f. reference level of 14 v peak-to-peak and has a linear response for input signal levels to 1·2 v. With no filtering, the r.f. leakage can be set to less than 0·01 v peak-to-peak and the d.c. offset to less than ±0·01 v with the 100 Ω balancing potentiometer.

The input and output impedances of the phase sensitive detector are about 50 Ω.

5 Conclusion

The probe and receiver system described above are free from many of the difficulties inherent in most other systems, as there are no balancing or damping elements. Also, since the entire system is impedance matched to the single coil probe without the use of transformers, there are no difficulties with transformer ringing, and there is optimum r.f. power transfer. The quarter wavelength cable and diode short effectively prevent the r.f. transmitter pulse from blocking the receiver.

Acknowledgments

We wish to thank D Douglas of the Bell Telephone Laboratories for suggesting the use of a quarter wavelength cable to isolate the receiver from the transmitter, and W Vollmers for building the r.f. probe.

It has been brought to our attention that a somewhat similar use of a quarter wavelength cable is described by U Häberlen (1967 Ph.D. Thesis, Tech. Hochshule, Stuttgart).

Research was sponsored by the Air Force Office of Scientific Research Office of Aerospace Research, United States Air Force under AFOSR grant no. 196–66.

References

Clark W G 1964 *Rev. Sci. Instrum.* **35** 316–33

Lowe I J and Barnaal D E 1963 *Rev. Sci. Instrum.* **34** 143–6

Mansfield P and Powles J G 1963 *J. Sci. Instrum.* **40** 232–8

McKay R A and Woessner D E 1966 *J. Sci. Instrum.* **43** 838–40

Trappaniers N J Gerritsma C J and Osting P H 1964 *Physica* **30** 997–1017

Quadrature Fourier NMR Detection: Simple Multiplex for Dual Detection and Discussion*

It is our experience that most users of Fourier transform NMR instruments are not fully aware of the utility of dual (quadrature) phase detection and how simple and trouble free it is in routine operation. There is also a widespread and incorrect belief that the technique is difficult and expensive to implement. We believe that this mode of detection should be viewed as a standard and well-integrated part of any spectrometer system, rather than as an expensive add-on feature. It is our purpose to summarize our experience and that of others, in support of these views.

We report, first, a simple method of adapting a dual (quadrature) phase detector system to most commercial Fourier NMR spectrometer computer systems in order to increase their sensitivity and their usefulness for observing weak proton signals. The advantage of the method is that absolutely no reprogramming of the computer need be done and only a single signal input is needed.

Our dual phase detector and filter operates on the 6.15-MHz intermediate-frequency signal and reference of a Bruker WH-90 spectrometer. It is carefully constructed along lines described previously (*1*) to avoid images of signals about the transmitter frequency. It is tedious but not difficult (*1*, *2*) to construct such a detector well enough to reduce images to a 1% level. Even better image rejection (or the use of less precise detectors) is possible using a method recently demonstrated by Stejskal and Schaefer (*3*). We will describe later how their method could certainly be combined with ours.

Our method differs from simple repeated alternating sampling of the two (real and imaginary) outputs of the phase detector (*1*) in that every two times we invert the incoming signal. Thus, the computer samples the complex signals in the four step sequence (Re, Im, -Re, -Im) which is repeated 2048 times in an 8 K sample of a single transient. The advance rate for each step is the usual sampling frequency ν_a(16 kHz for 8 kHz output spectral width) and the cycle time for the sequence is thus $4/\nu_a$. The dual phase detector can be said (*1*) to emit the real and imaginary parts of a complex audio NMR signal of form $\exp(-t/T_2)\exp(i\omega t)$, where ω is the difference (algebraic) between the nuclear and the rf carrier frequencies. In effect, our multiplexer converts this to a real signal of form $\exp(-t/T_2)\cos[(\omega + 2\pi\nu_a/4)t]$.

The only essential difference in operating procedure from that of the commercial instrument is that the observation frequency is set in the middle of the spectrum rather than at one end. Phasing is done manually in the usual way, by means of two controls on the Nicolet (BNC 12) computer. A spectrum obtained in this way as compared with one obtained with a standard single phase detector is shown in Fig. 1.

The multiplex circuitry is extremely simple, consisting of a single-pole double throw analog switch that selects between the two filter outputs, followed by an operational amplifier inverter and a second analog switch that alternates ±outputs. These are

* Partially supported by National Science Foundation Grants GU 3852 and GP 37156, and U.S. Public Health Service Grant GM 20168.

Reprinted with permission from Journal of Magnetic Resonance **19**, 250–254 (1975); ©Academic Press.

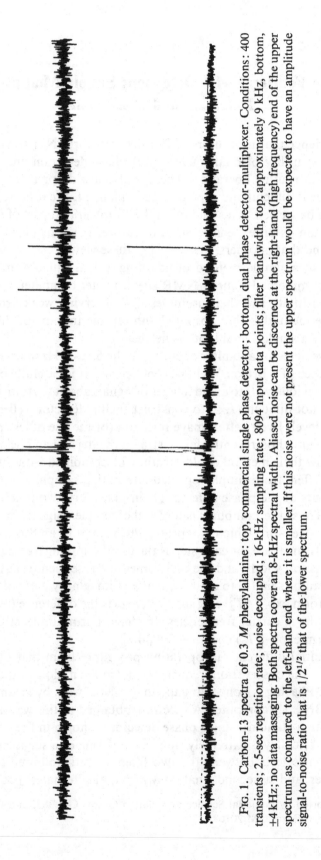

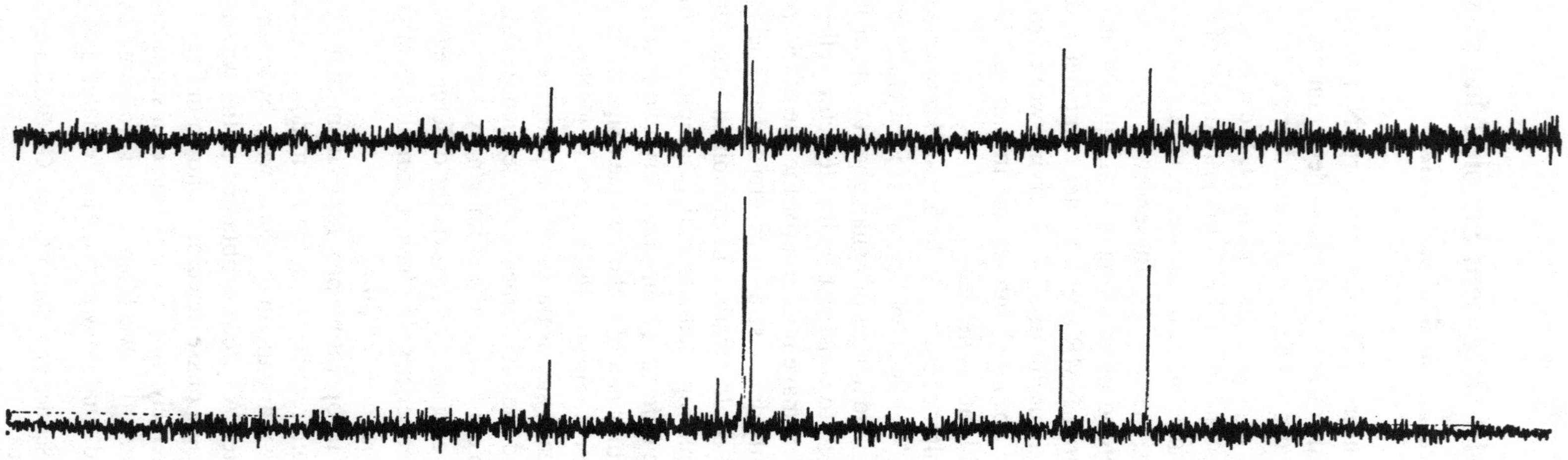

FIG. 1. Carbon-13 spectra of 0.3 *M* phenylalanine: top, commercial single phase detector; bottom, dual phase detector-multiplexer. Conditions: 400 transients; 2.5-sec repetition rate; noise decoupled; 16-kHz sampling rate; 8094 input data points; filter bandwidth, top, approximately 9 kHz, bottom, ±4 kHz; no data massaging. Both spectra cover an 8-kHz spectral width. Aliased noise can be discerned at the right-hand (high frequency) end of the upper spectrum as compared to the left-hand end where it is smaller. If this noise were not present the upper spectrum would be expected to have an amplitude signal-to-noise ratio that is $1/2^{1/2}$ that of the lower spectrum.

controlled by a two-stage binary counter, which is advanced by the address advance of the computer delayed 10–100 μsec by a variable length one-shot multivibrator. This is intended to avoid the possibility that the computer samples just as the multiplex is switching. The counter is cleared by the observation pulse, and its first and second stages run the Re/Im and the +/– switches, respectively.

The entire multiplexer consists of five integrated circuits. The analog switches are based on National Semiconductor AH0015 FET switches and standard operational amplifiers. The other components are standard 7400-series digital circuitry. The dual-detector/filter is similar to one described in great detail in (*1*). Circuit diagrams of these components and of the rf gating system (see below) are available from us on request.

The WH-90 spectrometer is equipped by the manufacturer to reverse the rf phase of the rf pulse every two transients, and the computer subtracts the signal when this is done. This is almost essential for implementation of our scheme, to avoid saturating the memory with a signal at a frequency $\nu_a/4$, which arises from unavoidable dc offset of the filter output. To implement the scheme of Stejskal and Schaefer (*2*) we would alternately shift the transmitter pulse phase by 90° and compensate by advancing the multiplexer phase +90° by, for example, initializing it at the +Im rather than the +Re position using straightforward logic. Data would have to be collected in blocks of four transients for complete cancellation

Some words of warning should be given to those interested in trying these techniques, especially for observation of weak proton signals in the presence of strong ones. Nearly perfect gating of the transmitter pulse is desirable; otherwise, a signal often appears at center band that looks like a misphased NMR peak. This problem disappeared after we rebuilt the WH-90 gating system along the lines described elsewhere (*1*). This is not a problem for ^{13}C gating, in the Bruker WH-90, because an extra conversion, from 90 to 22.5 MHz, improves the quality of gating.

Despite these improvements, usually there is still a single-address glitch in our proton spectra at exact center band. It is less serious, and can generally be eliminated, in ^{13}C spectra where the dynamic range is smaller than for most proton spectra. This glitch can be eliminated by adjusting the dc offset of the dual filter system to zero, and by empirically adjusting the delay between the address advance and the binary counter.

For proton NMR, excellent detector linearity may be important to avoid spurious audio beat signals between two strong NMR signals. We have always used very linear FET switch phase detectors operating at 50 or 100 kHz second intermediate frequency (*1*) and have not had such problems. High-frequency diode detectors may be inferior in this regard, but we have never tested such detectors in demanding applications.

Next we comment on the relative merits of various detection schemes. It is a minor scandal that commercial instruments have not, until recently, used either crystal-filter or quadrature single-sideband detectors despite the fact that they have been known to radio amateurs for years and appeared in the FT-NMR literature five years ago (*4*). The expected twofold decrease in running time, for the same signal-to-noise ratio, has been well-documented (*2*, *3*, *5*). The increase in cost is small compared to the cost of the instrument.

We have no experience with the crystal filter method of single-sideband detection. It is probably the method of choice for spectrometers dedicated to ^{13}C using a single spectral width (*5*). It is simpler in this case and, apparently, the rf gating requirements are not as stringent. It is necessary that the intermediate frequency (or the transmitter frequency, if direct detection is used) be constant, that is, the first local oscillator must track the transmitter frequency, if the latter is changed appreciably. This is relatively simple to do, and is also desirable for dual detection, to avoid the necessity of readjusting the phase whenever the rf frequency is shifted. Crystal filtering requires more rf power than does dual phase detection.

The dual-detector scheme seems to us to be advantageous for general purpose instruments where a variety of spectral widths are used for different nuclei. It is useful for observation of portions of a spectrum, either for observation of weak signals in the presence of a strong signal, or when computer capacity is inadequate to cover a spectrum at the necessary resolution. The latter point is especially relevant for fluorine spectroscopy, for example. The dual detector scheme filters at audio frequency using low-pass filters whose characteristics can be changed easily. Our dual filters are six-pole Butterworth types. They can be set with a bandwidth somewhat greater than the output spectral width, in which case the full output width is usable but some noise and any signals from outside the intended spectral width will be aliased onto the spectrum. Or, they can be set at somewhat less than the total possible spectral width for high-resolution searches of regions of the spectrum or for those cases where the computer has more than enough memory capacity to give the desired resolution. At this writing we do not have a dual filter whose bandwidth is suitable for ^{13}C work ($\sim \pm 3$kHz), and the experimental conditions used in Fig. 1 were chosen to match the available filter.

The question of phase correction of Fourier transform spectra deserves some discussion since the methods currently used, though adequate, are unnecessarily primitive and inconvenient. The manual phase correction used on most commercial spectrometers is satisfactory for less demanding applications using either quadrature or crystal filter detection. However, it is quite useful to have the computer automatically correct for amplitude and phase error produced by the filter. Such a correction program has been described in detail elsewhere (*1*). Its advantages are: (a) the filter can be set to nearly eliminate aliasing both noise and signals (aliasing is the spurious shifting of a signal by an amount equal to the digitizer sampling rate); (b) distortion in both the baseline (*3*) and the signal amplitude produced by the filter is precisely compensated; and (c) the maddening manual frequency-proportional phase correction is not needed. The computing requirements are modest: The filter characteristics can be stored in sampled form as relatively few (50) words; the correction program itself requires less than 300 words; and the number of usable output points is 10–20% less than in conventional Fourier transform NMR because this fraction of the spectrum may have excessive aliasing and digitizer noise. Automatic filter correction with variable spectral width is relatively easy with quadrature detection because the audio filter characteristics can be made to scale in proportion to the address advance frequency, so that one phase correction suffices for all spectral widths. It is our experience that the majority of potential users of automatic phase correction do not understand or believe how easy and useful it is to set up such a routine. It is a feature that is not essential, but it is so useful that it should eventually be incorporated into any spectrometer.

The recommended filter settings of commercial instruments are often such that there is a roughly twofold increase in noise power as a result of aliasing, in addition to the twofold increase from imaging. Thus, a fourfold decrease in noise, and increase in productivity, is often possible by means of quadrature detection combined with properly designed filtering.

For high-dynamic range applications there are several demonstrated rf excitation schemes that affect resonances in only a band of frequencies while not exciting some particular frequency or band thereof (*1*, *4*, *6–8*). It is desirable that the detection system sensitivity span roughly the same frequency band as the transmitter pulse, so that their effectiveness reinforces. Thus, for example, dual phase detection is almost essential for simple long pulse stimulation (*1*, *4*, *8*), while a crystal filter might be better for certain variants of correlation spectroscopy (*6*, *7*).

REFERENCES

1. A. G. Redfield and R. K. Gupta, *Adv. Magn. Resonance* **5,** 81 (1971).
2. D. M. Wilson, R. W. Olson, and A. L. Burlingame, *Rev. Sci. Instrum.* **45,** 1095 (1974).
3. E. O. Stejskal and J. Schaefer, *J. Magn. Resonance* **14,** 160 (1964).
4. A. G. Redfield and R. K. Gupta, *J. Chem. Phys.* **54,** 1418 (1971); first reported at 1970 E.N.C. Meeting, Pittsburgh, Pa.
5. A. Allerhand, R. F. Childers, and E. Oldfield, *J. Magn. Resonance* **11,** 272 (1973).
6. J. Dadok and J. Sprecher, *J. Magn. Resonance* **13,** 243 (1974).
7. B. L. Tomlinson and H. D. W. Hill, *J. Chem. Phys.* **59,** 1775 (1973).
8. A. G. Redfield, S. D. Kunz, and E. K. Ralph, *J. Magn. Resonance* (in press); and references therein.

A. G. Redfield
Sara D. Kunz

Department of Physics
Brandeis University
Waltham, Massachusetts 02154

Received December 18, 1974
Revised April 14, 1975

* Also at Department of Biochemistry and the Rosensteil Basic Medical Sciences Research Center, Brandeis University.

The Signal-to-Noise Ratio of the Nuclear Magnetic Resonance Experiment

D. I. HOULT AND R. E. RICHARDS

Department of Biochemistry, University of Oxford, South Parks Road, Oxford OX1 3QU, England

Received March 5, 1976

A fresh approach to the calculation of signal-to-noise ratio, using the Principle of Reciprocity, is formulated. The method is shown, for a solenoidal receiving coil, to give the same results as the traditional method of calculation, but its advantage lies in its ability to predict the ratio for other coil configurations. Particular attention is paid to the poor performance of a saddle-shaped (or Helmholtz) coil. Some of the practical problems involved are also discussed, including the error of matching the probe to the input impedance of the preamplifier.

INTRODUCTION

An important reason for the development of high-resolution NMR spectrometers employing superconducting, rather than iron, magnets has been the belief that the signal-to-noise ratio (S:N) available is proportional to the Larmor frequency to the three-halves power (*1*, *2*), and thus that an increase of, for example, three times in frequency from 90 to 270 MHz would bring as a return an improvement of a factor of 5.2 in S:N. It is always difficult accurately to compare the performances of different instruments as a variety of factors (to be discussed further) come into play; nevertheless, it is generally felt by those with experience of superconducting systems that the improvement obtained is disappointing—for example, about 2 for the frequencies quoted. As it is acknowledged that the electronics needed at very high and ultrahigh frequencies are "difficult," the accusing finger has tended to point in the engineers' direction. However, recent developments in the design of low-noise amplifiers (*3*) and the general improvement in frequency-changing techniques, due mainly to the increased use of hot carrier diodes, have caused us to consider afresh a fundamental dictum of NMR., the $\frac{3}{2}$ power law, and to examine anew the derivation of the formula, *which was derived prior to the advent of superconducting systems.*

PRIMARY AND SECONDARY CONSIDERATIONS

The usual formula for the signal-to-noise ratio available after a 90° pulse is given by (*1*, *2*)

$$\Psi_{\mathrm{rms}} = K\eta M_0 (\mu_0 Q\omega_0 V_c/4FkT_c\Delta f)^{1/2}, \qquad [1]$$

where K is a numerical factor (~1) dependent on the receiving coil geometry; η is the "filling factor," i.e., a measure of the fraction of the coil volume occupied by the sample; M_0 is the nuclear magnetization which is proportional to the field strength B_0; μ_0 is the permeability of free space; Q is the quality factor of the coil; ω_0 is the Larmor angular frequency; V_c is the volume of the coil; F is the noise figure of the preamplifier; k is Boltzmann's constant; T_c is the probe (as opposed to sample) temperature; and Δf is the bandwidth (in Hertz) of the receiver.

We may consider that the primary factors involved in any analysis of S:N are those contained within the equation. Secondary factors, such as whether or not quadrature detection is used, the availability of Fourier transform techniques, sweep rate or pulsing rate, the use of decoupling or Overhauser effect, though of great importance, are not of such a fundamental nature and will therefore not be considered further.

For all its usefulness, Eq. [1] is not a "fundamental" equation. It contains four unknowns, K, η, Q and F, only two of which (Q and F) are easily measurable. The definition of filling factor $\eta = V_s/2V_c$ (V_s is the sample volume) may well be satisfactory for a solenoidal coil, but its validity for other coil configurations must be questioned. Further, the equation contains little information as to the dependency of S:N on various physical parameters; for example, if we quadruple the coil volume while keeping η constant, do we obtain only a doubling of S:N, or does the change in the coil dimensions alter K and Q also? Table 1 shows how complex the use of Eq. [1] may be. The interactions between the various factors are manifold and mostly strong. This is partly a consequence of the method of calculation. The reader is referred to Ref. (*1*, *2*) for further details, but it is perhaps worth quoting Abragam who states that "the above calculation gives only an order of magnitude."

TABLE 1

INTERACTION OF TERMS IN THE TRADITIONAL EQUATION FOR SIGNAL-TO-NOISE RATIO AS GIVEN BY EQ. [1]

Temperature T					
Strong	Preamplifier noise figure F				
Strong		Coil Q			
		Weak	Filling factor and geometry $K\eta$		
		Strong	Strong for small samples	Volume V_c	
	Strong	Strong		Strong	Field strength B_0

THE PRINCIPLE OF RECIPROCITY

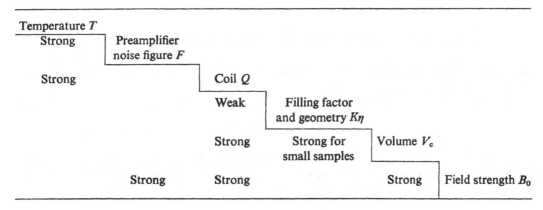

Clearly it would be of use to formulate a different method of calculation which gives a direct insight into the various factors involved and which removes some of the interactions inherent in Eq. [1]. This may be done to a reasonable extent by invoking the principle of reciprocity. Consider the induction field $\mathbf{B}_1$ produced by a coil C carrying unit current (See Fig. 1). Obviously, the field at point A is much stronger than at point B. Intuitively, one would expect therefore that if a magnetic dipole $\mathbf{m}$ were placed at point A and set rotating about the z axis, the alternating signal it induced in the coil would be much greater than that induced by the same dipole placed at point B. This is indeed the case, and it may easily be shown that the induced emf is given by

$$\xi = -(\partial/\partial t)\{\mathbf{B}_1 \cdot \mathbf{m}\}, \qquad [2]$$

where $\mathbf{B}_1$ is the field produced by the unit current at $\mathbf{m}$. It follows that for a sample of volume V_s, which has been recently subjected to a 90° pulse, we need only know the

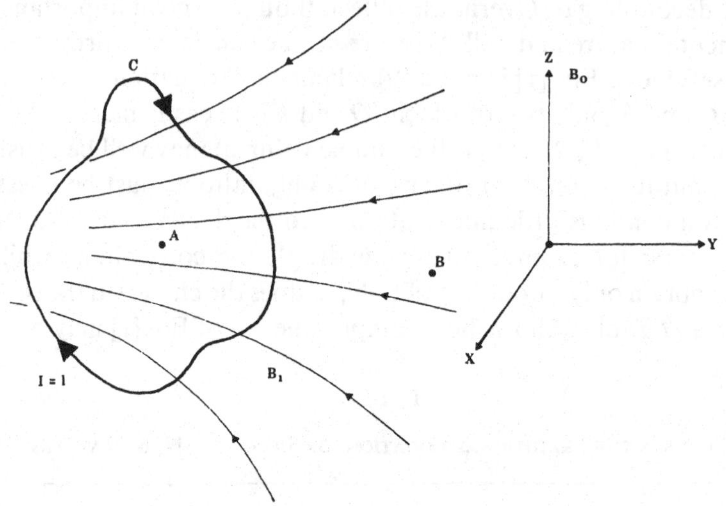

FIG. 1. The induction field $\mathbf{B}_1$ produced by a coil C carrying unit current.

value of $\mathbf{B}_1$ at all points in the sample to be able to calculate the emf induced in the coil. Thus, if $\mathbf{M}_0$ lies in the xy plane,

$$\xi = -\int_{\text{sample}} (\partial/\partial t)\{\mathbf{B}_1 \cdot \mathbf{M}_0\}\, dV_s. \qquad [3]$$

The calculation of $\mathbf{B}_1$ is feasible for most shapes of coil; of course, if $\mathbf{B}_1$ may be considered to be reasonably homogeneous over the sample volume, the calculation is considerably simplified as the integration of Eq. [3] becomes trivial, giving

$$\xi = K\omega_0 (B_1)_{xy} M_0 V_s \cos \omega_0 t, \qquad [4]$$

where K is an "inhomogeneity factor" which may if necessary be calculated, $(B_1)_{xy}$ is the component of $\mathbf{B}_1$ perpendicular to the main field B_0, and phase has been neglected. The magnetization M_0 is given by

$$M_0 = N\gamma^2 \hbar^2 I(I+1) B_0 / 3kT_s, \qquad [5]$$

where N is the number of spins at resonance per unit volume, γ is the magnetogyric ratio, and T_s is the sample temperature. As $\omega_0 = -\gamma B_0$ it follows from Eqs. [4] and [5] that the EMF induced in the coil is proportional to the square of the Larmor frequency.

THE NOISE

Having laid the basis of the calculation for the emf induced by the nuclear magnetization in the receiving coils, we now consider the noise. In a correctly designed system, this should originate solely from the resistance of the coil. As the dimensions of the coil are inevitably much less than the wavelength of the radiation involved, the radiation resistance is negligible, and it should therefore be possible to predict the thermal noise present purely on the basis of the equation

$$V = (4kT_c \Delta f R)^{1/2}. \qquad [6]$$

Here, T_c is the temperature of the coil and R its resistance. Unfortunately, the calculation of R may not be performed accurately, and it is solely this factor which leads to uncertainty in the theoretical prediction of signal-to-noise ratio. At the frequencies of most interest in NMR spectroscopy (say >5 MHz) the "skin effect" associated with the magnetic field generated by a current ensures that that current flows only in regions of the conductor where there is also a magnetic flux. Thus, for a long, straight cylindrical conductor, the current flows in a skin on the surface. This situation is easily amenable to calculation and yields the result that

$$R = (l/p)\,(\mu\mu_0\,\omega_0\,\rho(T_c)/2)^{1/2}, \qquad [7]$$

where l is the length of the conductor; p is its circumference, μ is the permeability of the wire; and $\rho(T_c)$ is the resistivity of the conductor, which is of course a function of temperature. However, the situation is considerably more complicated if the conductor is not cylindrical, and worse still, if there are many conductors in close proximity, as is, in effect, the case with a coil, the magnetic field created by the current of one conductor influences the distribution of current in another. This "proximity effect" (*4*), which is also manifest when conductors such as the silvering on a dewar, or even the sample itself, are close to the coil, normally tends to reduce the surface area over which current is flowing, and thus the resistance is increased from that calculated from Eq. [7] by a factor ζ of about 3. Attempts have been made to calculate ζ, but only for a single-layer solenoid may any confidence be placed in the results (*5*).

THE SIGNAL-TO-NOISE RATIO

By combining Eqs. [4] to [7] we may arrive at an equation for the signal-to-noise ratio,

$$\Psi_{\mathrm{rms}} = \frac{K(B_1)_{xy}\,V_s\,N\,\gamma\,\hbar^2\,I(I+1)}{7.12\,kT_s}\cdot\left(\frac{p}{FkT_c l\zeta\Delta f}\right)^{1/2}\cdot\frac{\omega_0^{7/4}}{[\mu\mu_0\,\rho(T_c)]^{1/4}}. \qquad [8]$$

Note that the proximity factor ζ and the noise figure F of the preamplifier have been included. At first sight, this equation appears unmanageable, but this is in fact not the case. First, the unknowns η and Q of Eq. [1] are absent; they have been replaced by a single function ζ, which, from experience, is reasonably well known, and second, the number of factors in Eq. [8] which are variable in a given experimental situation is small. These factors are:

(a)	$K(B_1)_{xy}$	the effective field over the sample volume produced by unit current flowing in the receiving coil.
(b)	p	the perimeter of the conductor.
(c)	l	the length of the conductor.

The above factors are dependent only on coil geometry.

(d)	T_c	the temperature of the coil.
(e)	$\rho(T_c)$	the material from which the coil is made.
(f)	F	the quality of the preamplifier.

It is of interest to note that the frequency dependence is to the power of 7/4. This is not a new conclusion; it has also been postulated by Soutif and Gabillard (*6*) and it must be stressed that, *for a solenoid*, in no way is Eq. [8] in contradiction with Eq. [1]. Figure 2 shows an experimental plot of Q versus frequency for a set of solenoids all wound in the same manner with the same overall dimensions. This plot clearly recon-

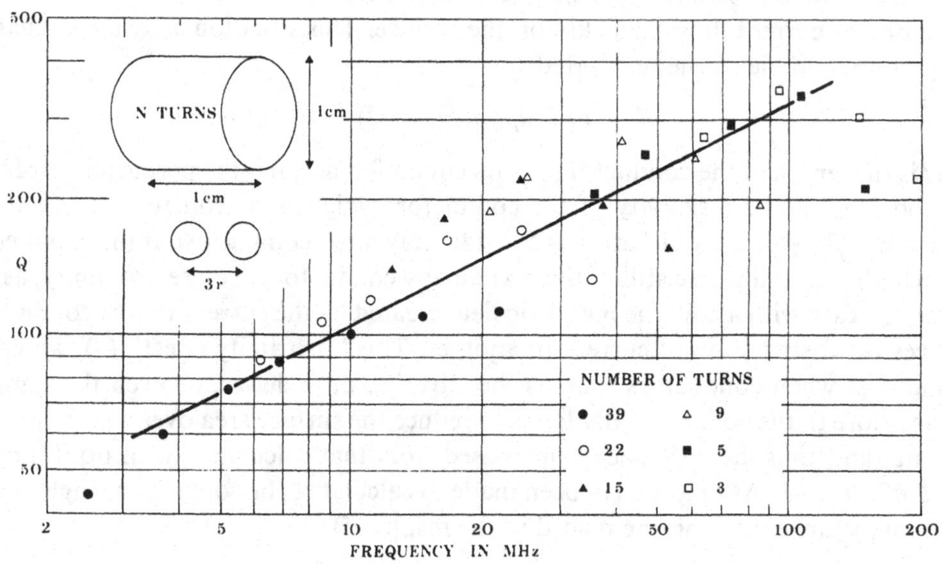

FIG. 2. The Q of a set of solenoids. The dimensions of each solenoid are the same, but the number of turns varies with frequency. The winding geometry is shown in the inset, and gives the optimum Q (*4*). For many turns, and well below the self-resonant frequency, $Q \propto f^{1/2}$.

ciles the two equations and the derivation of Eq. [1] from Eq. [8] is shown in Appendix 1. The major advantage of the calculation from first principles which results in Eq. [8] is that it is applicable to any coil geometry, for it is found that ζ changes but little with a change of configuration provided the separation of the windings is small in comparison with the overall dimensions of the coil. Of particular interest are saddle-shaped (or Helmholtz) coils used mainly with superconducting instruments and solenoidal coils used predominantly with conventional machines. Let us therefore compare the performance of these two configurations.

SADDLE-SHAPED AND SOLENOIDAL COILS

The factors of interest are (a) to (c) above, and so, assuming the same volume of sample is used in each case, the essential part of Eq. [8] is

$$\Psi \propto V_s (B_1)_{xy}/R^{1/2} \qquad \text{or} \qquad \Psi \propto V_s (B_1)_{xy} (p/l)^{1/2}. \qquad [9]$$

Our first task therefore is to calculate $(B_1)_{xy}$ for the two coils. While this is trivial for a many-turn solenoid, it is not for the saddle-shaped coils, and the essence of the calculation is indicated in Appendix 2. Let a be the radius of the coils, and $2g$ the lengths, as shown in Fig. 3.

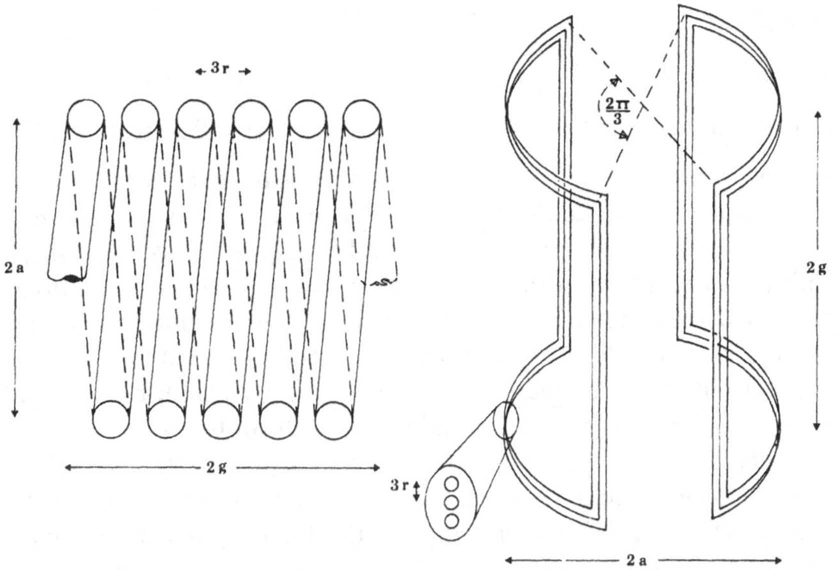

FIG. 3. The two winding geometries considered. The angular width of the saddle-shaped coils is 120°, as this value gives the best homogeneity, and the width of the windings is approximately $g/5$.

Saddle-shaped

$$(B_1)_{xy} = \frac{n\sqrt{3}\,\mu_0}{\pi}\left\{\frac{ag}{[a^2+g^2]^{3/2}} + \frac{g}{a[a^2+g^2]^{1/2}}\right\},$$

$$n \geqslant 1.$$

Solenoid

$$(B_1)_{xy} = \frac{\mu_0 n}{2}\frac{1}{[a^2+g^2]^{1/2}},$$

$$n \gg 1.$$

Typically, $a \simeq g$, and hence

$$(B_1)_{xy} = 0.585\, n\mu_0/a. \qquad (B_1)_{xy} = 0.354\, n\mu_0/a.$$

We must now calculate p and l. Let us assume that wire is used in both cases. We then have that

$$l \simeq 8na\{(g/a) + (\pi/3)\}. \qquad l \simeq 2\pi an.$$

When $a \simeq g$,

$$l \simeq 16.4\, na \qquad l \simeq 6.3\, na.$$

The length of wire per unit turn is thus greater for the saddle-shaped coil by a factor of about 2.6, a significant fact which will be considered later. The radius r of the wire used is dependent upon the manner in which the coils are wound, but assuming that a planar structure is used, the n turns must fit in a length $2g$ for the solenoid and in a width $g/5$ approximately for the saddle-shaped coils. To optimize the performance of each coil with respect to proximity effect, the distance between the centers of each turn should be roughly $3r$ (4), and so, assuming this value, we have that for $a \simeq g$,

$$3r(n-1) \simeq a/5, \qquad\qquad 3r(n-1) = 2a,$$

$$\therefore \quad p \simeq 2\pi a/15(n-1). \qquad\qquad \therefore \quad p = 4\pi a/3(n-1).$$

Thus, for $n \gg 1$,

$$p \simeq 0.42a/n, \qquad\qquad p \simeq 4.2\,a/n,$$

$$\therefore \quad \psi \propto 0.094\,\mu_0\,V_s/a. \qquad\qquad \therefore \quad \psi \propto 0.29\,\mu_0\,V_s/a. \qquad [10]$$

Thus the performance of the solenoid would appear to be approximately three times better than that of the saddle-shaped coils. It is difficult to explain this result on the basis of the traditional formula as represented by Eq. [1]. This is not surprising, as there is a critical assumption in the derivation (shown in Appendix 1) involving the energy stored by current flowing in the coil. Briefly, this assumption is that half the energy is stored, via the field $\mathbf{B}_1$, within the confines of the coil and that the field is homogeneous within those confines. While this is broadly true for a solenoid, due to the continuity and closed form of the winding, it is certainly not true for the saddle-shaped coils, where the open structure dictates that much of the magnetic energy is stored in flux which lies close to the wires and which does not pass through the sample. A further deficiency of Eq. [1] concerns the length of a 90° pulse when the probe is of the single-coil variety. One would be led to believe that the length was dependent only on $Q^{-1/2}$. Typically a saddle-shaped coil has a lower Q than a solenoid, and so the 90° pulse length should be a little longer (up to, say, 60%) but Eq. [10] show the inadequacy of this statement. This may be seen from the following argument.

90° PULSE LENGTHS

When the probe is matched to the transmitter (a point to be considered later), the power supplied W is dissipated entirely in the resistance R of the coil. The current flowing through the coil is thus given by

$$I = (W/R)^{1/2}.$$

Now the $\mathbf{B}_1$ field employed in the calculations to date has been derived for unit current. It follows that when the transmitter is on, the irradiating field $(B_1^*)_{xy}$ is given by

$$(B_1^*)_{xy} = I(B_1)_{xy} = (B_1)_{xy}(W/R)^{1/2}.$$

Thus, from Eq. [9],

$$(B_1^*)_{xy} \propto \mathrm{W}^{1/2}\,\psi/V_s. \qquad [11]$$

The 90° pulse length is thus a direct measure of the S:N obtainable from a single coil system and if the preceding calculations are correct, one would expect the length to be about three times longer for the saddle-shaped coils. To check the results obtained, experiments were performed at two frequencies: 20 MHz, where the condition $n \gg 1$ holds, and 129 MHz, where it does not. Table 2 shows the data collected, which are in good agreement with the theory. Finally, it is of interest to note that the results of Eq. [10] are independent of the number of turns on the coil. This is equivalent to saying that

TABLE 2

EXPERIMENTAL RESULTS[a] SHOWING THE SUPERIOR PERFORMANCE OF A SOLENOIDAL RECEIVING COIL AS COMPARED TO A SADDLE-SHAPED COIL WHEN USING THE SAME SAMPLE: A SPHERE 7.5 MM IN DIAMETER

Frequency	Coil type	Number of turns	Q	Height (mm)	Radius (mm)	90° Pulse[b] (μsec)	Signal[c] (volts)
129 MHz, ^{31}P	Saddle-shaped	2	210	7	4	27	1.0
	Solenoid	4	300	6	4	9	2.6
			Ratio 1.4			Ratio 3	Ratio 2.6
20 MHz, ^{1}H	Saddle-shaped	6	80	10	5	46	0.78
	Solenoid	18	208	10	5	18	2.20
			Ratio 2.5			Ratio 2.6	Ratio 2.8

[a] All values obtained are accurate to better than 10%.

[b] No comparison is intended between the performances at the two frequencies as the two transmitters are of different powers.

[c] The signals are measured relative to a constant noise background.

the Q of a coil is mainly dependent on the overall dimensions of that coil rather than the number of turns within those dimensions. That this is broadly true may be seen from Fig. 2.

DEPENDENCIES

From the various equations derived, it is possible to determine how the signal-to-noise ratio varies as parameters are altered. With regard to coil geometry, the latter is usually determined by the homogeneity of the main magnetic field B_0. Thus, for example, a superconducting magnet with a field of 11 tesla might have a homogeneity of one part in 10^9 over 1 ml, whereas with a 4 tesla magnet the equivalent volume might be 25 ml. For simplicity, we shall assume that coil length and diameter are approximately equal, though it should be stressed that this ratio is not necessarily optimal for the best utilization of the main field homogeneity nor, in the case of a saddle-shaped coil, does it produce the best B_1 homogeneity (7). Thus, from Eq. [10] we can see that if *all* linear dimensions are scaled in the same manner, the S:N varies as a^2 or as $V^{2/3}$. Assuming a frequency dependence of $\omega_0^{7/4}$, it is thus easy to show that the low-field magnet gives 25% *better* signal to noise than its high-field counterpart. It has of course been assumed that 25 ml of sample is available. In most biochemical applications of NMR, such extravagence would be unthinkable and so, for higher sensitivity, a higher field is still called for.

The major variable in Eq. [8] not yet considered is temperature. While the sample temperature T_s may well be fixed by the chemistry of the system under observation, there is no such limitation on the temperature T_c of the probe, and if the latter is cooled, significant improvements may be obtained. From Eq. [10], *for a constant sample volume*, the signal-to-noise ratio varies only as a^{-1}, and so the insertion of a dewar vessel between the coil and the sample may well only decrease ψ by 30%. On the other hand, at liquid nitrogen temperature (77 K), T_c is a quarter of room temperature and thus, if the preamplifier has an excellent noise figure, the noise is reduced by a factor of 2 on cooling. However it must not be overlooked that the conductivity of the conductor is temperature dependent, and for copper, $\rho(T_c)$ is one-tenth of its room-temperature value at 77 K. The net gain in signal-to-noise ratio obtainable by cooling the probe is therefore about 2.5 (Note, however, that the 90° pulse length should decrease by only about 25%.) Of course, if the sample may also be cooled, further improvement is possible.

LIMITATIONS OF THE FORMULA

So far, little indication has been given of the limitations of Eq. [8]. These are predominantly twofold. First, the calculation breaks down below about 5 MHz, where, for an average size of sample, the radius of the wire used becomes comparable with the skin depth. Second, the calculation breaks down when the distributed capacitance between the turns on the coil is sufficiently large at the frequency of interest to change the phase of the emf induced in one part of the coil relative to another. In its extreme manifestation, this effect causes self-resonance, a condition whereby the coil resonates without an external tuning capacitor. This should be avoided, and the onset of self-resonance may be seen in each of the curves of Fig. 2, where, with increasing frequency, the Q of each coil begins to drop away from the $f^{1/2}$ line. To avoid this effect, the number of turns on the coil must be decreased with increasing frequency. There is a limit of course to this procedure; for the saddle-shaped coil, the limit is a single turn. In this case, the formula still holds, provided attention is paid to the fact that the leads from the coil and the links between the two sections contribute appreciable resistance. A further modification may be to connect the two halves in parallel rather than series. This situation may also be accounted for and allows satisfactory use up to about 300 MHz. Also, it is usual for the conductor to be foil rather than wire with such a configuration.

With a solenoid, the calculation breaks down when the condition $n \gg 1$ is not obeyed and this may be seen in Fig. 2 to occur for frequencies in excess of 150 MHz. However, a single-turn solenoid fabricated from foil is a special case for which provision can be made, and whereas a single-turn saddle-shaped coil is useful up to about 200 MHz, a single-turn solenoid is useful to 600 MHz. The reason for this behavior lies in the fact, mentioned earlier, that a saddle-shaped coil requires 2.6 times the length of conductor per unit turn of a solenoid, and thus has greater inductance and self-capacitance per unit turn. This is ironic when it is remembered that it is only at the high frequencies made available by superconducting systems that the saddle-shaped configuration is required. It is the authors' opinion that the disappointing signal-to-noise ratio experienced with superconducting systems is a direct consequence of the use of saddle-shaped coils, and the construction of a spectrometer to work at a frequency of 470 MHz being undertaken in this laboratory presents a considerable challenge, as the only satisfactory coil configuration so far found is solenoidal.

A further limitation is the amount of space available for the construction of the coil. If there is appreciable coupling, magnetic or electrostatic, between the coil and, say, a shield or a strongly conducting sample (for example, a saline solution), then in general, the S:N ratio will be degraded. The mechanism by which the degradation takes place is dependent upon geometry and frequency; the field generated by unit current may be lessened by induction, proximity effect may change the coil resistance, and there may be resistive losses in the coupled element. The simplest way of monitoring these effects is to measure the Q of the coil in free air and then to observe the change when the coil is in place in the probe. It should preferably be less than 10%, and a good rule of thumb for obtaining this value is that no conductor should be closer to the coil than the largest dimension of the latter. The effects of coupling are particularly noticeable at low temperatures. It is quite easy to obtain Q's of over 600 at 77 K in free liquid nitrogen, but another matter in the confines of the probe. Of particular importance is coupling to conductors at room temperature. The component of resistance introduced into Eq. [6] by this coupling carries with it a temperature four times greater than that of liquid nitrogen and its noise contribution is thus disproportionately large. It might be added that the construction of a low-temperature probe with a room-temperature sample is no easy matter, and that to date, we do not have a reliable system.

THE PREAMPLIFIER AND TRANSMITTER

The only factor not so far considered is the noise figure of the preamplifier. For the frequency range 50 to 500 MHz the best semiconductor now available is probably a gallium arsenide field effect transistor. The authors have described elsewhere the design and construction of a preamplifier with a noise figure of 0.3 db at 129 MHz (*3*), and it

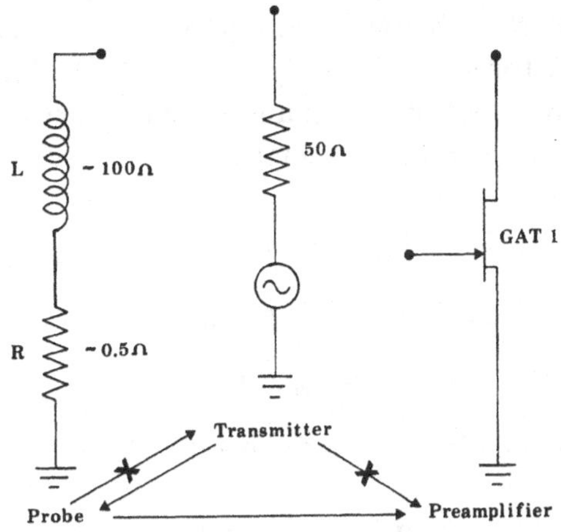

FIG. 4. The three elements probe, transmitter, and transistor must be interfaced in such a way that the signal-to-noise ratio is not degraded and the transistor is not damaged.

remains therefore to consider in what manner the probe coil and the preamplifier can be interfaced in order to obtain the best noise performance. If a single-coil probe is used, there is also the problem of interfacing to the transmitter, while protecting the preamplifier from the pulses and conserving the noise performance. As this is the most difficult situation likely to be encountered, it is to this that we turn our attention. The problem is illustrated in Fig. 4. Considering first the interface between probe coil and transmitter, it is obvious that at the frequency of interest, the impedance of the coil $Z \sim 0.5 + j\ 100$ ohms $[j = (-1)^{1/2}]$ must be transformed in such a manner as to power match the transmitter. As is well known, an essentially lossless transformation may be effected with the aid of the circuit of Fig. 5, provided high-Q capacitors are used. However, a word of warning is required here. It is desirable to keep the leads from the coil

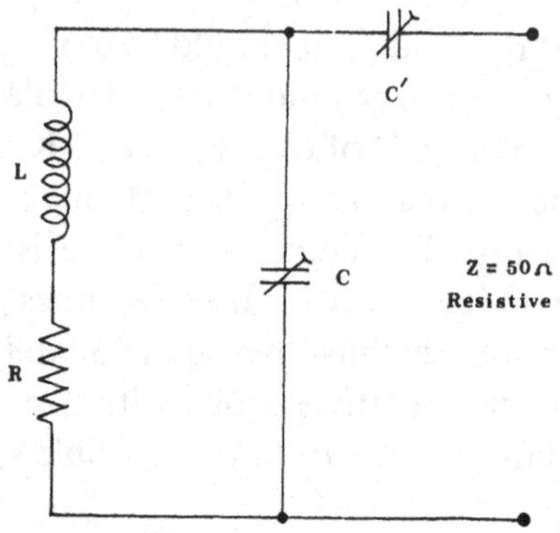

FIG. 5. Impedance transformation using a tuned circuit. The value of C is slightly less than that required to tune to resonance and the matching capacitance C' is given by $C' \simeq (C/50Q\ \omega_0)^{1/2}$.

to the capacitors as short as possible to minimize the resistance R. In this situation, the capacitor C is physically close to the coil and unfortunately, many high-Q variable capacitors are ferromagnetic. The main field homogeneity is thus disturbed.

Turning now to the interface between the coil and the F.E.T. one must ask under what conditions the transistor gives its best performance. Robinson (*8*) has considered this problem and has shown that the optimum noise figure is obtained when the input capacitance is almost tuned out and the signal source has a source impedance which is resistive and given by

$$R_{SO} \simeq 1.6 f_T / f g_m, \qquad [12]$$

f_T is the figure of merit for the F.E.T. and is given by

$$f_T = g_m / 2\pi C_j, \qquad [13]$$

where g_m is the transconductance of the device and C_j is the junction capacitance. Alternatively,

$$R_{SO} \simeq 1.6 / 2\pi f C_j. \qquad [14]$$

With a junction capacitance of 2 pF and a frequency of $f = 129$ MHz, $R_{SO} \simeq 980\ \Omega$. A practical value obtained with the amplifier of Ref. (*3*) gave $R_{SO} = 800\ \Omega$. On the other hand, the input impedance of the F.E.T. with its junction capacitance tuned out is resistive, and given by

$$R_{in} \simeq g_m / (2\pi f C_j)^2. \qquad [15]$$

At 129 MHz, and taking the values $C_j = 2$ pF, $g_m = 15$ mA/V, we have $R_{in} \simeq 5.7$ kΩ. Obviously, the source and the F.E.T. are grossly mismatched powerwise when they are noise matched, and it follows that one should never tune a probe by looking for the maximum signal from the receiver. The signal may well be a maximum; the signal-to-noise ratio most certainly will not be.

The probe has been matched to 50 Ω resistance in order to power match to the transmitter. It follows, to make a noise match to the F.E.T., that 50 Ω must be transformed to R_{SO} within the preamplifier. This, and the tuning out of the gate capacitance C_j, is easily accomplished using the transformation properties of tuned circuits and the reader is referred to Ref. (*3*) for further details. In general, the greater the ratio f_T/f, the better the noise performance. However, in the pursuit of excellence one must beware, particularly at low frequencies, of making the optimum source impedance higher than is practical. Above a value of about 2 kΩ, losses in the transforming device must also be taken into account. Hence to obtain very low noise figures, it may be necessary to cool the preamplifier.

Finally, the preamplifier must be protected from the potentially destructive transmitter pulses, and if a class A transmitter is used, there must be a gate of some sort which prevents the injection of noise from transmitter to receiver. Nor must the protection or the gate degrade the noise performance of the system. As one progresses to higher frequencies, crossed diodes (*9*) become increasingly ineffective due to their junction capacitance—typically 4 pF. Not only does such a large value allow noise to pass from the transmitter; it also ensures that any diodes used to protect the F.E.T. form, above say 50 MHz, a major part of the tuning capacitance. This may, depending on the diode, be disastrous, for after passing heavy current, many diodes exhibit a change of capacitance which lasts many milliseconds. This change can ruin the noise performance of a tuned amplifier and even cause it to oscillate. A far more elegant way to protect the preamplifier is to use PIN diodes (*10*) but unfortunately, PIN diodes have a low resistance to radio frequencies when they are passing a heavy direct current (say, 2 Ω at 40 mA). A PIN diode in the line from the probe to the preamplifier therefore introduces shot noise and can degrade the noise figure of the receiver by as much as 4 db. Fortunately, it is possible to construct a PIN diode circuit which not only protects the preamplifier (60 db isolation) from the transmitter, but which also has all the diodes "off" when the spectrometer is receiving signal. Details may be found in Ref. (*11*).

CONCLUSION

The authors have attempted to provide a direct physical picture of the factors governing the signal-to-noise ratio in an NMR experiment. It has been shown that the signal received from a sample by a set of coils is directly proportional to the magnetic field that would be created at the sample if unit current were passed through the coils, while the noise present in the coils has been shown to be purely a function of the coil resistance. This simple argument allows a direct comparison of the efficiency of different coil configurations to be made. For example, while unit currents flowing through saddle-shaped and solenoidal coils create similar B_1 fields, and the coils thus receive similar signals from the sample, the resistance of a saddle-shaped coil is considerably larger than that of a solenoid and so the signal-to-noise ratio is much less. The correct manner of interfacing between the probe, the transmitter and the preamplifier has been discussed, and attention has been drawn to the importance of noise matching the probe to the amplifying device used, and the distinction between noise and power matching. Finally, the problems of protecting the receiver from the transmitter pulses and noise have been considered, and the use of PIN diodes advocated as a solution.

APPENDIX 1: THE EQUIVALENCE OF THE PRESENT AND THE TRADITIONAL FORMULATIONS

From Eqs. [4] and [6], the signal-to-noise ratio may be written as

$$\Psi_{\mathrm{rms}} = K\omega_0 (B_1)_{xy} M_0 V_s/(8kT_c R\Delta f)^{1/2}. \qquad [16]$$

To convert to the traditional formula of Eq. [1], we must find a relationship between the energy stored in the B_1 field, which is a measure of the coil inductance, and the value of $(B_1)_{xy}$ at the sample. If over the sample volume, the B_1 field is predominantly homogeneous and in the xy plane, we may say that $(B_1)_{xy} \simeq B_1$ for the sample. The energy stored in the sample volume is given by

$$E = \frac{1}{2\mu_0} \int_{\mathrm{sample}} B_1^2\, dV \simeq (B_1)_{xy}^2 (V_s/2\mu_0). \qquad [17]$$

The inductance of the coil is given by

$$L = (1/\mu_0) \int_{\mathrm{all\ space}} B_1^2\, dV. \qquad [18]$$

If, following Hill and Richards (*2*), we define the filling factor as

$$\eta = \int_{\mathrm{sample}} B_1^2\, dV \Big/ \int_{\mathrm{all\ space}} B_1^2\, dV, \qquad [19]$$

then from Eqs. [17] to [19],

$$(B_1)_{xy} \simeq (\mu_0 \eta L/V_s)^{1/2}$$

or [20]

$$K(B_1)_{xy} = K(\mu_0 \eta L/V_s)^{1/2},$$

where mean and root mean square inhomogeneity factors have been introduced, $K \simeq K \simeq 1$.

Substituting in Eq. [16] and adding the noise figure of the preamplifier we thus obtain

$$\Psi_{\mathrm{rms}} = KM_0 \left[\left(\frac{\omega_0 L}{R} \right) \frac{\eta \mu_0 V_s}{8kT_c F\Delta f} \right]^{1/2}. \qquad [21]$$

For the special case of a solenoid, it may be shown that

$$\int_{\mathrm{coil\ volume}\ V_c} B_1^2\, dV \simeq \tfrac{1}{2} \int_{\mathrm{all\ space}} B_1^2\, dV.$$

Thus if the field within the solenoid is homogeneous

$$\eta \simeq V_s/2V_c. \qquad [22]$$

Substitution in Eq. [21] gives Eq. [1],

$$\Psi_{\mathrm{rms}} = K\eta M_0 (\mu_0 Q\omega_0 V_c/4FkT_c \Delta f)^{1/2},$$

which is valid only for a solenoid.

APPENDIX 2: THE FIELD AT THE CENTER OF A SADDLE-SHAPED COIL

The vector magnetic potential **A** at point P due to an element of arc **ds** is given by

$$d\mathbf{A} = (\mu\mu_0 I/4\pi)\,(d\mathbf{s}/v), \tag{23}$$

where $v = |\mathbf{p} - \mathbf{a}|$ is the distance of P from **ds** (see Fig. 6). For the special case of P at

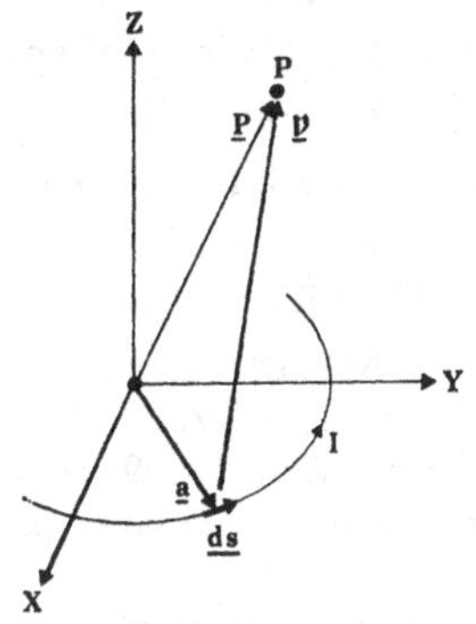

FIG. 6. The coordinate system.

the center of the coil of Fig. 3, we have that

$$v = (a^2 + g^2)^{1/2}$$

and further, the contributions to $(B_1)_{xy}$ from all four arcs add. Thus we have that

$$(\mathbf{B}_1)_{\text{arcs}} = \text{curl}\left\{\frac{\mu\mu_0 I}{\pi}\int_{-\pi/3}^{+\pi/3}\frac{(-a\sin\phi)\,\mathbf{i} + (a\cos\phi)\,\mathbf{j}}{(a^2+g^2)^{1/2}}\,d\phi\right\}$$

where **i** and **j** are unit vectors in the x and y directions.

$$\therefore\quad (\mathbf{B}_1)_{\text{arcs}} = \text{curl}\left\{\frac{\sqrt{3}\,\mu\mu_0 Ia}{\pi(a^2+g^2)^{1/2}}\,\mathbf{j}\right\};$$

$$\therefore\quad (\mathbf{B}_1)_{\text{arcs}} = -\frac{\sqrt{3}\,\mu\mu_0 I}{\pi}\,\frac{ag}{[a^2+g^2)^{3/2}}\,\mathbf{i}. \tag{24}$$

The potential due to one of the four verticals of the coil is given by

$$\mathbf{A} = \int_{-g}^{+g}\frac{\mu\mu_0 I}{4\pi(a^2+z^2)^{1/2}}\,d\mathbf{z}$$

and the field produced by them, which is parallel to that produced by the arcs, is given by

$$(\mathbf{B}_1)_{\text{verticals}} = \frac{\sqrt{3}\,\mu\mu_0 I}{2\pi}\,\frac{\partial}{\partial a}\left\{\int_{-g}^{+g}\frac{dz}{(a^2+z^2)^{1/2}}\right\}\mathbf{i}$$

$$= \frac{\sqrt{3}\,\mu\mu_0 I}{\pi}\,\frac{\partial}{\partial a}\left\{\sinh^{-1}\left(\frac{g}{a}\right)\right\}\mathbf{i}; \tag{25}$$

$$\therefore\quad (\mathbf{B}_1)_{\text{verticals}} = -\frac{\sqrt{3}\,\mu\mu_0 I}{\pi}\,\frac{g}{a(a^2+g^2)^{1/2}}\,\mathbf{i}.$$

Hence, from Eqs. [24] and [25], the field at the center of a saddle-shaped coil which is passing unit current is given by

$$(B_1)_{xy} = \frac{\sqrt{3}\,\mu\mu_0}{\pi}\left\{\frac{ag}{(a^2+g^2)^{3/2}} + \frac{g}{a(a^2+g^2)^{1/2}}\right\}. \qquad [26]$$

By using the type of analysis, briefly indicated above, for a point *P* which is off-center, the total magnetic field can be analyzed in a series of spherical harmonics, from which it may be shown that the optimum homogeneity is obtained when the angular width of the coil is 120°, as shown in Fig. 3, and the length is twice the diameter.

REFERENCES

1. A. Abragam, "The Principles of Nuclear Magnetism," pp. 82–83, Clarendon Press, Oxford, 1961.
2. H. D. W. Hill and R. E. Richards, *J. Phys. E, Ser. 2* **1,** 977 (1968).
3. D. I. Hoult and R. E. Richards, *Electron. Lett.* **11,** 596 (1975).
4. F. E. Terman, "Radio Engineer's Handbook," 1st ed., pp. 77–85, McGraw–Hill, New York, 1943.
5. A summary of Butterworth's extensive work on this subject is given by B. B. Austin, *Wireless Eng. Exp. Wireless* **11,** 12 (1934).
6. M. Soutif and R. Gabillard, "La Résonance paramagnétique nucléaire" (P. Grivet, Ed.), 1st ed., pp. 149–161, Centre National de la Recherche Scientifique, Paris, 1955.
7. D. I. Hoult, D. Phil. Thesis, Oxford, 1973.
8. F. N. H. Robinson, "Noise and Fluctuations in Electronic Devices and Circuits," Chaps. 11, 12, 13, Clarendon Press, Oxford, 1974.
9. I. J. Lowe and C. E. Tarr, *J. Phys. E, Ser. 2* **1,** 320 (1968).
10. K. E. Kisman and R. L. Armstrong, *Rev. Sci. Instrum.* **45,** 1159 (1974).
11. D. I. Hoult and R. E. Richards, *J. Magn. Resonance* **22,** 561 (1976).

Slotted tube resonator: A new NMR probe head at high observing frequencies

Hans Jürgen Schneider and Peter Dullenkopf

Institut für Elektronik, Ruhr-Universität Bochum, 4630 Bochum, Germany

(Received 24 June 1976; in final form, 12 August 1976)

Superconducting magnetic systems permit increasing observing frequencies for NMR experiments. The sensitivity and range of application of the single-layer detection system is evaluated. High measuring frequencies demand new detection systems because the conventional detection systems lose their good features above about 100 MHz. In this paper a new detection system is presented. The slotted tube resonator (STR) is suitable for all intensities of magnetic field which will be produced in the near future. The characteristic parameters of the STR are calculated and the construction is described in detail. Theoretical sensitivity considerations are compared with practical measurements.

I. INTRODUCTION

The capability of a nuclear magnetic resonance (NMR) spectrometer is mainly determined by its resolution and sensitivity. Both properties are improved by increasing the measuring frequency f_0 and the induction B_0 of the magnetic field. Superconducting coils are able to produce inductions up to $B_0 \approx 8.5$ T with proper field homogeneities. At frequencies $f_0 \approx 360$ MHz, traditional resonance detection circuits with concentrated capacitors and measuring coils can hardly be realized. Resonance circuits designed by hollow waveguide elements are excluded because of their huge dimensions. Possible solutions are hybrids of concentrated elements and coaxial line elements like the slotted tube resonator (STR) described in this publication.

II. SENSITIVITY CONSIDERATIONS

Under the common conditions of NMR experiments using the modulation or time-sharing method, the sensitivity[1] is given by

$$\left(\frac{S}{N}\right)_{\text{NMR}} = 2.5\,\frac{U_S}{U_{N\text{pp}}} = \frac{1}{2\sqrt{2}}\,\frac{U_S}{U_{N\text{rms}}} = \frac{1}{2\sqrt{2}}\left(\frac{S}{N}\right), \qquad (1)$$

with U_S = peak value of signal voltage, $U_{N\text{pp}}$ = p-t-p value of noise voltage, and $U_{N\text{rms}}$ = rms value of noise voltage.

When a well-matched resonant circuit is employed appropriately as a detection system, the sensitivity S/N becomes[2]

$$\left(\frac{S}{N}\right) = \frac{1}{8}\left(\frac{\omega_0^3 \eta Q V_S}{2\mu_0 kTFb_R}\,\frac{T_2}{T_1}\right)^{1/2}\frac{\chi_0}{\gamma}, \qquad (2)$$

with η = filling factor (magnetic energy stored in the sample)/(magnetic energy stored in the detection system), Q = quality factor, V_S = volume of the sample, T_1, T_2 = relaxation time constants, μ_0 = permeability, k = Boltzmann constant, T = absolute temperature, F = noise figure for the preamplifier, b_R = receiver bandwidth, χ_0 = magnetic susceptibility, and γ = gyromagnetic ratio.

A 5-mm sample tube filled with a 1% solution of ethylbenzene ($C_6H_5C_2H_5$) in $DDCl_3$ constitutes the reference for sensitivity measurement. The magnitude of signal voltage U_S is given by the amplitude of the two inner lines of the CH_2 quartet. If the sample tube is filled to the height of l_S = 10 mm (sample diameter D_S = 4 mm), the sensitivity at room temperature will be described by

$$\left(\frac{S}{N}\right)_{\text{NMR}} = \frac{59\times 10^{-3}}{2\sqrt{2}}\left(\frac{T_2/T_1}{Fb_R/\text{Hz}}\right)^{1/2} \times \left[\eta Q\left(\frac{f}{\text{MHz}}\right)^3\right]^{1/2}. \qquad (3)$$

Sensitivity optimization demands well-defined characteristics from the receiver: noise figure and bandwidth must be kept as small as possible (see first square root). The second part of Eq. (3) provides information for the design of the actual detection system: the observing frequency affects the sensitivity with the power 1.5. The filling factor and the quality factor also enhance the sensitivity. Both filling and quality factor depend on the particular configuration of the detection system, and therefore the product ηQ is decisive for sensitivity considerations.

III. CONVENTIONAL DETECTION SYSTEM

The detection system most often used is the single-layer solenoid (Fig. 1) which, together with a parallel capacitor C, forms a resonant circuit. Used in resonance experiments, the usual dimensions of the solenoids (coil

Reprinted from Review of Scientific Instruments **48**, 68–73 (1977); ©American Institute of Physics.

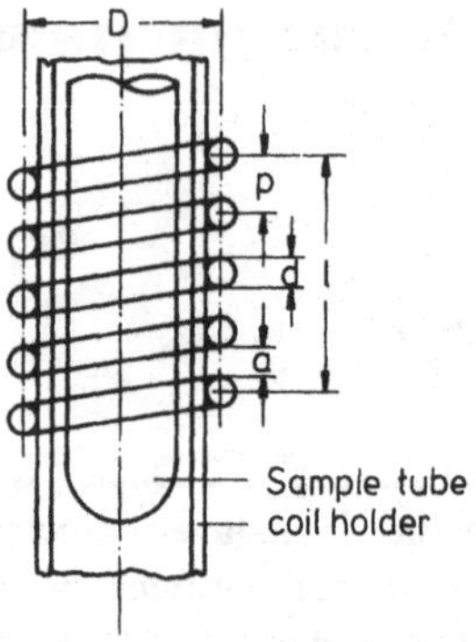

FIG. 1. Single-layer solenoid detection system, 5-mm sample tube: $D = 8$ mm, $l = 10$ mm.

diameter $D = 8$ mm, coil length $l = 10$ mm) dictate a number of turns N of[2]

$$N = \frac{2334}{(f/\mathrm{MHz})(C/\mathrm{pF})^{1/2}}\,. \tag{4}$$

The range of application is limited by practical conditions, demonstrated in Fig. 2.

Tunability of the detection system demands a minimum capacity of $C \approx 5$ pF (boundary ①); the necessary space required for high quality capacitors excludes the use of capacitors of above about $C \approx 100$ pF (boundary ②). Common rf coil wire diameters of $d = 0.2$–2 mm lead to turn numbers of $N = 3$–20, determining a coil length of $l = 10$ mm and $a \geqslant d$ (boundaries ③ and ④). This range of application is determined by the ability to resonate and the mechanical limitations. Obviously, the detection system solenoid falls just within the range of convenient application at a Larmor frequency $f_0 = 270$ MHz. Higher observing frequencies (e.g., $f_0 = 360$ MHz) would almost certainly cause design problems.

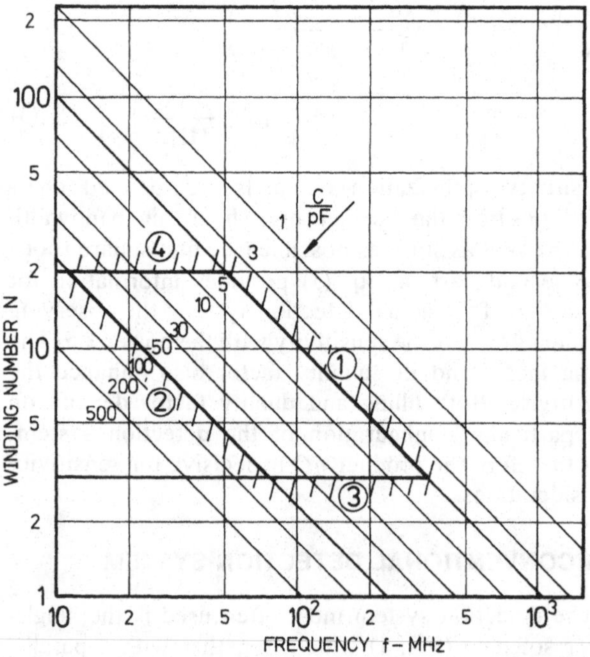

FIG. 2. Range of application of the single-layer solenoid detection system.

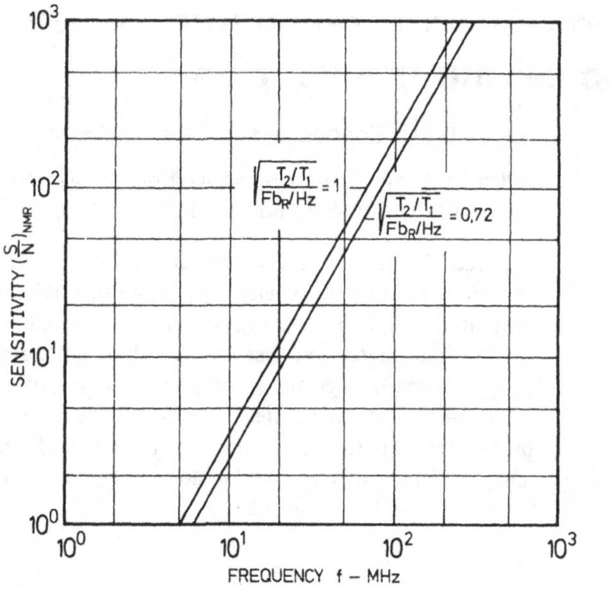

FIG. 3. Sensitivity of the single-layer solenoid detection system versus Larmor frequency.

When the required coil volume (standard sample) is stipulated, optimum coil sizes[3] are determined according to the inductivity,

$$d/p \approx 0.52 + 0.6/N, \quad l/D = 0.7. \tag{5}$$

For optimally dimensioned coils both the quality factor[3] and the filling factor[2] reach maximum values. The decisive product ηQ will be given by

$$\begin{aligned}\eta Q &= 0.77(D_s^2 l_s/D^3) \times 4.1(D/\mathrm{mm})(f/\mathrm{MHz})^{1/2} \\ &= 3.16(D_s/D)^2(l_s/\mathrm{mm})(f/\mathrm{MHz})^{1/2}.\end{aligned} \tag{6}$$

With this result the maximum attainable sensitivity of the solenoid detection system for standard sample dimensions can be calculated:

$$\left(\frac{S}{N}\right)_{\mathrm{NMR}} = 5.86 \times 10^{-6}\left(\frac{T_2/T_1}{Fb_R/\mathrm{Hz}}\right)^{1/2}\left(\frac{f}{\mathrm{MHz}}\right)^{7/4}. \tag{7}$$

The relationship between the sensitivity and the observing frequency is demonstrated in Fig. 3.

In practice the values $T_1 = T_2$, $F = 1.8$ dB, and $\tau = 2/\pi b_R = 500$ msec determine the square root, $[(T_2/T_1)/Fb_R/\mathrm{Hz})]^{1/2} = 0.72$. Up to observing frequencies of about 100 MHz the above computation of sensitivity is in accordance with the practically obtainable values. At increasing frequencies, however, the realized sensitivity of commercial spectrometers falls behind the calculated values derived from Eq. (7). There are essentially two reasons for this. Above 100 MHz, proton resonance experiments must be performed in superconducting magnetic solenoids. The orthogonality of static and rf magnetic field demands a modified detection system which can only roughly be described by a solenoid, particularly because the number of turns goes down to $N = 2$. Moreover the quality factor responds insufficiently to the supposed equation because high

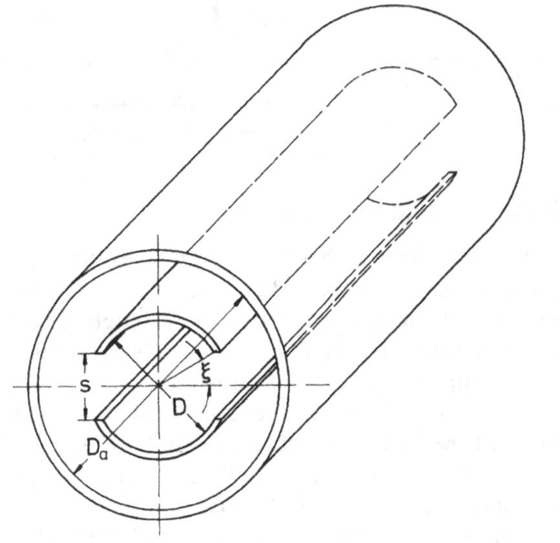

FIG. 4. Shielded symmetrical slotted tube line STL.

frequencies give rise to radiation losses which are not taken into account.

In conclusion, it can be stated that above about 100 MHz the NMR detection system solenoid loses its good features due to mechanical and electronic factors. Therefore a new, suitable detection system must be sought.

IV. THE SLOTTED TUBE RESONATOR (STR)

A. Description of the system

For purposes of sample changing and sample spinning a detection system is preferred where the symmetry axis coincides with the axis of the solenoid magnet. At the same time the irradiating field and the magnetic steady field have to be perpendicular to each other. In first approximation a Lecher system fulfills these conditions. The homogeneity of the rf field in the sample region increases if a strip line is used. In view of the cylindrical symmetry of the solenoid magnet, best adaptation will be obtained by a transmission line formed like a tube.

When a tube is cut lengthwise, one has two symmetrical tube sections which form a strip line consisting of arched conductors. This transmission line, hereafter called the slotted tube line (STL), can constitute a resonant line, if it has the appropriate dimensions. A resonant circuit like this is suitable for NMR detection and will be called the slotted tube resonator (STR system). The simple unsymmetrical STR is of an open type and is therefore susceptible to interference. Furthermore, the magnetic rf field will be deformed when the unsymmetrical STR is inserted into the inner channel of the cryostat and the sample volume will not sit in a homogenous rf field. With reference to these considerations the STR has been constructed as a three-wire system which renders a shielded resonant line possible, as shown in Fig. 4.

A coaxial tube shields the intrinsic STL, which will be fed symmetrically. This configuration eliminates external influences to a large extent. A short circuit near a quarter wavelength converts this transmission line into a resonant line. The coaxially inserted sample tube meets the appropriate conditions for NMR experiments. All components, the solenoid magnet, the detection system, and the sample tube, are fitted into one another coaxially, providing the orthogonality between the magnetic rf and steady fields.

B. Optimizing geometrical dimensions

The rf field configuration in the detection system described is a function of the geometrical dimensions of the slotted tube line. It will be determined by the shielding diameter D_a, the slotted tube diameter D, and the slot width s. This dependence has been investigated by the electrostatic method.[4] The shielding diameter D_a was selected so wide ($D_a \approx 2D$) that it does not greatly influence the intrinsic field configuration of the STL. Increasing the slot width stepwise it was found that an unambiguous optimum of field homogeneity exists.

The most homogeneous field was obtained with a width–diameter ratio of $s/D = 0.77$, corresponding to an opening angle of

$$\xi = \arcsin(s/D) = 50.35°. \tag{8}$$

The corresponding field configuration is shown in Fig. 5.

For $D_s < D$ the sample will be irradiated by a homogeneous magnetic rf field perpendicular to the steady field. In the direction of the sample axis the rf-field follows the current distribution along the STL. The relative field deviation is

$$\Delta B_s/B_{s\,\max} \leq 1 - \cos[2\pi(l_s/\lambda)]. \tag{9}$$

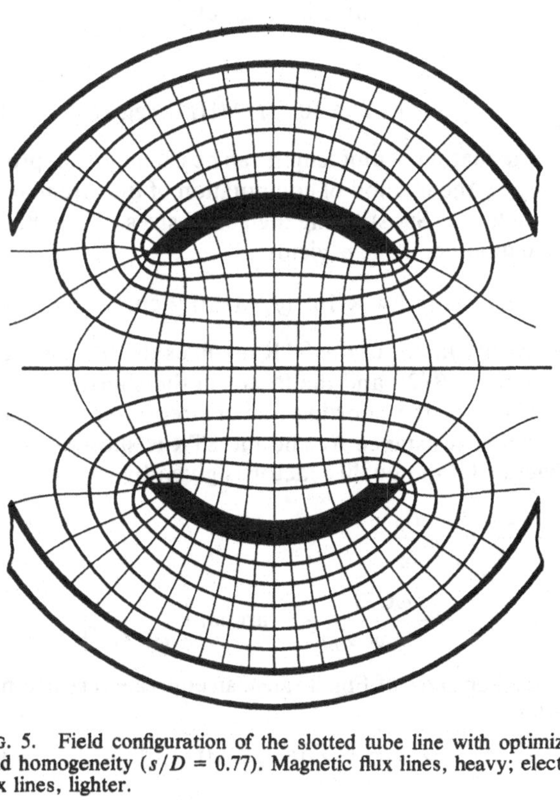

FIG. 5. Field configuration of the slotted tube line with optimized field homogeneity ($s/D = 0.77$). Magnetic flux lines, heavy; electric flux lines, lighter.

With a sample length of $l_s = 10$ mm at a measuring frequency of 360 MHz, the maximum deviation will be smaller than 3×10^{-3}.

C. Characteristic parameters of the STR

The field configuration (Fig. 5) is suitable for determination of the capacitance C' and inductance L' per unit length of the STL. By counting the lines of force the characteristic parameters can be calculated:

$$C' = 0.28 \text{ pF cm}^{-1},$$
$$L' = 3.96 \text{ nH cm}^{-1}, \qquad (10)$$
$$Z_L = 118.9\ \Omega.$$

If the warm channel of the cryostat has a sufficiently wide diameter for maintaining the condition $D_a \gg D$, the characteristic parameters are obtainable by analytical computation using the conformal representation[2,5]

$$C' = \epsilon[K'(k)/K(k)],$$
$$L' = \mu[K(k)/K'(k)], \qquad (11)$$
$$Z_L = (\mu/\epsilon)^{1/2}[K(k)/K'(k)],$$

with

$$K(k) = \text{complete elliptic integral}$$
$$= \int_0^{\pi/2} \frac{d\psi}{(1 - k^2 \sin^2\psi)^{1/2}},$$

$$K'(k) = K(k') \quad \text{with} \quad k' = (1 - k^2)^{1/2},$$
$$k = \tan^2(\xi/2).$$

The phase constant β is given by

$$\beta = \omega(L'C')^{1/2}$$
$$= [1.2 \times 10^{-2}(f/\text{MHz})] \text{ deg cm}^{-1}. \qquad (12)$$

The attenuation of unit length can be calculated approximately by the method of computation of the attenuation of strip lines.[6] Substituting the actual sizes of Fig. 8, the attenuation constant α yields

$$\alpha = 1.4 \times 10^{-6}(f/\text{MHz})^{1/2} \text{ cm}^{-1}. \qquad (13)$$

The quality factor of the STR involves the basic quality factor $Q_0 = \beta/2\alpha$ and the losses in the shorting bar. If the losses in the shorting bar are represented by the equivalent resistance R_B and the STR is shorted by the capacitor C, the quality factor is given by

$$\frac{1}{Q} = \frac{2\alpha}{\beta} + 2\frac{R_B}{Z_L}$$
$$\times \left[\frac{\omega C Z_L}{1 + (\omega C Z_L)^2} + \arctan\left(\frac{1}{\omega C Z_L}\right)\right]^{-1}. \qquad (14)$$

The chosen sizes of Fig. 8 cause an equivalent resistance of about

$$R_B \approx 3.5 \times 10^{-3}(f/\text{MHz})^{1/2}. \qquad (15)$$

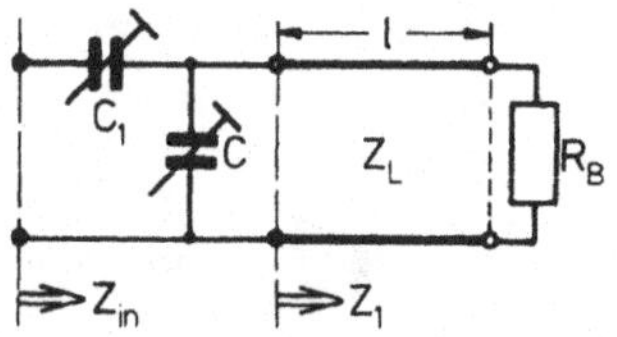

FIG. 6. Matching and tuning of the STR.

D. Matching and tuning network

Shorting the STR with a capacitor causes the resonant impedance to decrease. Matching the characteristic impedance of the feeder needs a STR length of only a few millimeters. By application of a series resonant circuit the appropriate series capacitor C_1 reduces to a fractional part of 1 pF. The two matching methods mentioned are unfit for practical application. A combination of both of them is a suitable arrangement for the tuning and matching network (Fig. 6).

A shunting capacitor C shortens the slotted tube line to a practicable length l and a series capacitor C_1 matches the resonant impedance to the characteristic impedance $R_i = 50\ \Omega$ of the feeder. The components C and C_1 may be determined by mathematical analysis:

$$\omega C Z_L = \frac{x}{x^2 + r^2} - \left[\frac{x^2}{(x^2 + r^2)^2} - \frac{1 - r/r_i)}{x^2 + r^2}\right]^{1/2}$$
$$\omega C_1 Z_L = \frac{(r/r_i)}{x} \frac{x^2 + r^2 + r_i r[1 - 2(r/r_i)^{1/2}]}{x^2 + r^2 - (r_i r)^{1/2}} \qquad (16)$$

with

$$x = \tan(\beta l)\frac{1 - (R_B/Z_L)^2}{1 + (R_B/Z_L)^2 \tan^2(\beta l)},$$
$$r = \frac{R_B}{Z_L} \frac{1 + \tan^2(\beta l)}{1 + (R_B/Z_L)^2 \tan^2(\beta l)},$$
$$r_i = R_i/Z_L.$$

When $R_B \ll Z_L$, R_i and $l \leq \lambda/8$, the equations are approximately

$$\omega C Z_L \approx 1/x \approx \cot(\beta l),$$
$$\omega C_1 Z_L \approx \frac{(r/r_i)^{1/2}}{x} \approx \frac{(R_B/Z_L)^{1/2}}{\sin(\beta l)}. \qquad (17)$$

It is remarkable that the capacitors C and C_1 fulfill the two functions almost independently from each other. The shunting capacitor C represents the tuning element and the series capacitor C_1 matches the resonant circuit without notable variation of the resonant frequency. The ratio of the two adjusting capacitors is

$$\frac{C_1}{C} \approx \left(\frac{r}{r_i}\right)^{1/2} \approx \frac{(R_B/R_i)^{1/2}}{\cos(\beta l)}. \qquad (18)$$

E. Construction of the STR

High-resolution experiments require high stability of the magnet system, necessitating a stabilizing channel. Therefore the detection system must be able to resonate at the two frequencies, the observing frequency f_0 and

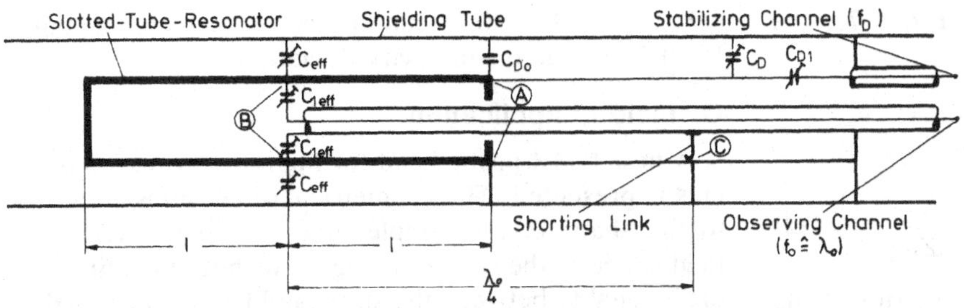

FIG. 7. Detection system: slotted tube resonator (double resonator for stabilizing application).

the stabilizing frequency f_D (generally D_2O resonance: $f_D < f_0$) simultaneously. Operation in the stabilizing channel should not affect the main resonance. Consequently the coupling point of the stabilizing channel at the STR must be of extremely low impedance for the observing frequency. The most appropriate location is the short circuit of the STR, but this place is occupied by the sample and the rf field adjacent to this should not be disturbed. An additional identical STR fixed homologously to the first STR creates an additional short circuit (Fig. 7). This configuration represents a capacitively shortened $\lambda_0/2$ resonator which has the same quality specifications as the $\lambda_0/4$ resonator. Computation of the corresponding effective capacitors yields

$$C_{eff} = 4C = 32 \text{ pF}, \qquad C_{1eff} = 2\sqrt{2}C_1 = 0.9 \text{ pF}. \tag{19}$$

The observing channel will be coupled to the detection system in the center of the STR (point B). By rotating the semirigid cable, the capacitors C_{1eff} may be adjusted to matching the resonant circuit. C_1 is split into two capacitors, C_{1eff}, which feed the STR symmetrically. If a shorting link (point C) is inserted at a distance of a quarter wavelength, the sheath current of the unbalanced feeder will be suppressed. The tuning capacitors C_{eff} are formed by adjustable arched electrodes which are placed near to the outside of the STL. This assembly occupies a small space and represents a well-defined feeding point, necessary for calculation of the resonant line length.

A very small slot at the new short circuit (point A) will not disturb the short circuit properties as far as the high observing frequency is concerned. However, the remaining bowl-shaped electrodes can be used as go and return lines for the low stabilizing frequency. With reference to the stabilizing frequency f_D the STR forms only one loop, the inductance of which constitutes a resonant circuit together with the capacitor C_{D0} and the slot capacitor C_S. This parallel resonant circuit is tuned by the tuning capacitor C_D and matched to the stabilizing channel by the capacitor C_{D1}. An overall capacity of $C_p = C_S + C_{D0} + C_D \approx 260$ pF is suitable for tuning the stabilizing frequency $f_D = 41.5$ MHz. When the stabilizing resonant circuit is determined by the quality factor Q, the parallel resistance R_p, and the input impedance R_i, then the capacitor ratio can be calculated:

$$\frac{C_{D1}}{C_p} = \frac{1}{2}\frac{R_p/R_i}{Q^2}\left[1 + \left(1 + \frac{4Q^2}{R_p/R_i - 1}\right)^{1/2}\right]. \tag{20}$$

This equation simplifies to

$$C_{D1}/C_p \approx (1/Q)(R_p/R_i)^{1/2},$$

where

$$C_{D1} \approx (C_p/Q\omega R_i)^{1/2} \tag{21}$$

for $Q \gg (\omega C_p R_i)^{-1}$. With a quality factor $Q \approx 100$ a series capacitor $C_{D1} \approx 14$ pF will be appropriate for matching. Figure 8 shows the sample head with optimized dimensions for 10-mm sample tubes.

F. Sensitivity of the STR

The filling factor η and the quality factor Q constitute the decisive product which is suitable for sensitivity

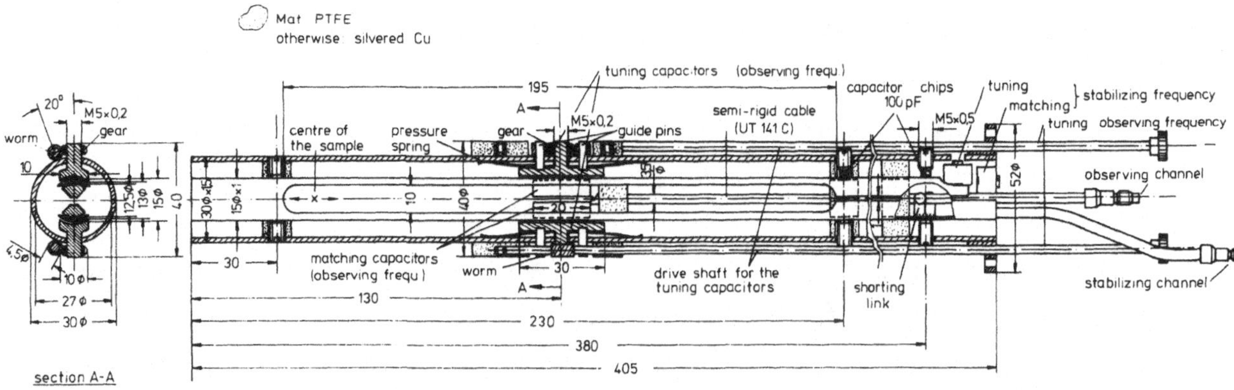

FIG. 8. NMR probe head with the detection system: slotted tube resonator (STR) optimized for 10-mm sample tubes.

evaluation. Calculating η and Q from Eq. (17), the product will be obtained as

$$\eta Q = \frac{w_i}{w}\frac{A_s}{A_i}\left[4\frac{l}{\lambda_0} + \frac{1}{\pi}\sin\left(4\pi\frac{l_s}{\lambda_0}\right)\right] \times \left\{\frac{2\alpha}{\beta}\left[4\frac{l}{\lambda_0} + \frac{1}{\pi}\sin\left(4\pi\frac{l}{\lambda_0}\right)\right] + \frac{4}{\pi}\frac{R_B}{Z_L}\right\}^{-1}, \quad (22)$$

with w_i = magnetic energy stored in the interior of the STR, w = total magnetic energy stored in the STR, A_s = cross-sectional area of the sample, and A_i = cross-sectional area of the interior of the STR. The sensitivity increases with decreasing length l of the resonator caused by capacitive shortening. With w_i/w taken as ~50% and considering the dimensions of the STR (Fig. 8), $\eta Q = 15$ for an observing frequency of $f_0 = 270$ MHz. If 10-mm sample tubes are employed, the sensitivity can be calculated:

$$\left(\frac{S}{N}\right)_{\text{NMR }10} = \left(\frac{S}{N}\right)_{\text{NMR}}\left(\frac{V_{s\,10}}{V_{s\,5}}\right)^{1/2} = 33{,}2 \times 10^{-3} \times \left(\frac{T_2/T_1}{Fb_R/\text{Hz}}\right)^{1/2}\left[\eta Q\left(\frac{b}{\text{MHz}}\right)^3\right]^{1/2} = 410. \quad (23)$$

Up to now field homogeneity factors prevented the use of 10-mm sample tubes. If a 5-mm sample tube is inserted into the 10-mm detection system, the theoretical sensitivity decreases to

$$\left(\frac{S}{N}\right)_{\text{NMR }5/10} = \left(\frac{V_{s\,5}\eta_5}{V_{s\,10}\eta_{10}}\right)^{1/2}\left(\frac{S}{N}\right)_{\text{NMR }10} = \left(\frac{D_{s\,5}}{D_{s\,10}}\right)^{2}\left(\frac{S}{N}\right)_{\text{NMR }10} = 81. \quad (24)$$

Practical measurements with a 5-mm sample tube reveal $(S/N)_{\text{NMR}5/10} = 58$. The divergence between calculated and measured sensitivity is caused by the usual difference between the theoretical and real quality factor. From quality measurements a correction factor of $(Q_{\text{real}}/Q_{\text{theor}})^{1/2} = 0.63$ can be derived, which accounts for the measured sensitivity deviation.

G. Field of application

A new NMR probe head for high observing frequencies is presented. The geometrical dimensions conform to the shape of the sample tube and to the available field space of the solenoid magnet without violating the orthogonality between the static and RF magnetic field. The high product ηQ guarantees high sensitivity of the detection system. The STR is appropriate for double resonance experiments using a crossed slotted tube line resonator (CSTR).[7]

All tuning and matching elements are externally adjustable and capacitive, thus favoring a high filling factor. They fit closely together, thereby concentrating the feeding point. In contrast to all conventional detection systems, the STR tuning range is within 200–500 MHz, permitting the observation of different nuclei without changing the probe head. In addition to the normal sensitivity enhancement, with increasing observing frequencies the sensitivity will further increase as ηQ increases.

With these properties the STR detection system presented is suitable for all intensities of magnetic field which will be produced in the near future.

ACKNOWLEDGMENTS

The authors are very much obliged to Prof. Dr. G. Laukien and to Spektrospin AG, Zürich for supporting this work.

[1] R. R. Ernst, *Advances in Magnetic Resonance, Vol. 2* (Academic, New York, 1966).
[2] H.-J. Schneider, thesis, Ruhr-Universität Bochum, 1974.
[3] W. Lorenz, Frequenz **24**, 20 (1970).
[4] F. Wolf, Studienarbeit S 3, Institut für Elektronik, Ruhr-Universität Bochum, 1971.
[5] H. Buchholz, *Elektrische und magnetische Pontentialfelder* (Springer, Berlin, 1957).
[6] M. V. Schneider, Bell Syst. Tech. J. **48**, 1421 (1969).
[7] H.-J. Schneider and P. Dullenkopf (unpublished).

RF Magnetic Field Penetration, Phase Shift and Power Dissipation in Biological Tissue: Implications for NMR Imaging

P. A. BOTTOMLEY and E. R. ANDREW

Department of Physics, University of Nottingham, University Park, Nottingham NG7 2RD, U.K.

Received 22 *December* 1977, *in final form* 23 *February* 1978.

ABSTRACT. The magnetic field penetration, phase shift and power deposition in planar and cylindrical models of biological tissue exposed to a sinusoidal time-dependent magnetic field have been investigated theoretically over the frequency range 1 to 100 MHz. The results are based on measurements of the relative permittivity and resistivity dispersions of a variety of freshly excised rat tissue at 37 and 25 °C, and are analysed in terms of their implications for human body nuclear magnetic resonance (NMR) imaging.

The results indicate that at NMR operating frequencies much greater than about 30 MHz, magnetic field amplitude and phase variations experienced by the nuclei may cause serious distortions in an image of a human torso. The maximum power deposition envisaged during an NMR imaging experiment on a human torso is likely to be comparable to existing long-term safe exposure levels, and will depend ultimately on the imaging technique and NMR frequency employed.

1. Introduction

A recent advance in the nuclear magnetic resonance (NMR) technique has been to produce cross-sectional images which represent the spatial distribution of mobile protons in heterogeneous systems (Lauterbur 1974, Kumar, Welti and Ernst 1975, Hinshaw 1976, Andrew, Bottomley, Hinshaw, Holland and Simaroj 1977, Mansfield and Maudsley 1977b). This new development, combined with the observation that excised biological tissues exhibit different NMR properties depending on their health and origin (Hazelwood, Cleveland and Medina 1974, Hollis, Saryan, Eggleston and Morris 1975, Kiricuta and Simplaceanu 1975), has led to the suggestion that the method may prove a useful tool in medical diagnosis (Grannell and Mansfield 1975, Damadian, Goldsmith and Minkoff 1977, Hinshaw, Bottomley and Holland, 1977, Holland, Bottomley and Hinshaw 1977, Mansfield and Maudsley 1977a) and several laboratories are at present engaged in the development of NMR imaging systems capable of scanning the whole human body (Andrew 1977).

Since objects of such size are of comparable dimension to the electromagnetic skin depth in biological tissue at radiofrequencies (RF) (Johnson and Guy 1972), skin effects arising from the resonant RF magnetic field cannot be neglected and any consequent amplitude or phase variation of this field towards the centre of the object will have a two-fold effect on the medical viability of NMR imaging. Firstly, it will tend to degrade the image quality by altering the nature of the NMR signal derived from this region, and, secondly, the power loss associated with the induced fields produced by the RF pulse will result in heating of the specimen. In medical applications this heating should be kept below recognised danger levels.

Although the effects of RF electromagnetic (EM) radiation on biological tissue are fairly well documented (Schwan 1965, Johnson and Guy 1972, Michaelson 1972, Lin, Guy and Johnson 1973), the NMR imaging situation differs in that the resonant applied alternating magnetic field has only a very small associated electric field so that the electrically induced component of the absorbed power is negligible in comparison with the magnetically induced component. In addition, variations in phase, as well as amplitude, are important where phase-sensitive detection is employed (Hinshaw 1976, Mansfield, Maudsley and Baines 1976).

The present paper represents a first attempt at quantifying the extent of RF magnetic field penetration, phase shift and power deposition in biological tissue samples during an NMR imaging experiment. Calculations showing the behaviour of the magnetic field within samples assume semi-infinite planar (Bottomley 1977) and infinitely long cylindrical models of homogeneous biological tissue. The latter model is particularly relevant to thin section NMR imaging (Hinshaw *et al.* 1977, Mansfield and Maudsley 1977b, Hinshaw, Andrew, Bottomley, Holland, Moore and Worthington 1978) where end effects can be ignored in all but the extremities of a specimen. This is because only the NMR signal derived from a well defined cross-section is of interest. The power deposition estimates also assume a cylindrical homogeneous tissue model. All of these calculations assume a uniform applied RF field, and refer to the distortions and power deposition of the RF field prior to NMR absorption. The calculations are based on measurements of the resistivity and relative permittivity of freshly excised rat lung, brain, liver, hepatoma D23, kidney, heart and abdominal wall muscle tissues at 37 and 25 °C over the frequency range 1 to 100 MHz using a technique developed in this laboratory (Bottomley 1978).

2. Theory

2.1. *Semi infinite planar model*

The magnetic induction $\boldsymbol{B}$, in a linear, homogeneous and isotropic medium must satisfy the wave equation

$$\nabla^2 \boldsymbol{B} - \varepsilon\mu \frac{\partial^2 \boldsymbol{B}}{\partial t^2} - \frac{\mu}{\rho}\frac{\partial \boldsymbol{B}}{\partial t} = 0 \qquad (1)$$

where ε is the permittivity, μ the permeability and ρ the resistivity of the medium. The solution of eqn (1) which describes the behaviour of a sinusoidal time-dependent magnetic induction field in a semi-infinite plane of conducting material is the same as that which describes the propagation of plane electromagnetic waves in conducting media (Lorrain and Corson 1970). Thus if the conducting material which occupies the region $x \geqslant 0$ (fig. 1(a)) is subjected to the magnetic induction field

$$\boldsymbol{B}(x \geqslant 0) = B_0 \exp(\mathrm{j}\omega t)\,\hat{\mathbf{z}} \tag{2}$$

then within the medium

$$\boldsymbol{B}(x \geqslant 0) = B_0 \exp[-K_\mathrm{I} x + \mathrm{j}(\omega t - K_\mathrm{R} x)]\,\hat{\mathbf{z}} \tag{3}$$

where

$$K_\mathrm{I} = \omega(\tfrac{1}{2}\varepsilon\mu\{[1 + (1/\rho^2 \varepsilon^2 \omega^2)]^{\frac{1}{2}} - 1\})^{\frac{1}{2}} \tag{4}$$

$$K_\mathrm{R} = \omega(\tfrac{1}{2}\varepsilon\mu\{[1 + (1/\rho^2 \varepsilon^2 \omega^2)]^{\frac{1}{2}} + 1\})^{\frac{1}{2}} \tag{5}$$

and ω is the angular frequency, and where it is understood that we are considering only the real parts of eqns (2) and (3).

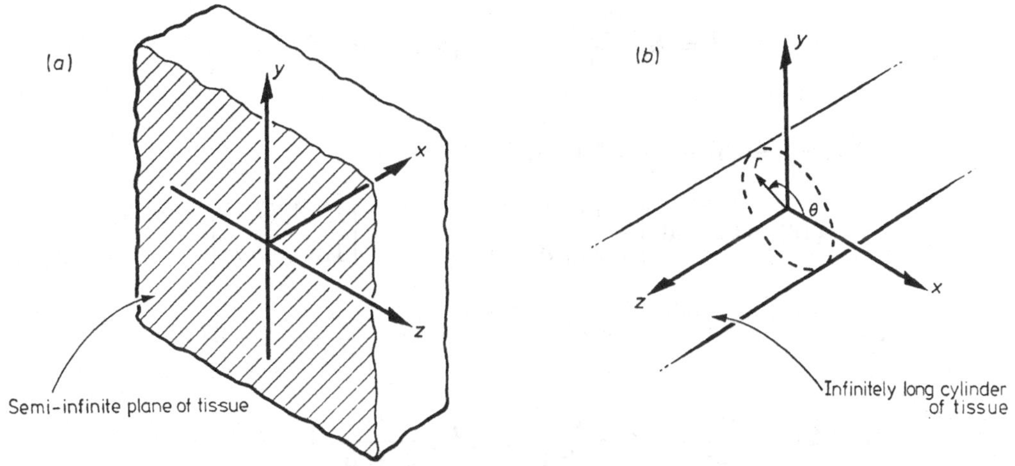

Fig. 1. Cartesian (*a*) and cylindrical (*b*) coordinate systems used for the models. The corresponding unit vectors are indicated by circumflex's in the text.

The reciprocal of the imaginary part K_I, of the wave number, is the distance over which the amplitude of the field is attenuated by a factor $1/e$ and is the skin depth of the medium. The real part, K_R ($= 2\pi/\lambda$, where λ is the wavelength in the medium), gives rise to a change in phase with position in the material.

2.2. *Cylindrical model*

Consider an infinitely long cylinder of radius r_0 of the conducting material coaxial with the z-axis and immersed in the alternating magnetic field described by eqn (2). Cylindrical polar coordinates r, θ and z are employed (fig. 1(*b*)). By symmetry, the field inside the cylinder will be a function of the radial position coordinate only, and if we assume a solution of (1) of the form

$$\boldsymbol{B} = B_0 R(r) \exp(\mathrm{j}\omega t)\,\hat{\mathbf{z}} \tag{6}$$

then the function $R(r)$ must satisfy the modified Bessel equation

$$\frac{\partial^2 R}{\partial r^2} + \frac{1}{r}\frac{\partial R}{\partial r} - K^2 R = 0 \tag{7}$$

of order zero and independent variable Kr, where $K = K_\mathrm{I} + \mathrm{j}K_\mathrm{R}$ is the complex wave number. The solution of (7) that remains finite for all $r \leqslant r_0$ is

$$R(r) = \frac{\mathrm{I}_0(Kr)}{\mathrm{I}_0(Kr_0)} \tag{8}$$

where I_0 is a modified Bessel function of the first kind of order zero, and the integration constant, $[\mathrm{I}_0(Kr_0)]^{-1}$ has been chosen so that $R(r) = 1$ at $r = r_0$. This result is similar to that which describes uniform cylindrical electromagnetic waves with the $\boldsymbol{E}$ or $\boldsymbol{B}$ vectors parallel to the axis (Schelkunoff 1943).

Eqn (8) can be expressed in polar form by expanding the modified Bessel function into a power series (Spiegel 1968) and invoking De Moivre's formula. Thus the solution (6) can be written

$$\boldsymbol{B} = B_0 \frac{|\mathrm{I}_0(Kr)|}{|\mathrm{I}_0(Kr_0)|} \exp \mathrm{j}(\omega t - \xi(r))\, \hat{\boldsymbol{z}} \tag{9}$$

where

$$|\mathrm{I}_0(Kr)| = [\mathrm{Re}\,\mathrm{I}_0(Kr)^2 + \mathrm{Im}\,\mathrm{I}_0(Kr)^2]^{\frac{1}{2}}$$

$$\xi(r) = \arctan\left[\frac{\mathrm{Im}\,\mathrm{I}_0(Kr_0)}{\mathrm{Re}\,\mathrm{I}_0(Kr_0)}\right] - \arctan\left[\frac{\mathrm{Im}\,\mathrm{I}_0(Kr)}{\mathrm{Re}\,\mathrm{I}_0(Kr)}\right]$$

$$\mathrm{Re}\,\mathrm{I}_0(Kr) = 1 + \frac{s^2 r^2}{2^2}\cos 2\phi + \frac{s^4 r^4}{2^2 4^2}\cos 4\phi + \frac{s^6 r^6}{2^2 4^2 6^2}\cos 6\phi + \ldots$$

$$\mathrm{Im}\,\mathrm{I}_0(Kr) = \frac{s^2 r^2}{2^2}\sin 2\phi + \frac{s^4 r^4}{2^2 4^2}\sin 4\phi + \frac{s^6 r^6}{2^2 4^2 6^2}\sin 6\phi + \ldots$$

$$s = (K_{\mathrm{R}}{}^2 + K_{\mathrm{I}}{}^2)^{\frac{1}{2}} \quad \text{and} \quad \phi = \arctan\left(\frac{K_{\mathrm{R}}}{K_{\mathrm{I}}}\right).$$

Here the attenuation in amplitude of the magnetic field is described by the ratio $|\mathrm{I}_0(Kr)|/|\mathrm{I}_0(Kr_0)|$ and the phase variation is described by $\xi(r)$. The infinite series $\mathrm{Re}\,\mathrm{I}_0(Kr)$ and $\mathrm{Im}\,\mathrm{I}_0(Kr)$ converge rapidly in practice.

2.3. *Power deposition*

Power losses in the form of joule heating within the specimen will derive from eddy currents induced by the alternating magnetic field. In the above configuration, the induced electric field $\boldsymbol{E}$, as deduced from Faraday's Law, is

$$\boldsymbol{E} = -\frac{\mathrm{j}\omega B_0 \mathrm{I}_1(Kr)\exp(\mathrm{j}\omega t)}{K\mathrm{I}_0(Kr_0)}\,\hat{\boldsymbol{\theta}} \tag{10}$$

where I_1 is a modified Bessel function of the first kind of order one (Schelkunoff 1943). The absorbed power density resulting from both conduction and displacement current losses is given by (Johnson and Guy 1972)

$$P = \tfrac{1}{2}\sigma|E|^2$$

where $\sigma = 1/\rho$ is the effective (measured) conductivity of the tissue, and

$$|E| = \frac{|\omega B_0 \mathrm{I}_1(Kr)|}{|K\mathrm{I}_0(Kr_0)|}$$

is the amplitude of the real part of (10).

If the field is pulsed, as in an NMR imaging experiment, the power density is averaged over a complete pulse cycle (Schwan 1965). Thus, if T and τ are respectively the NMR pulse length and the pulse repetition period, the time averaged absorbed power density is

$$P = \sigma \frac{T}{\tau} |E|^2. \tag{11}$$

It may be noted that the amplitude of the RF field, B_0, is also related to the pulse length, T, via the Larmor equation, and for 90° pulses at the proton resonance, $B_0 = 5{\cdot}87 \times 10^{-9}/T$ tesla, T in seconds.

3. Experiment

The values of the permittivity and resistivity of biological tissue used in the present investigation were obtained from measurements of the tissue impedance using a type 4815A Hewlett–Packard RF vector impedance meter (Bottomley 1978). The measurements were performed at frequencies of 1, 2, 3, 4, 5, 10, 15, 20, 25, 30, 40, 50, 60, 70, 80, 90 and 100 MHz at 37 °C and 25 °C. The temperature was controlled to within ±1 °C during measurements by blowing dry heated air over the meter probe and specimen cell assembly.

The resistivity and relative permittivity dispersions of seven different tissue types at 37 °C are shown in figs 2(*a*) and 2(*b*). All of the tissue samples were taken from freshly killed white Wistar rats and all measurements were completed within six hours of death. The curves shown for each tissue type represent an average from measurements taken from six tissue samples. Individual samples exhibited curves which deviated from these mean curves by less than 5%. This can be attributed to a combination of experimental error and the natural heterogeneity of biological specimens. Temperature dependence over the range 25 °C to 37 °C is approximately −1% per °C for the resistivity and approximately +1·2% per °C at 1 MHz decreasing to approximately 0·3% per °C at 100 MHz for the relative permittivity of the tissues.

These experimental results are in good agreement with results measured at 25, 50 and 100 MHz by Osswald (1937) and with data collected by Schwan (1965). A possible cause of the difference between the results for healthy and tumorous liver tissues is the increase in water content and change in the ratio of free to bound water of tumorous tissue relative to its host (Kiricuta and Simplaceanu 1975).

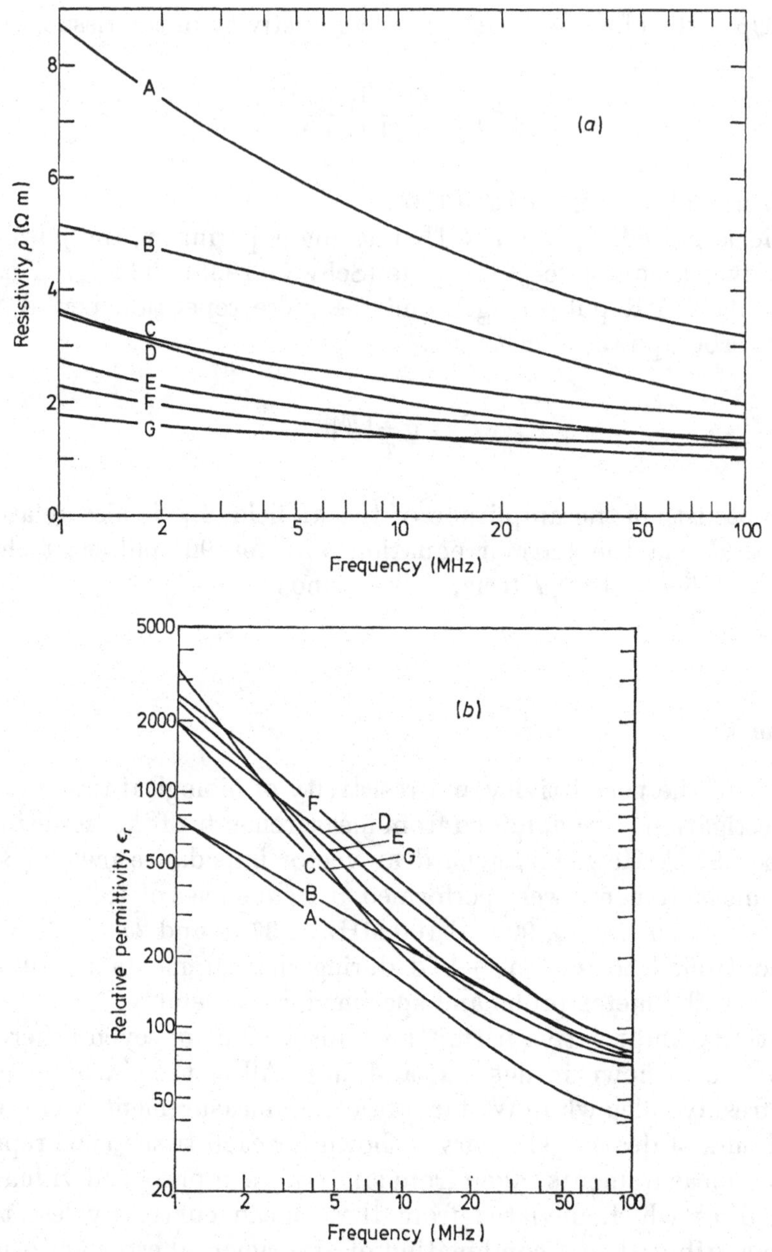

Fig. 2. Resistivity (*a*) and relative permittivity (*b*) of rat lung (A), brain (B), liver (C), kidney (D), heart (E), liver hepatoma D23 tumour (F) and abdominal wall muscle (G) tissues at 37 °C as a function of frequency.

4. Results

Curves showing the variation of the skin-depth of penetration of the RF magnetic field from 1 to 100 MHz in the seven tissue types at 37 °C are shown in fig. 3(*a*). The corresponding phase shifts at one skin depth and 37 °C are shown in fig. 3(*b*) for the planar tissue model. The curves are calculated from eqns (3), (4) and (5) with the data shown in fig. 2 and assuming a tissue magnetic permeability equivalent to that of free space. The skin depths of abdominal

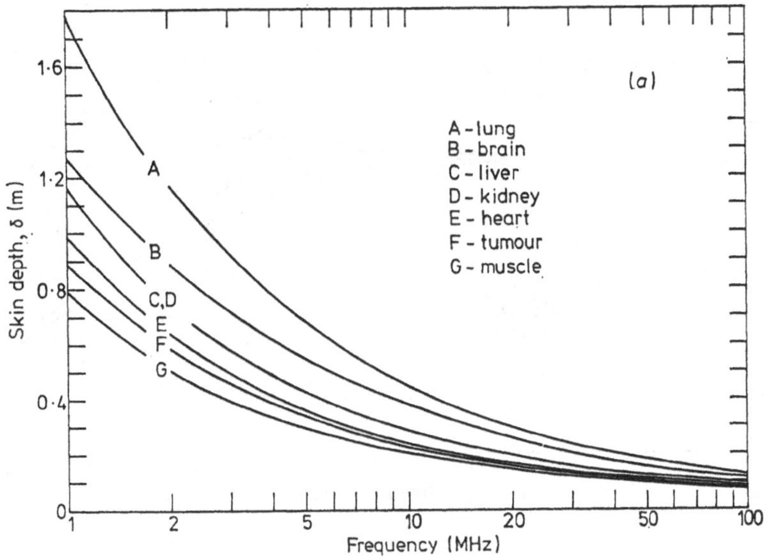

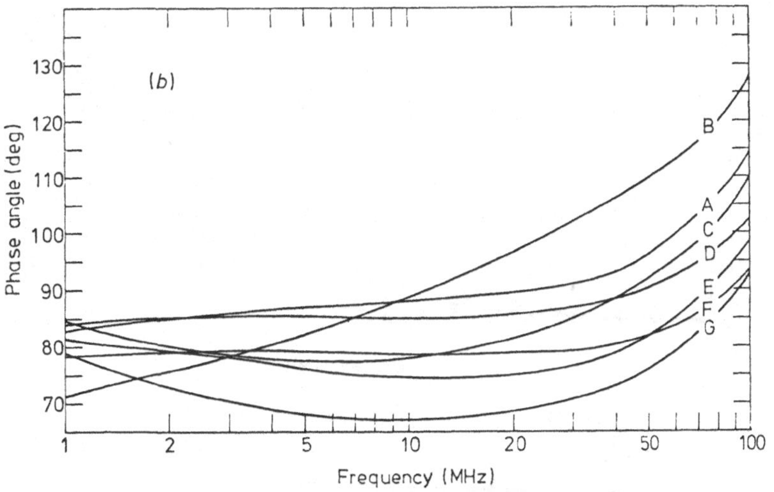

Fig. 3. Skin depths (a) and corresponding phase shifts at one skin depth (b) of the tissues as a function of frequency at 37 °C.

wall muscle, tumour, heart, liver and kidney at 1, 10, 27, 41 and 100 MHz are in tolerable agreement with values characteristic of tissue of high water content published by Johnson and Guy (1972). However, the large variations in skin depth with tissue type observed here emphasise the difficulty in assigning a single meaningful representative value to such a broad range of tissues.

The behaviour of the RF magnetic field within a cylinder of tissue at 37 °C, as described by eqn (9), is illustrated in figs 4 and 5. In figs 4(a), 4(b) and 4(c) the field amplitude, phase shift and combined amplitude–phase variation at the axis of tissue cylinders of radius 20 cm, as a function of frequency, are shown. Cylinders of this size give a fair approximation to an average human torso. Fig. 5(a) shows the amplitude–phase variation at the axis of a muscle cylinder

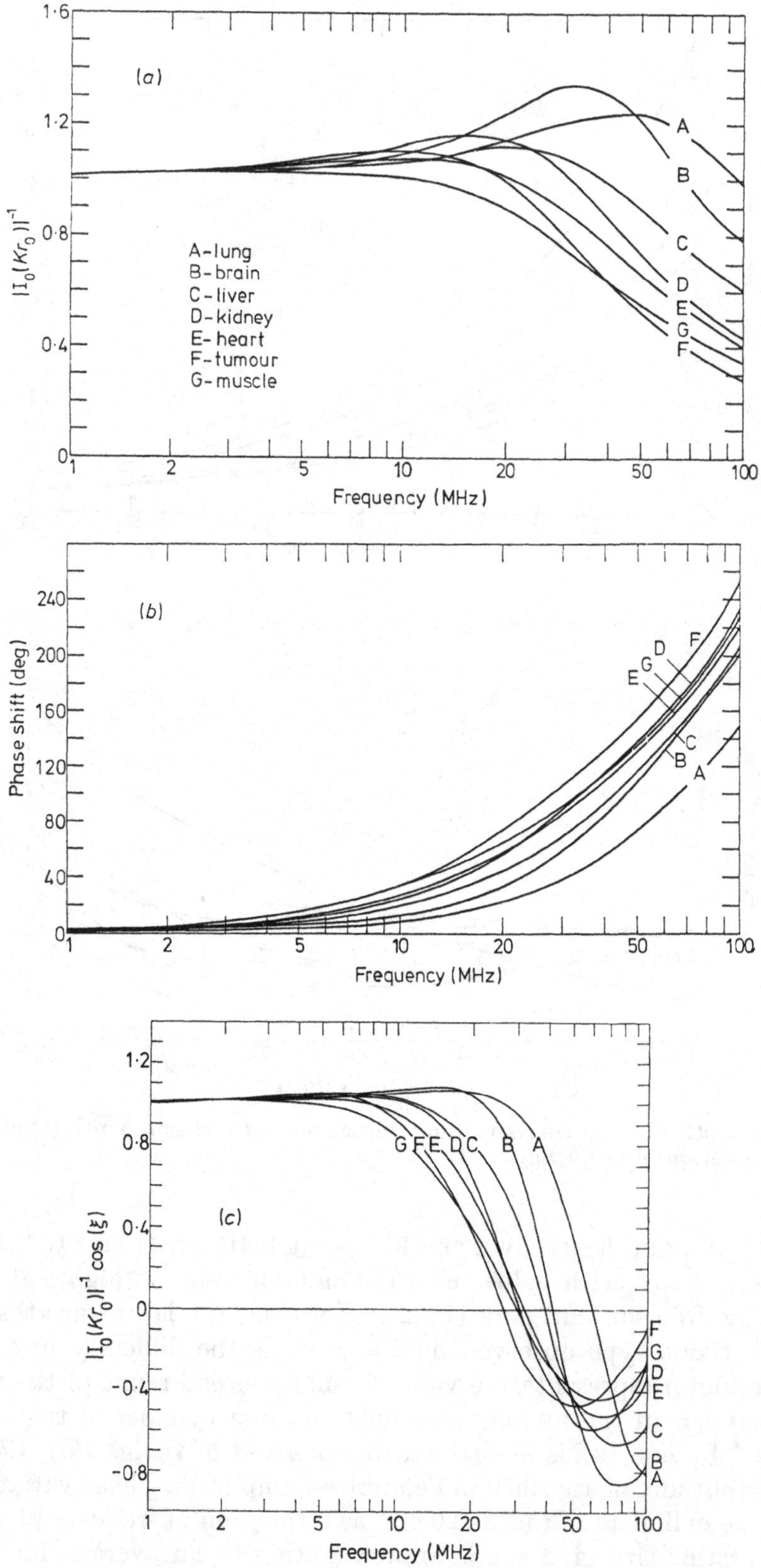

Fig. 4. Magnetic field amplitude (*a*), phase (*b*) and amplitude–phase (*c*) variation at the axis of tissue cylinders of radius 20 cm as a function of frequency. These are represented by the functions $|I_0(Kr_0)|^{-1}$, $\xi(0)$ and $|I_0(Kr_0)|^{-1} \cos(\xi)$ respectively.

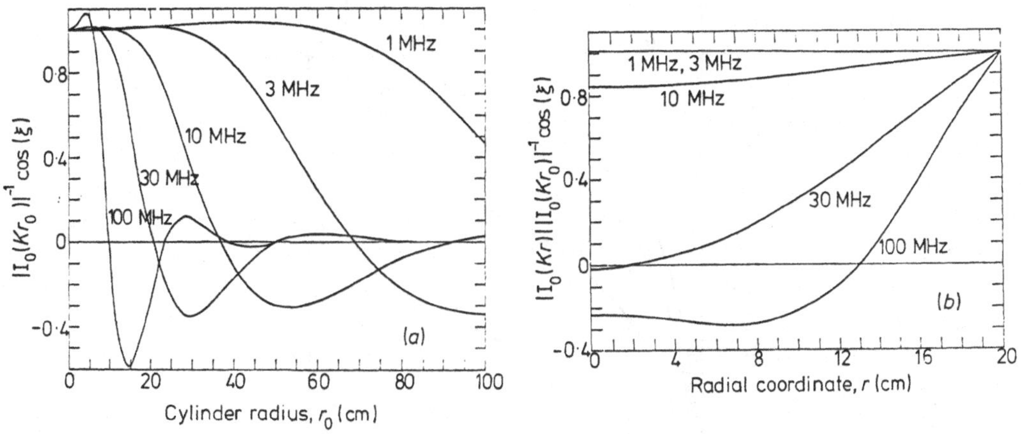

Fig. 5. Part (*a*) shows the amplitude–phase variation ($|I_0(Kr_0)|^{-1}\cos(\xi)$) at the axis of a muscle cylinder as a function of the cylinder radius for different frequencies. Part (*b*) shows the amplitude–phase variation ($|I_0(Kr_0)||I_0(Kr_0)|^{-1}\cos(\xi)$) along the radius of a muscle cylinder of fixed radius 20 cm for different frequencies.

as a function of the cylinder radius for different frequencies. Fig. 5(*b*) shows the amplitude–phase variation along the radius of a muscle cylinder of fixed radius 20 cm for different frequencies.

The results for the cylindrical tissue model differ significantly from those for the planar tissue model. Within the cylinder, the variation in field amplitude with position is far from the exponential decay exhibited in the planar model (eqn 3) and may even increase towards the centre of the cylinder. This increase is due to a combination of the reduced wavelength in the medium and the radius of curvature which can produce a focussing of the RF field that more than compensates for the conduction and displacement current losses. The zero's in the curves of figs 4(*c*), 5(*a*) and 5(*b*) occur at phase shifts of $n\pi/2$, where n is an odd positive integer.

The absorbed power density associated with the induced eddy current losses in a 20 cm radius tissue cylinder is shown in fig. 6. The calculated curves (eqn (11)) assume that the cylinder is subjected to a string of 10 μs 90° RF pulses at the proton resonance, spaced at intervals of 10 ms. This is comparable to our existing NMR image operating conditions (Hinshaw *et al.* 1978). Power absorption estimates for different T or τ values can be obtained from fig. (6) by appropriate scaling as indicated by eqn (11).

Greatest power absorption occurs at the surface of the cylinder where the eddy current loops are linked by the greatest flux. Within the cylinder, the power absorption density decreases monotonically to zero at the axis where the flux linkage is zero. For muscle tissue this is shown in fig 6(*a*) for different frequencies. The surface absorbed power density for all of the tissues as a function of frequency is shown in fig. 6(*b*). At the higher frequencies the surface power absorption is less than would be expected from a low frequency extrapolation because the magnetic field penetration is reduced, thus giving rise to a decreased flux linkage.

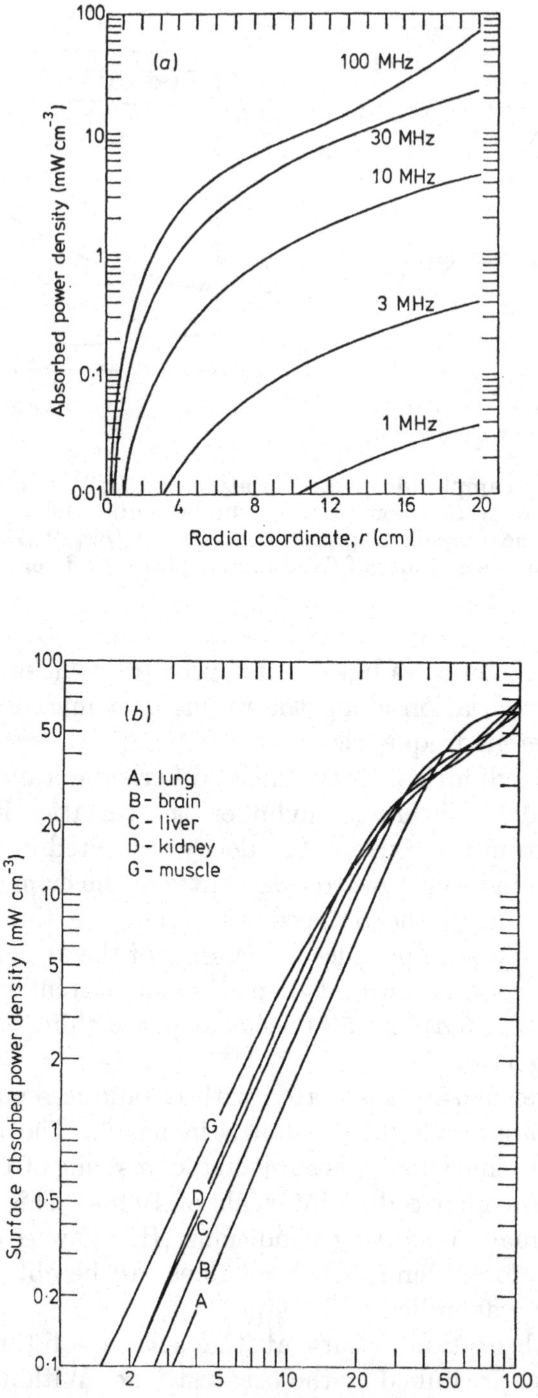

Fig. 6. The absorbed power density along the radius of a muscle cylinder of fixed radius 20 cm for different frequencies is shown in part (*a*). Part (*b*) shows the absorbed power density at the surface of the 20 cm radius cylinder for most of the tissues as a function of frequency. The curve for heart falls midway between those of muscle and kidney and the curve for tumour lies midway between that of muscle and heart. The calculated curves assume T and τ are 10 μs and 10 ms respectively.

5. Discussion

In the planar tissue model, fig. 3 indicates that a skin depth of 20 cm or less is approached by most tissues at frequencies of approximately 20 MHz, with corresponding phase shifts of around 80°. Remembering that the skin depth represents an attenuation in amplitude to 37% of that at the surface, it would appear that we might expect serious RF field attenuation and phase problems to occur at frequencies of 20 MHz and higher in human body sized NMR imaging experiments. The results from the cylindrical tissue model lend qualitative support to this conclusion although the RF field attenuation is less severe owing to the surface curvature.

From the NMR imaging viewpoint, an attenuation in amplitude to 37% at the centre of the specimen will mean that an NMR pulse which is 90° at the surface will be reduced to a 33° pulse at the centre resulting in a 45% reduction in the NMR signal. Similarly, an attenuation of 50% results in a 30% reduction in the NMR signal at the centre. For a 20 cm radius muscle cylinder, the latter condition is reached at a frequency of nearly 60 MHz (fig. 4(*a*)). The effect on an NMR image would be to produce 'fading' or loss of sensitivity in this region.

Phase variations in the RF field throughout a specimen will produce distortions in the NMR signal and hence the resultant NMR image. For example, if the NMR signal is pure absorption mode at the cylinder surface, it will be pure dispersion mode at regions where the phase shift is 90°, and inverted at regions where the phase shift is 180°, thus giving rise to a large negative signal in these regions. The extent of the image deterioration caused by these phase variations will depend upon the imaging technique employed. A possible solution to this problem would be to use the power spectrum of the NMR signal. For the cylindrical model, phase shifts of $\pi/4$, $\pi/2$ and π are reached at frequencies of 14, 30 and 74 MHz respectively, within the 20 cm muscle cylinder (fig. 4(*b*)).

Present power density safety standards are all expressed in terms of flux levels. The maximum recommended safe power density flux for long term human exposure in the United Kingdom is 10 mW cm^{-2} (Ministry of Technology 1968). This corresponds to an absorbed power density of order 1 mW cm^{-3} or a current density of approximately 3 mA cm^{-2} (Schwan 1972). The surface power absorption in an NMR imaging experiment operating under the specified conditions exceeds this figure at frequencies greater than about 5 MHz (fig. 6(*a*)) for the 20 cm radius tissue cylinders. The averaged power absorption throughout the cylinders exceeds this figure at frequencies greater than about 10 MHz. Different imaging techniques may increase or reduce power absorption levels. For example, with T and τ values of 60 and 800 μs respectively as employed in the technique of Damadian *et al.* (1977), the above absorbed power level estimates would be doubled, and the surface power absorption would exceed 1 mW cm^{-3} at operating frequencies greater than about 3·5 MHz. It should be noted however that power levels much greater than the recommended value are used in clinical diathermy treatment of many areas of the human body (Schwan 1965, Johnson and Guy 1972).

Finally, there are several additional considerations which must be taken into account when applying the above results to a real human body sized NMR imaging situation. Firstly, there are the problems which arise from the assumptions made in the derivation of the simplified theory presented. These are: (i) that the applied RF field is perfectly uniform and unaffected by the specimen and (ii) that the specimen is composed of a single homogeneous tissue infinite in extent and either of planar or cylindrical geometry. In heterogeneously composed specimens, reflections and refractions at tissue interfaces should also be taken into account (e.g. Kritikos and Schwan 1976). Secondly, we have considered only the behaviour of the RF field at the resonating nuclei. The NMR signal itself may undergo additional distortions prior to detection. There will also be an additional power absorption component resulting from nuclear magnetic relaxation processes, but this is likely to be several orders of magnitude below that due to eddy–current losses (Grannell and Mansfield 1975).

6. Conclusion

Evidence based on simple planar and cylindrical tissue models indicates that the distortion in both amplitude and phase of the resonant magnetic field is sufficient to cause serious problems in NMR imaging of the human body at frequencies greater than 30 MHz. Should it be desired to operate at such frequencies in order to improve the signal-to-noise ratio for example (Hinshaw 1976), it may be possible to correct for these distortions using the theory and results presented.

Power deposition during an NMR imaging experiment on a human torso will depend on the imaging technique employed. Although the power deposition estimates presented relate to our existing NMR image operating conditions, with a simple modification they can be made to apply to other NMR imaging systems. With the operating conditions specified, the maximum recommended safe power density for long term human exposure is exceeded at NMR imaging frequencies greater than 5 MHz. However, the power levels achieved below 100 MHz are no greater than those used in clinical diathermy treatment, and with the relatively short exposure times envisaged for a completed image (Andrew *et al.* 1977, Hinshaw *et al.* 1977), it is improbable that RF heating will occur at dangerous levels.

We would like to thank Dr. M. R. Price of the Nottingham University Cancer Research Laboratory for providing the tissue samples. The work described forms part of an MRC funded project on NMR imaging undertaken by a group consisting of E. R. Andrew, P. A. Bottomley, W. S. Hinshaw, G. N. Holland and W. S. Moore.

Résumé

Pénétration à haute fréquence de champ magnétique, déphasage et dissipation de puissance dans les tissus biologiques: implications pour la représentation de la RMN

La pénétration d'un champ magnétique, le déphasage et le dépôt de puissance dans des modèles planaires et cylindriques de tissus biologiques exposés à un champ magnétique sinusoïdal à dépendance temporelle ont été étudiés sur le plan théorique sur une gamme de fréquences de 1 à 100 MHz. Les résultats sont fondés sur des mesures des dispersions relatives de la constante diélectrique et de la résistivité sur divers tissus de rats récemment excisés, à 37 et 25°C, et leurs implications pour la représentation de la résonance magnétique nucléaire (RMN) du corps humain sont analysées.

Les résultats indiquent qu'à des fréquences de RMN beaucoup plus élevées que celles se situant aux alentours de 30 MHz, l'amplitude du champ magnétique et les variations de phase subies par les noyaux sont susceptibles de provoquer de graves distorsions sur une image du torse humain. Il est probable que le dépôt maximum de puissance envisagé pendant une expérience de représentation de RMN sur un torse humain est comparable aux niveaux d'exposition existants sans danger à long terme, et qu'il dépendra en définitive de la technique de représentation et de la fréquence de RMN utilisée.

Zusammenfassung

Durchdringung des Magnetfeldes im Hochfrequenzbereich, Phasenverschiebung und Energiezerstreuung in biologischen Geweben: Bedeutung für die kernmagnetische Resonanzabbildung

Theoretisch untersucht wurde die Durchdringung des magnetischen Feldes, die Phasenverschiebung und Energieablagerung in ebenen und zylindrischen Modellen biologischer Gewebe, die einem sinusförmigen, zeitabhängigen Magnetfeld ausgesetzt wurden. Der bei dieser Untersuchung zugrundegelegte Frequenzbereich war 1 bis 100 mHz. Die Ergebnisse stützen sich auf Messungen der relativen Permissivität und der Resistivitäten bzw. deren Verteilung an einer Vielzahl von Geweben, die unmittelbar vor der Messung bei 37 and 25°C an Ratten ausgeschnitten wurden. Deren Analyse erfolgte mit Hinsicht auf die Bedeutung, die für die kernmagnetische Abbildung am menschlichen Körper besteht.

Die Ergebnisse deuten an, dass bei kernmagnetischen Resonanzfreuqenzen, die wesentlich über 30 mHz liegen, die magnetischen Feldamplituden und die Phasenunterschiede, die die Kerne auf weisen, zu schwerwiegenden Verzerrungen bei der Abbildung des menschlichen Körpers führen können. Die maximal zu veranschlagende Energieabladung bei der kernmagnetischen Resonanzabbildung am menschlichen Körper dürfte denjenigen Werten nahekommen, die für die auf lange Dauer vom Menschen ohne Schaden zu ertragenden Werte der Energieeinwirkung gelten. Letztlich hängen diese Werte wohl von der verwandten Abbildungsmethode und der kernmagnetischen Resonanzfrequenz ab.

References

Andrew, E. R., 1977, *Phys. Bull.*, **28**, 323–324.

Andrew, E. R., Bottomley, P. A., Hinshaw, W. S., Holland, G. N., and Simaroj, C., 1977, *Phys. Med. Biol.*, **22**, 971–974.

Bottomley, P. A., 1977, presented at the *British Radio Spectroscopy Group Meeting on Biological Applications of NMR, Dundee, U.K.*, 14–15 *September.*

Bottomley, P. A., 1978, *J. Phys. E: Sci. Instrum.*, **11**, 413–414.

Damadian, R., Goldsmith, M., and Minkoff, L. 1977, *Physiol. Chem. Phys.*, **9**, 97–100.

Grannell, P. K., and Mansfield, P., 1975, *Phys. Med. Biol.*, **20**, 477–482.

Hazelwood, C. F., Cleveland, G., and Medina, D., 1974, *J. Nat. Cancer Inst.*, **52**, 1849–1853.

Hinshaw, W. S., 1976, *J. Appl. Phys.*, **47**, 3709–3721.

Hinshaw, W. S., Andrew, E. R., Bottomley, P. A., Holland, G. N., Moore, W. S., and Worthington, B. S., 1978, *Br. J. Radiol.*, **51**, 273–280.

Hinshaw, W. S., Bottomley, P. A., and Holland, G. N., 1977, *Nature, Lond.*, **270**, 722–723.

Holland, G. N., Bottomley, P. A., and Hinshaw, W. S., 1977, *J. Mag. Res.*, **28**, 133–136.

Hollis, D. P., Saryan, L. A., Eggleston, J. C., and Morris, H. P., 1975, *J. Nat. Cancer. Inst.*, **54**, 1469–1472.

Johnson, C. C., and Guy, A. W., 1972, *Proc. IEEE*, **60**, 692–718.

Kiricuta, I. C., and Simplaceanu, V., 1975, *Cancer Res.*, **35**, 1164–1167.

Kumar, A., Welti, D., and Ernst, R. R., 1975, *J. Mag. Res.*, **18**, 69–83.

Kritikos, H. N., and Schwan, H. P., 1976, *IEEE Trans. Biomed. Eng.*, **BME-23**, 168–172.

Lauterbur, P. C., 1974, *Pure Appl. Chem.*, **40**, 149–157.

Lin, J. C., Guy, A. W., and Johnson, C. C., 1973, *IEEE Trans. Microw. Theory Tech.*, **MTT-21**, 791–797.

LORRAIN, P., and CORSON, D., 1970, *Electromagnetic Fields and Waves*, 2nd edn (San Francisco: W. H. Freeman).
MANSFIELD, P., and MAUDSLEY, A. A., 1977a, *Br. J. Radiol.*, **50**, 188–194.
MANSFIELD, P., and MAUDSLEY, A. A., 1977b, *J. Mag. Res.*, **27**, 101–119.
MANSFIELD, P., MAUDSLEY, A. A., and BAINES, T., 1976, *J. Phys. E: Sci. Instrum.*, **9**, 271–278.
MICHAELSON, S. M., 1972, *Proc. IEEE*, **60**, 389–421.
OSSWALD, V. K., 1937, *Hochfreq. Tech Elektroakust.*, **49**, 40–49.
MINISTRY OF TECHNOLOGY, 1968, *Intense Radio Frequency Radiation Code of Practice*, MoT Report BR 19945 (H.M.S.O.).
SCHELKUNOFF, S. A., 1943, *Electromagnetic Waves* (New York: Van Nostrand).
SCHWAN, H. P., 1965, in *Therapeutic Heat and Cold*, Ed. S. Licht (New Haven, Conn.: Licht) 63–125.
SCHWAN, H. P., 1972, *IEEE Trans. Biomed. Eng.*, **BME-19**, 304–312.
SPIEGEL, M. R., 1968, *Mathematical Handbook of Formulas and Tables* (New York: McGraw-Hill).

The Sensitivity of the Zeugmatographic Experiment Involving Human Samples

D. I. HOULT*

Department of Biochemistry, University of Oxford, South Parks Road, Oxford OX1 3QU, United Kingdom

and

PAUL C. LAUTERBUR

Department of Chemistry, State University of New York at Stony Brook, Stony Brook, New York 11794

Received August 26, 1978

An attempt is made to remove some of the uncertainty surrounding the sensitivity of an NMR experiment involving human samples. It is shown that noise may be associated not only with the receiving coil resistance, but also with dielectric and inductive losses in the sample. Although steps may be taken to minimize the dielectric losses, this is not the case for the magnetic losses, and an estimate is made of their effects upon the signal-to-noise ratio. Approximate values of the latter are calculated for the head and torso and some experimental constraints briefly discussed.

INTRODUCTION

There exists a need for reasonable estimates of the sensitivity of the zeugmatographic experiment involving human subjects (*1*), based upon an appreciation of some fundamental limitations of such experiments, to ensure that instrument design and specifications are realistic. The paper of Hoult and Richards (*2*) provides a starting point for the required calculations, but a major uncertainty is introduced into results obtained by their method, for the sample (a human being) is electrically lossy. These losses are associated, when signal is being received, with the induction of noise from the sample in the receiving coils, thus reducing the sensitivity. We therefore define two limiting cases—the absence of sample losses giving an upper limit to the sensitivity, and the presence of losses (assuming that the body is a homogeneous, uniform saline solution of physiological concentration) giving a lower limit.

THE LOSSLESS CASE

It may be shown (*2*) that if a conductor carrying unit current produces a field $\mathbf{B}_1$ at point P, then a rotating magnetic dipole $\mathbf{m}$ placed at point P introduces an EMF in

* Present address: National Institutes of Health, Biomedical Engineering and Instrumentation Branch, Building 13, Room 3W13, Bethesda, Md. 20014.

Reprinted with permission from Journal of Magnetic Resonance **34**, 425–433 (1979); ©Academic Press.

that conductor given by

$$\xi = -\frac{\partial}{\partial t}\{\mathbf{B}_1 \cdot \mathbf{m}\} \qquad [1]$$

For the NMR case, **m** may be considered to be the nuclear magnetic moment flipped in the usual way from the z direction into the xy plane by the application of a 90° pulse. Thus, for an elementary volume dV of sample,

$$\mathbf{m} = jM_0\, dV \exp(j\omega_0 - 1/T_2)t,$$

where M_0 is the nuclear magnetization, ω_0 is the Larmor frequency, T_2 is the spin–spin relaxation time, t is time, and $j = -1^{1/2}$. Thus:

$$\xi \simeq \omega_0 B_1 M_0 \exp(j\omega_0 - 1/T_2)t\, dV, \qquad [2]$$

where it is assumed that B_1 is in the xy plane. The nuclear magnetization is given (in SI units) by

$$M_0 = N\gamma^2\hbar^2 I(I+1)B_0/3kT_s \qquad [3]$$

where N is the number of spins at resonance per unit volume, γ is the magnetogyric ratio, and T_s is the sample temperature. For protons in water, $M_0 = 3.25 \times 10^{-3} B_0\ \mathrm{A\ m^{-1}}$, assuming a sample temperature of 310 K.

The EMF induced in the coil is given by Eq. [2]; the incremental mean square thermal noise in the coil is given by

$$\frac{d\overline{N^2}}{d\nu} = 4kT_cR, \qquad [4]$$

where T_c is the coil temperature, R is its resistance and ν is frequency. To calculate the sensitivity, i.e., the signal-to-noise ratio, we therefore need to know B_0 (and hence ω_0), B_1 and R. The ratio $B_1/R^{1/2}$ should be maximised in accordance with the geometry of the experimental situation in order to obtain the best signal-to-noise ratio. A solenoidal configuration gives the best ratio, and the optimal winding geometry (*3, 4*) is shown in Fig. 1. The field due to such a coil is given by, for unit current,

$$B_1 \simeq \frac{n\mu_0}{2}\frac{1}{[a^2+g^2]^{1/2}}, \qquad n \gg 1, \qquad [5]$$

while the resistance is given by

$$R \simeq \frac{3\sigma\rho n^2 a}{2\delta g}, \qquad n \gg 1, \qquad [6]$$

a is the radius of the coil, g its half height, n the number of turns, ρ the resistivity of the conductor, δ the radio frequency skin (or penetration) depth (*5*) and σ the proximity effect factor (*4*). The penetration depth is given by

$$\delta = \left[\frac{2\rho}{\mu\mu_0\omega_0}\right]^{1/2}. \qquad [7]$$

At 1 MHz and at room temperature, the resistivity ρ of annealed copper is 1.69 ×

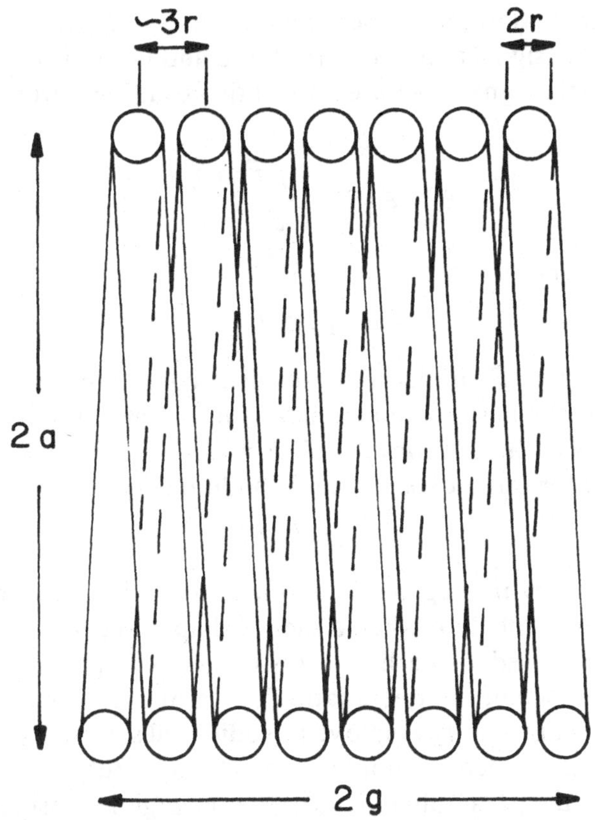

FIG. 1. The optimal winding geometry for a solenoidal receiving coil. The coil length (2g) should be somewhat less than the diameter (2a), for example, 70% thereof, and the distance between turns should be three to four times the wire radius.

10^{-8} Ωm and thus the skin depth δ is 6.6×10^{-5} m. The proximity factor is the main unknown in Eq. [6], but the proximity effect usually increases resistance by a factor of between three and six for the coil geometries encountered in NMR. A coil was therefore constructed, according to the dictates of Fig. 1, with a diameter of 0.25 m, a length of 0.2 m, and six turns. The conductor was copper tubing 0.02 m in diameter, and the coil fitted comfortably over a person's head. At 1 MHz, the inductance was 9 μH and the unloaded quality factor Q was 650. Thus the resistance was 87 mΩ, and from Eq. [6], the proximity factor was approximately equal to 5.

We are now in a position to calculate, for the coil described above, the sensitivity of the experiment and for convenience we consider a water sample of volume 1 ml at a Larmor frequency of 1 MHz. Thus $B_0 = 0.0235$ T, and $M_0 dV = 7.63 \times 10^{-11}$. From Eq. [5] $B_1 = 2.36 \times 10^{-5}$ T, and this value was checked with the aid of a small search coil and a known current through the main coil. The experimental value was 2.27×10^{-5} T. Thus from Eq. [2], the induced EMF in the coil is given by

$$\xi \simeq 1.13 \times 10^{-8} \exp(j\omega_0 - 1/T_2)t, \qquad [8]$$

while the mean square noise per unit bandwidth is, from Eq. [4],

$$\frac{d\overline{N^2}}{d\nu} = 1.41 \times 10^{-21}; \qquad T_c = 293 \text{ K}. \qquad [9]$$

For both the above equations, the experimental values of B_1 and R were taken. Let us now assume that the signal and noise are stored and Fourier transformed with an optimum filter. Then the signal-to-noise ratio of the resulting Lorentzian line is given (*6*) by:

$$\psi = |\xi| \left\{ T_2/2 \frac{d\overline{N^2}}{d\nu} \right\}^{1/2}. \qquad [10]$$

Hence from Eqs. [8] and [9],

$$\psi \simeq 213\, T_2^{1/2}.$$

A reasonable value for T_2 under physiological conditions is 10^{-1} sec and so, following a 90° pulse, the best possible signal-to-noise ratio obtainable from a 1-ml sample at 1 MHz is, with the coil considered, $\psi = 67.4$. Following Ref. (*2*), this result may be extrapolated for other experimental conditions giving

$$\psi \simeq 8.4\nu^{7/4}/a, \qquad [11]$$

where ν is the frequency in megahertz, and a is the coil radius in meters. When condition $n \gg 1$ is not fulfilled, or when a saddle-shaped receiving coil is employed, the ratio $B_1/R^{1/2}$ referred to earlier is no longer an optimum, and a smaller sensitivity may be expected, the degradation being by a factor of up to three (*2*), depending on the exact geometry employed. Additional sensitivity may be obtained if the receiving coils are cooled to liquid nitrogen temperature (77 K). Assuming (as was done above) that the preamplifier contributes negligible noise, an improvement in sensitivity of up to three may be expected (*2*). An additional factor of $2^{1/2}$ may be obtained with receiving coils in quadrature.

SAMPLE LOSSES

(*a*) *Dielectric Loss*

A major objection to the above calculation is the presence of dielectric and magnetic losses within the sample. Every coil has distributed capacitance associated with its turns, and while this is difficult to calculate, it may be shown (*3*) that its value is approximately proportional to the coil diameter, decreases slowly with increasing coil length, and is essentially independent of the number of turns. For the solenoidal form already discussed, a rough estimate of the capacitance is 2 pF per centimeter diameter, and the capacitance may be considered to be between the two ends of the coil. Electric lines of force associated with this capacitance pass through the sample, and it may be shown (*3*) that the equivalent series resistance associated with the distributed dielectric loss is given by

$$R_e = \tau \omega_0^3 L^2 C_d, \qquad [12]$$

where τ is the loss factor, L is the coil inductance and C_d is the distributed capacitance. To minimize the dielectric loss, the ratio R_e/R must be minimized also. Now the coil resistance $R = \omega_0 L/Q$, and Q is essentially independent of the coil inductance, being more a function of the coil dimensions than the number of turns.

Thus

$$\frac{R_e}{R} \simeq \tau Q \frac{C_d}{C}, \qquad [13]$$

where C is the tuning capacitor placed across the coil. This equation may be interpreted as stating that most of the dielectrical energy stored in the tuned circuit comprising L, C, and C_d must be in C rather than C_d. Loss factors range from 10^{-4} for ceramics and quartz to 10^{-1} for plastics, and if we, very arbitrarily, assume for the human body a value $\tau = 0.1$ and a 30% dielectrical filling factor, we find, for the coil already considered, that $R_e/R \sim 20C_d/C$. Experiments with loaded coils, with and without Faraday shields interposed between human extremities and a surrounding coil, have revealed the presence of a dielectric loss, but it is difficult to estimate the exact size of the loss without knowledge of the efficiency of the shield. It would appear to be considerably less than the estimate above, and so a wise precaution is to insert a Faraday shield and to assume that $C_d = 50$ pF, that C should therefore be greater than 1 nF, and that L should be adjusted accordingly for the frequency of interest. At frequencies above about 3 MHz with human samples, this constraint may result in the number of turns on the coil tending to unity—a situation previously only encountered in high field, high resolution spectrometers.

(*b*) *Inductive Loss*

A far more serious loss mechanism is that associated with the conductivity of the sample. This loss cannot be avoided and it is therefore essential to obtain some

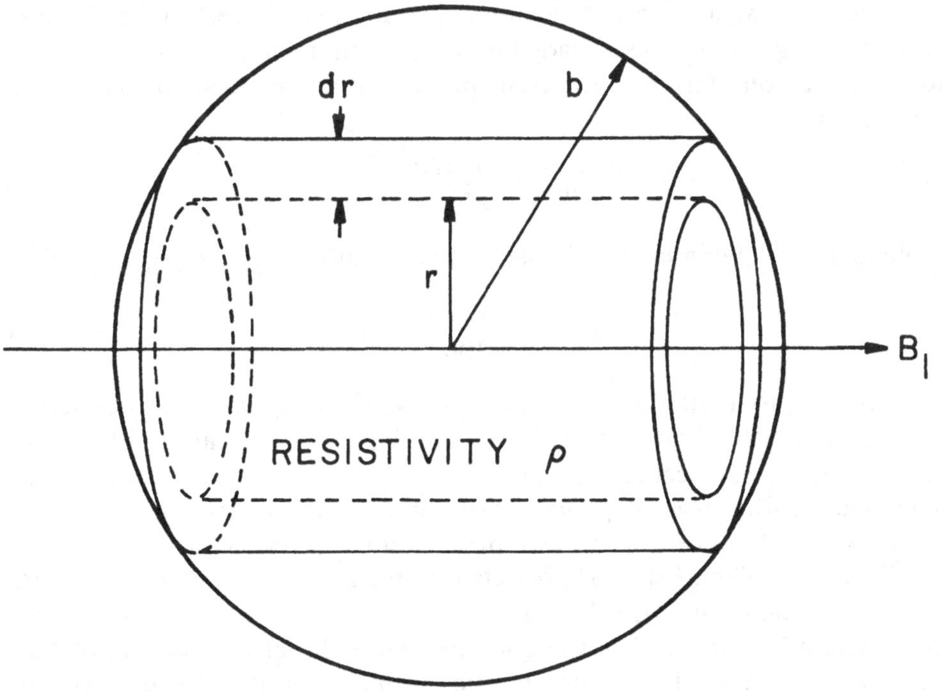

FIG. 2. A model for the calculation of losses in the sample due to induction.

estimate of its dependence upon physical dimensions and frequency, and a measure of its value. Associated with this loss is the depth of penetration of the B_1 field into the conducting sample. The penetration depth must ordinarily be far greater than the sample dimensions if the zeugmatographic experiment is to be feasible. Let us assume that this is indeed the case, and as a model for the calculation of the power dissipated in the sample, consider a homogeneously conducting sphere, as shown in Fig. 2. Let the sphere have a radius b, and consider the conductance of an elementary cylinder, coaxial with the direction of B_1, of radius r and width dr. The conductance is given by

$$dG = \frac{2(b^2 - r^2)^{1/2}\, dr}{2\pi\rho r}, \qquad [14]$$

where ρ is the specific resistivity of the sphere. The EMF induced around the cylinder by the alternation of B_1 is given by

$$V = -\pi r^2 \frac{\partial B_1}{\partial t} = \pi r^2 \omega_0 B_1 \sin \omega_0 t. \qquad [15]$$

Thus the power dissipated in the sphere is

$$W = \int_0^b \left(\frac{\pi \omega_0^2 B_1^2}{2\rho} \right) r^3 (b^2 - r^2)^{1/2}\, dr$$

or

$$W = \frac{\pi \omega_0^2 B_1^2 b^5}{15\rho}. \qquad [16]$$

Now this power dissipation can be expressed as an effective resistance R_m in series with the receiving coil, for we already know B_1 for unit current from Eq. [5]. Thus, because we are considering the current *amplitude* as opposed to the rms value, $W = R_m/2$, and hence

$$R_m = \frac{\pi \omega_0^2 \mu_0^2 n^2 b^5}{30\rho[a^2 + g^2]} \qquad [17]$$

for a solenoidal coil. For a saddle-shaped coil, the result is slightly larger, typically (*2*)

$$R_m = \frac{1.37\pi \omega_0^2 \mu_0^2 n^2 b^5}{30\rho a^2}, \qquad [18]$$

where a is the radius of the coil. The validity of the calculation was tested by monitoring the Q of a saddle-shaped coil containing a spherical sample of sodium chloride solution. An accurate linear dependence of R_m on conductivity was obtained and the fifth-order dependence on radius roughly verified. Using a 2-liter sample of 100 mM concentration (a typical average physiological value (*7*)) and $\omega_0 = 4$ MHz, $a = 0.13$ m, Eq. [18] predicts that for $\rho = 0.85\ \Omega$m, $R_m = 29$ mΩ. An experimental value of 30 m$\Omega \pm 10\%$ was obtained. A similar resistance value was recorded with a human head in the coil, and while direct comparison of the two results is not meaningful due to uncertainties in the distribution of resistivity and in the effective volume and shape of the head, at least one should be aware that inductive losses are present, that they are not negligible, and that the dependencies of Eq. [16] must be considered. One may also tentatively estimate the penetration depth on the basis of the results obtained. For 100 mM NaCl, from Eq. [7]

$$\delta = \frac{0.6}{\nu^{1/2}} \text{ meters,}$$

where ν is the frequency in megahertz, and thus it may be possible, remembering that penetration occurs from all sides of a sample, to work at frequencies as high at 10 MHz, provided dielectric losses can still be neglected.

We are now in a position to estimate the effects of the magnetic losses upon the signal-to-noise ratio. From Eqs. [2] and [3], the induced EMF is proportional to $\nu^2 B_1$. From Eqs. [6] and [7], the resistance of the coil is proportional to $n^2\nu^{1/2}$ while from Eq. [16] the "magnetic" resistance R_m is proportional to $\nu^2 B_1^2 b^5$. Thus for Eqs. [10] and [4], the signal-to-noise ratio may be represented by an equation of the form

$$\psi = \frac{\nu^2 B_1}{[\varepsilon n^2 \nu^{1/2} + \beta \nu^2 B_1^2 b^5]^{1/2}},$$

and from Eq. [5], for a given coil geometry,

$$\psi = \frac{\nu^2}{[\alpha a^2 \nu^{1/2} + \beta \nu^2 b^5]^{1/2}}, \qquad [19]$$

where α, β, and ε are constants. The numerator represents the signal, the first term in the denominator the coil resistance, and the second term the magnetic losses. Thus from Eq. [11], $\alpha \simeq 1.4 \times 10^{-2}$, and assuming a 100 mM NaCl solution at 310 K, $\beta \simeq 24$. If a saddle-shaped coil is used, the value of α should be increased by a factor of between 4 and 9, depending on the exact construction employed, and a reasonable value is $\alpha = 8 \times 10^{-2}$. The quantity ν is in megahertz.

Hence, assuming once again a 1-ml element of water within the sample, with $T_2 = 10^{-1}$ sec, a frequency of 1 MHz, a solenoidal coil of diameter 0.25 m, and an effective total sample radius b of 0.08 m we have, for a 100 mM solution, $\psi \simeq 58$. Taking the same sample, but at 4 MHz with a saddle-shaped coil, $\psi \simeq 250$, as opposed to $\psi \simeq 305$ if the magnetic losses are neglected. Because of the fifth-order dependence on sample radius (Eq. [14]), inductive losses will probably be greatest when obtaining zeugmatograms of the torso. Clearly, the size, shape, and composition of the body have a bearing upon the results. For example, one expects that the losses will be greater if the subject is in an advanced state of pregnancy, as a consequence of the continuous conducting nature of the amniotic fluid. For example, for $\nu = 4$ MHz, $a = 0.25$ m, $b = 0.2$ m, the magnetic losses of Eq. [19] are an order of magnitude larger than the resistive losses, giving a *lower limit* to the sensitivity of $\psi \simeq 45$. Without magnetic losses, the sensitivity would have been of the order of 150.

EXPERIMENTAL CONSTRAINTS

Attention has been paid above to the sensitivity achievable from an element of sample which possesses a spread of frequencies only of the order of T_2^{-1}. However if signal is to be obtained simultaneously from a range of frequencies dictated by the spatial volume of interest and the required resolution, it must be borne in mind that the probe coil, if tuned, has a limited bandwidth. For example, the solenoidal coil mentioned earlier for use at 1 MHz had a Q value of over 600, giving a usable frequency range of only about 700 Hz (to level −1 db). Even at 4 MHz with a loaded saddle-shaped coil, the bandwidth available is still only about 7 kHz, and concomitant with the bandwidth is a receiver recovery time which may be as long as 1 msec, due to ringing of the probe. If a single coil probe is used, one must also beware of the transient response of the probe to the transmitter pulses—the time constant of the probe is $Q/\pi\omega$ μsec, or about 200 μsec for the 1-MHz coil. Thus the transmitter pulse must be considerably longer than this value if serious phase glitch is not to occur. The solutions to these problems lie outside the scope of the present article, but attention is called to them, as they influence the extension of the results (obtained for a *point* in frequency space) to a line or plane of frequencies.

CONCLUSIONS

The conclusions to be drawn from the above calculations are that:

(1) Effects associated with the radiofrequency penetration depth suggest that the frequency of operation of the spectrometer with human samples should be less than about 10 MHz.

(2) Precautions against dielectric losses in the sample must be taken, but with suitable care in design these losses can be rendered negligible.

(3) Losses due to the conductivity of the sample are not negligible and can be the dominant source of noise as one approaches 10 MHz, or at frequencies above 1 MHz if the receiving coils are cooled or an exceptionally large subject is used.

(4) Sensitivity increases with frequency to the 7/4 power at frequencies below 1 MHz, but only linearly when sample losses are predominant.

(5) For a 1-ml element of sample in the head, the signal-to-noise ratio following Fourier transformation of the 90° pulse response can be expected to be in the region of 50 at 1 MHz with a solenoidal receiving coil, and of the order of 250 at 4 MHz with a saddle-shaped receiving coil. For the torso, the expected sensitivity will be somewhat lower depending on the structure and composition of the body.

Note added in proof. Eq. [12] assumes that the coil is immersed in the dielectric. Also, a closely-related discussion has been published since the completion of this paper: P. A. Bottomley and E. R. Andrew, *Phys. Med. Biol.* **23,** 630 (1978).

ACKNOWLEDGMENTS

This investigation was partially supported by Grant CA-153000, awarded by the National Cancer Institute, DHEW, and by Contract N01-HV-5-2970, awarded by the National Heart, Lung and Blood Institute, DHEW. Grateful acknowledgements are also made to the Gulf Oil Foundation and the General Electric Company for partial support.

REFERENCES

1. For recent references and information on new techniques, see P. C. LAUTERBUR, "NMR in Biology" (R. A. Dwek, I. D. Campbell, R. E. Richards, and R. J. P. Williams, Eds.), p. 323, Academic Press, London, 1977; W. S. HINSHAW, P. A. BOTTOMLEY, AND G. N. HOLLAND, *Nature* **270,** 722 (1977); R. DAMADIAN, M. GOLDSMITH, AND L. MINKOFF, *Physiol. Chem. and Phys.* **9,** 97 (1977); P. MANSFIELD AND A. A. MAUDSLEY, *Brit. J. Radiol.* **50,** 188 (1977).
2. D. I. HOULT AND R. E. RICHARDS, *J. Magn. Reson.* **24,** 71 (1976).
3. F. E. TERMAN, "Radio Engineer's Handbook," 1st. ed., pp. 73–90, McGraw-Hill, New York, 1943, and references therein.
4. B. B. AUSTIN, *Wireless Eng. Exp. Wireless* **11,** 12 (1934).
5. B. I. BLEANEY AND B. BLEANEY, "Electricity and Magnetism," 3rd ed., Chap. 8, Oxford Univ. Press, London, 1976.
6. R. R. ERNST AND W. A. ANDERSON, *Rev. Sci. Instrum.* **37,** 93 (1966).
7. R. M. DOWBEN, "General Physiology," p. 358, Harper & Row, New York, 1969.

Radiofrequency Losses in NMR Experiments on Electrically Conducting Samples

In recent years, NMR has been used to study metabolism in intact biological tissue, such as skeletal muscle (*1–3*), heart (*4, 5*), and kidney (*6*), and in whole cells (*7–10*). The intracellular concentration of potassium ions within tissues is typically about 150 m*M*, while the extracellular fluid contains about 150 m*M* sodium chloride, and the conductivity of these samples will cause radiofrequency losses which may adversely affect the signal-to-noise ratio of the NMR spectra.

Hoult and Lauterbur (*11*) have discussed the dielectric and inductive losses that one would expect from human samples in zeugmatography experiments. Here, we discuss in more general terms these two types of loss, and draw particular attention to the different dependences they have on sample conductivity. We also describe a simple method of determining their relative and absolute magnitudes. In subsequent papers, we hope to discuss the influence that these losses have upon rf coil and sample design.

Inductive Losses

The alternating field B_1 from the rf coil induces currents in a conducting sample which dissipate power. The power dissipation can be expressed as an effective resistance R_m in series with the receiving coil, and Hoult and Lauterbur (*11*) showed that, for a spherical sample,

$$R_m = \frac{\pi \omega_0^2 \mu_0^2 n^2 b^5 \sigma}{30(a^2+g^2)} \qquad [1]$$

for a solenoidal coil of radius a, length $2g$, and n ($\gg 1$) turns. The quantity ω_0 is the resonant frequency, b is the sample radius, and σ is the specific conductivity of the sample. In the case of a cylindrical sample of length $2g$, and radius b, it may also be shown that

$$R_m = \frac{\pi \omega_0^2 \mu_0^2 n^2 b^4 g \sigma}{16(a^2+g^2)}. \qquad [2]$$

Dielectric Losses

Electrical lines of force associated with the distributed capacitance of the rf coil pass through the sample, and Hoult and Lauterbur (*11*) quote the following formula for the equivalent series resistance R_e resulting from dielectric losses:

$$R_e = \tau \omega_0^3 L^2 C_d. \qquad [3]$$

Reprinted with permission from Journal of Magnetic Resonance **34**, 449–455 (1979); @ Academic Press.

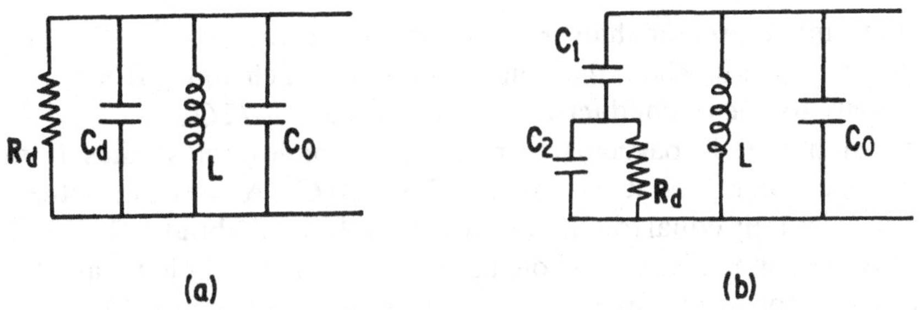

FIG. 1. Representations of tuned circuits (a) with rf coil immersed in lossy sample, (b) with rf coil insulated from lossy sample.

The term τ is the loss factor of the sample, L is the coil inductance, and C_d is the distributed capacitance. This formula is derived by assuming that the rf coil is actually located within the dielectric; the equivalent circuit is shown in Fig. 1a, where R_d is the parallel resistance associated with the lossy capacitor, and C_0 is the tuning capacitance.

However, in NMR experiments, the lossy dielectric is contained within a sample tube, and is insulated from the rf coil. As a result, the equivalent circuit is not that shown in Fig. 1a, but rather the circuit of Fig. 1b, where C_1 is the (lossless) capacitance from coil to sample, and C_2 and R_d now represent the lossy sample. The admittance Y of the circuit element comprising C_1, C_2, and R_d is

$$Y = Y_{\text{real}} + Y_{\text{imag}},$$

where

$$Y_{\text{real}} = \frac{\omega_0^2 R_d C_1^2}{1+\omega_0^2 R_d^2 (C_1+C_2)^2} \tag{4}$$

and

$$Y_{\text{imag}} = \frac{\omega_0 C_1 - \omega_0^3 R_d^2 C_1 C_2 (C_1+C_2)}{1+\omega_0^2 R_d^2 (C_1+C_2)^2}. \tag{5}$$

Note that the circuit element is lossless both for $R_d = 0$ and for $R_d = \infty$, and that Y_{real} goes through a maximum at $R_d = 1/\omega_0(C_1+C_2)$. This formula, together with the expression $1/R_d = \sigma C_2/\varepsilon\varepsilon_0$, where σ is the specific conductivity of the dielectric, leads to the conclusion that Y_{real} is maximal for a conductivity σ_0 given by

$$\sigma_0 = \frac{\omega_0 \varepsilon\varepsilon_0 (C_1+C_2)}{C_2}. \tag{6}$$

Further straightforward algebra leads to the following important conclusions: (i) If the maximum dielectric losses, which occur at $\sigma_0 = \omega_0\varepsilon\varepsilon_0(C_1+C_2)/C_2$, are expressed as an effective series resistance R_0, then

$$Q_0 = \frac{\omega_0 L}{R_0} = \frac{2C_0(C_1+C_2)}{C_1^2} \quad \text{(assuming } C_0 \gg C_1, C_2\text{)}.$$

(ii) The resonant frequency ω_0 decreases as the conductivity of the sample increases, and the reduction in frequency in changing from a sample of zero conductivity to one of infinite (i.e., much greater than σ_0) conductivity is $\Delta\omega_\infty = C_1^2\omega_0/2C_0(C_1+C_2)$. Thus $\Delta\omega_\infty/\omega_0 = 1/q_0$. (iii) The reduction in frequency in changing from a sample of zero conductivity to one of conductivity $\sigma = \sigma_0$ is $\Delta\omega = \omega_0/2Q_0$.

The conclusion that is particularly relevant to biological studies is that the dielectric losses are maximal at $\sigma_0 = \omega_0\varepsilon\varepsilon_0(C_1+C_2)/C_2$. Assuming for the present that C_1 is approximately equal to C_2, then at 100 MHz, σ_0 is about 1 Ω^{-1} m^{-1}. This is approximately the conductivity of whole tissues and cells (see below), and therefore tissue and cell preparations may generate significant dielectric losses in many high-resolution NMR experiments.

Results

The quality factors of rf coils were measured using the method described by Hoult (*12*), and in a few cases a *Q* meter was also used.

A seven-turn solenoid of length 20 mm and radius 8 mm wound from 16-s.w.g. copper wire was tuned with a high-*Q* air-spaced capacitor. A sample tube of inner diameter 11.5 mm was held within the solenoid and the *Q* of the circuit was measured using samples of varying sodium chloride concentrations. An electrical screen consisting of parallel axial strips of wire, 0.5 mm wide and 1.5 mm apart joined at one end, was then wrapped around the sample tube between the sample and rf coil and the experiment was repeated. Figure 2 shows the increase in effective resistance generated by the samples in the presence and absence of the screen. The screen prevents electric lines of force associated with the distributed capacitance from passing through the sample, and therefore eliminates the dielectric losses, although, of course, its presence near the rf coil does introduce a small additional

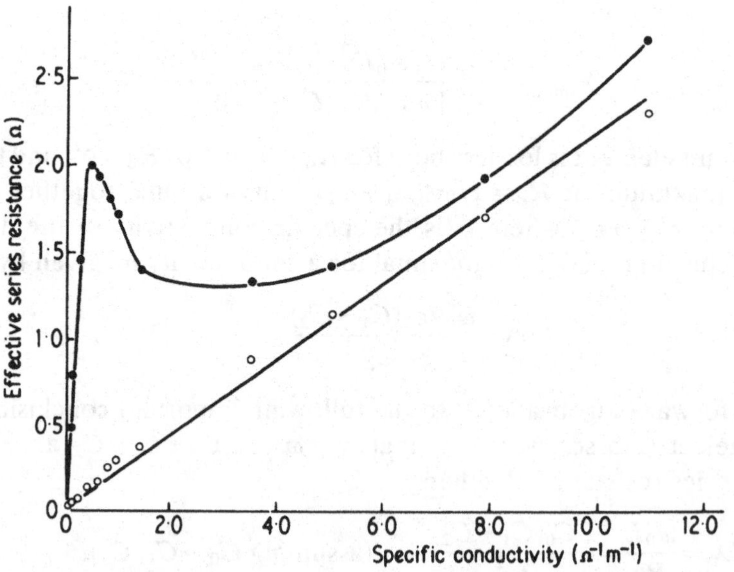

FIG. 2. The effective series resistance generated by samples of sodium chloride plotted as a function of sample conductivity, (●) in the absence of an electrical screen, and (○) in the presence of an electrical screen. The experimental conditions are described in the text.

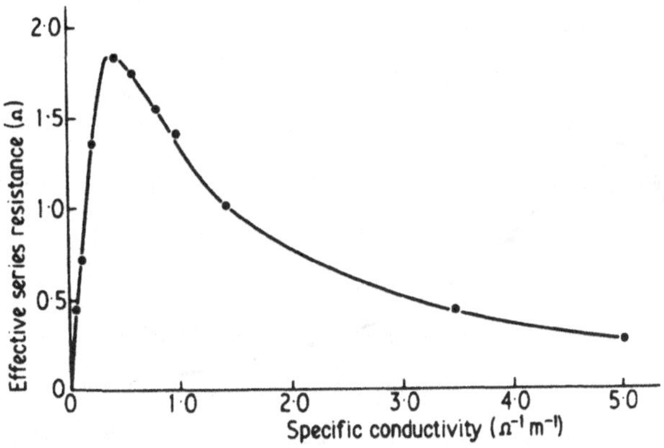

FIG. 3. The effective series resistance due to dielectric losses alone, plotted as a function of sample conductivity. The experimental conditions are described in the text.

loss. Thus the lower trace in Fig. 2 represents the magnetic losses, and the difference between the two sets of data, plotted in Fig. 3, represents the dielectric losses.

In further experiments, the sample was placed in a tube between the plates of the tuning capacitor of a tuned circuit. The functional dependence of sample losses on sodium chloride concentration was similar to the dependence shown in Fig. 3. This provides confirmation that the resistive losses plotted in Fig. 3 are indeed of dielectrical origin.

Experiments using the seven-turn solenoid were performed at 27, 70, and 120 MHz, and all the data agreed with the theory presented above. In particular, (i) there was a linear relationship between R_m and sample conductivity, and the absolute value of R_m was in fairly good agreement with the value predicted from Eq. [2]; (ii) the value of σ_0 increased linearly with ω_0, as predicted from Eq. [6]; (iii) the value of Q_0 was the same whether obtained directly from experiment, or more indirectly from the relationship $\Delta w_\infty/\omega_0 = 1/Q_0$; and (iv) the resonant frequency at $\sigma = \sigma_0$ was midway between that at zero and high conductivity. From the data, C_1 was found to be 0.6 pF, and C_2 was 1.5 pF.

Results obtained with a 14-turn solenoid of diameter 7 mm and length 18 mm are shown in Fig. 4a, while those obtained with a single-turn saddle coil of length and diameter 35 mm are shown in Fig. 4b. As expected, dielectric losses are far more significant for the smaller multiturn coil than for the larger single-turn coil.

Of particular relevance to studies of living tissue was the observation that a whole kidney placed in a 35-mm saddle coil produced about 70% of the losses generated by 150 m*M* potassium chloride. For such a coil, inductive losses dominate, as Fig. 4b clearly shows, and we can therefore conclude that the effective specific conductivity of the kidney under these conditions is about 1.2 Ω^{-1} m^{-1}.

A possible alternative explanation for the dielectric losses could be that there is a resonant dielectric loss associated with some form of molecular motion within the sample which has a characteristic time scale of about 10^{-9} sec, and which varies with salt concentration. However, it is difficult to imagine any form of molecular motion that would have the required concentration dependence. Also, experiments using a *Q* meter on samples of hydrogen chloride, potassium chloride, and sodium chloride

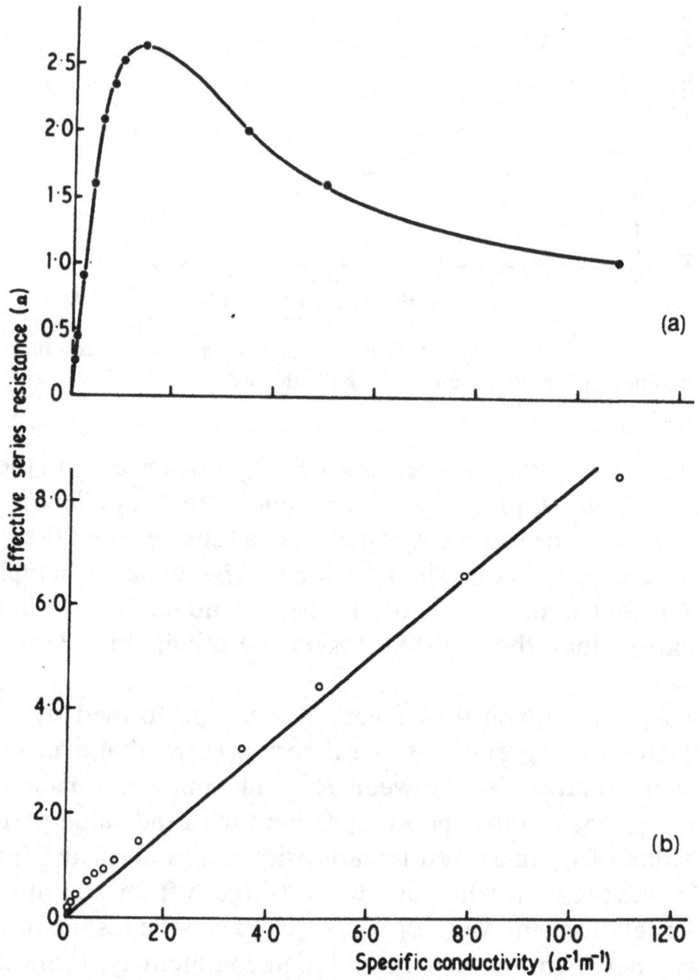

FIG. 4. The effective series resistance of samples of sodium chloride plotted as a function of sample conductivity, (a) using a 14-turn solenoid of diameter 7 mm and length 18 mm, and (b) using a single-turn saddle coil of length and diameter 35 mm.

showed that Q was dependent upon the conductivity rather than the concentration of the sample. Furthermore, it was found that the concentration of sodium chloride at which the maximum losses occurred varied from one tuned circuit to another, presumably because of differences in the ratios of C_1 to C_2 (see Eq. [6]). Thus we can conclude that the observed effects reflect the conductivity of the sample, rather than its chemical composition.

Conclusions

The observed inductive losses agree fairly well with the values predicted by Eq. [2]. This agreement, together with that found by Hoult and Lauterbur (*11*) for much larger samples, provides confirmation that inductive losses can in principle be predicted from the known geometry of the coil and sample. The dielectric losses are far more difficult to predict, because of the unknown values of the capacitances C_1

and C_2. However, they may be determined experimentally, in ways that we now summarize.

First, the losses measured as a function of sample conductivity can be fitted to the sum of one loss which is proportional to conductivity, and another corresponding to the dielectric loss. Sècond, if the inductive losses can be estimated from the geometry of coil and sample, then the difference between these and the observed losses can be assumed to be of dielectrical origin. Third, sample losses can be measured with and without an electrical screen, and the difference corresponds to the dielectric losses.

One can also deduce the maximal sample losses (i.e., at $\sigma = \sigma_0$) simply from the decrease in resonant frequency that is observed on changing from a sample of zero to one of high conductivity. This single experiment provides an indication of the dielectric losses that might be expected.

In conclusion, therefore, we have shown how a few rapid and straightforward experiments enable one to determine both the inductive and dielectric losses generated by a conducting sample. These losses can be particularly important in studies of living tissues and cells, for which conductivities can be expected to be about $1.2\ \Omega^{-1}\ m^{-1}$. In subsequent papers, we hope to discuss the effects that these losses have upon signal-to-noise and upon rf coil design.

ACKNOWLEDGMENTS

The support of Dr. David Hoult, Sir Rex Richards, and Mr. Peter Styles is gratefully acknowledged.

REFERENCES

1. D. I. Hoult, S. J. W. Busby, D. G. Gadian, G. K. Radda, R. E. Richards, and P. J. Seeley, *Nature (London)* **252,** 285 (1974).
2. C. T. Burt, T. Glonek, and M. Barany, *J. Biol. Chem.* **251,** 2584 (1976).
3. M. J. Dawson, D. G. Gadian, and D. R. Wilkie, *J. Physiol.* **267,** 703 (1977).
4. P. B. Garlick, G. K. Radda, P. J. Seeley, and B. Chance, *Biochem. Biophys. Res. Commun.* **74,** 1256 (1977).
5. W. E. Jacobus, G. J. Taylor, D. P. Hollis, and R. L. Nunnally, *Nature (London)* **265,** 756 (1977).
6. P. A. Sehr, G. K. Radda, P. J. Bore, and R. A. Sells, *Biochem. Biophys. Res. Commun.* **77,** 195 (1977).
7. R. B. Moon and J. H. Richards, *J. Biol. Chem.* **248,** 7276 (1973).
8. J. M. Salhany, T. Yamane, R. G. Shulman, and S. Ogawa, *Proc. Nat. Acad. Sci. USA* **72,** 4966 (1975).
9. G. Navon, S. Ogawa, R. G. Shulman, and T. Yamane, *Proc. Nat. Acad. Sci. USA* **74,** 87 (1977).
10. F. F. Brown, I. D. Campbell, P. W. Kuchel, and D. C. Rubenstein, *FEBS Lett.* **82,** 12 (1977).
11. D. I. Hoult and P. C. Lauterbur, *J. Magn. Reson.* **34,** 425 (1979).
12. D. I. Hoult, *Progr. NMR Spectrosc.* **12,** 41 (1978).

Department of Biochemistry
University of Oxford
South Parks Road
Oxford OX1 3QU, U.K.

D. G. Gadian

Clarendon Laboratory
University of Oxford
Parks Road
Oxford OX1 3PU, U.K.

F. N. H. Robinson

Received October 23, 1979

An Efficient Decoupler Coil Design which Reduces Heating in Conductive Samples in Superconducting Spectrometers

The heating of conductive samples, particularly aqueous solutions of biological molecules and buffer ions, by high-power wide-band proton decoupling is a vexing problem which constantly confronts experimenters applying high-field superconducting NMR spectroscopy of the less receptive nuclei such as carbon-13. Led and Petersen (*1*) have studied the problem and recommend that it be alleviated by gating off the decoupler, increasing the variable-temperature gas flow, and carefully controlling the ionic content, and thus the conductivity, of the sample. Grutzner and Santini (*2*) have pointed out that the nature of the modulation used to spread the decoupler power over the proton chemical shift range is also crucial. Often these techniques are inadequate or inapplicable, however, and the only alternative is to settle for less decoupling field than is necessary to fully collapse and narrow the lines. The result is that, though a nominally decoupled spectrum may be obtained, the full sensitivity of the instrument is not exploited because the lines remain wide.

Faced with this problem we have turned to the design of the decoupler coil to further reduce unnecessary heating of the sample. Since the penetration of the decoupling field itself into a conductive sample induces eddy currents which dissipate energy in the sample, it is not possible to eliminate all heating, but design criteria can be enunciated which will minimize it. Hoult and Lauterbur (*3*) and Gadian and Robinson (*4*) have recently pointed out that electric fields which penetrate the sample can be significant loss mechanisms which lower the Q of receiver coils. For decoupler coils driven at high powers these same electric field loss mechanisms produce significant energy dissipation in the sample. As a consequence it is important to minimize electric fields in the sample. This can be done by employing low-voltage, high-current (i.e., low inductance) coils. Further, the sample should be shielded insofar as is possible from electric fields produced by voltages across the coil. In addition, the decoupling field B_2 should be made as uniform as possible over that part of the sample to which the receiver coil is sensitive. Since energy dissipation goes as B_2^2, a nonuniform field can produce unnecessary hot spots in the sample without significantly improving the overall decoupling effect. Finally, the decoupling field in those parts of the sample to which the receiver coil is not sensitive should be minimized so as to reduce the total energy dissipation in the sample.

Figure 1 is a drawing of a decoupler coil, or structure, designed after the above criteria for use in the Varian SC-300 superconducting spectrometer at the University of Utah. Some of the origins of this design are in the work of Kan *et al.* (*5*) and the later slotted tube resonator work of Schneider and Dullenkopf (*6*). Certain features of the system are also similar to a structure used by J. Dadok (private communication). The wide "vertical bands" which run down the sides of the cylinder conform to the 80° specification given by Schneider and Dullenkopf for optimum

Reprinted with permission from Journal of Magnetic Resonance **36**, 447–451 (1979); ©Academic Press.

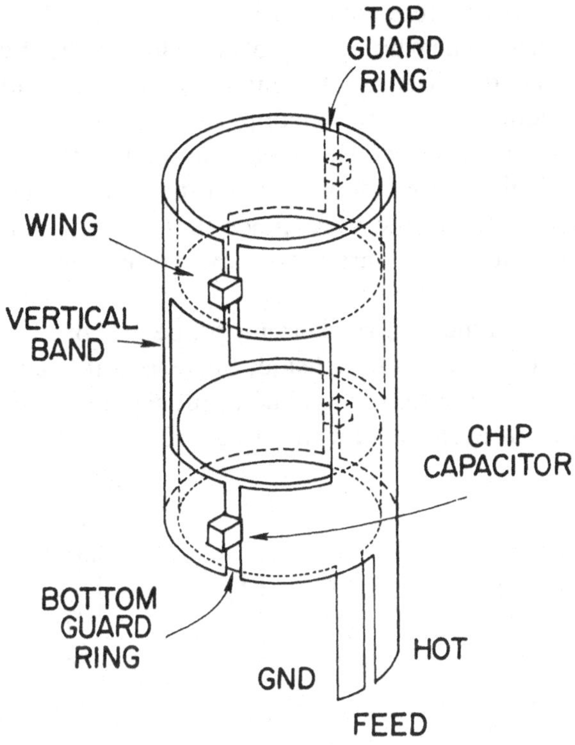

FIG. 1. Decoupler structure. Not shown is a layer of 0.005-in. Teflon dielectric between the guard rings and the wing-band pieces.

field homogeneity in their slotted tube resonator system. A significant difference is that the present structure has been made quite short instead of being a quarter wavelength long. The very low inductance of this short section of slotted tube line has been resonated with lumped capacity. The capacity has been distributed symmetrically around the structure to obtain a uniformity of current flow, thereby retaining the good field homogeneity characteristics of the slotted tube line. Part of the resonating capacity is that formed by the proximity of the "wings" of the vertical bands to the "guard rings." The additional capacity necessary to resonate the structure at 300 MHz was determined empirically and then added by soldering small procelain chip capacitors (American Technical Ceramics, Huntington Station, N.Y.) between the extremities of the wings. The guard rings are not essential for such a structure to resonate, but were incorporated in order to shield the sample from the electric fields generated by the rf voltage present between the wings. Only the bottom guard ring is grounded, but the symmetry of the structure keeps the top ring at rf ground. We believe that the guard rings are the principal reason for the superior performance of the structure.

The structure is built as a three-layer laminate. The innermost layer is the guard rings, which are constructed of 0.005-in. copper sheet soldered into the ring shape. Over the guard rings is a layer of 0.005-in. Teflon dielectric, which is not shown in Fig. 1. The Teflon is a single sheet which covers the region from the bottom of the bottom guard ring to the top of the top guard ring. Over the Teflon are the two

H-shaped pieces which constitute the bands and the wings. They too are constructed of 0.005-in. copper sheet. The whole assembly is held together by the chip capacitors which join the H-shaped pieces and by linen cord tied around the assembly. We intend to make subsequent structures of copper–Teflon bonded laminate, which will greatly simplify their construction. The inner diameter of the guard rings is 0.72 in. and they are 0.4 in. high. The window between the guard rings is 0.6 in. high. The whole structure is mounted in the same place that the original decoupler coil was mounted, on the outside of the variable-temperature Dewar incorporated in the SC-300 probe.

In order to understand the electrical circuit properties of the decoupler structure and to see how it is matched to 50 Ω, consider the resonant circuits of Fig. 2a through i, all of which resonate at the same angular frequency, $\omega = (LC)^{-1/2}$. In a stepwise manner this series of figures shows how the circuit of Fig. 2i, which is a close analog of

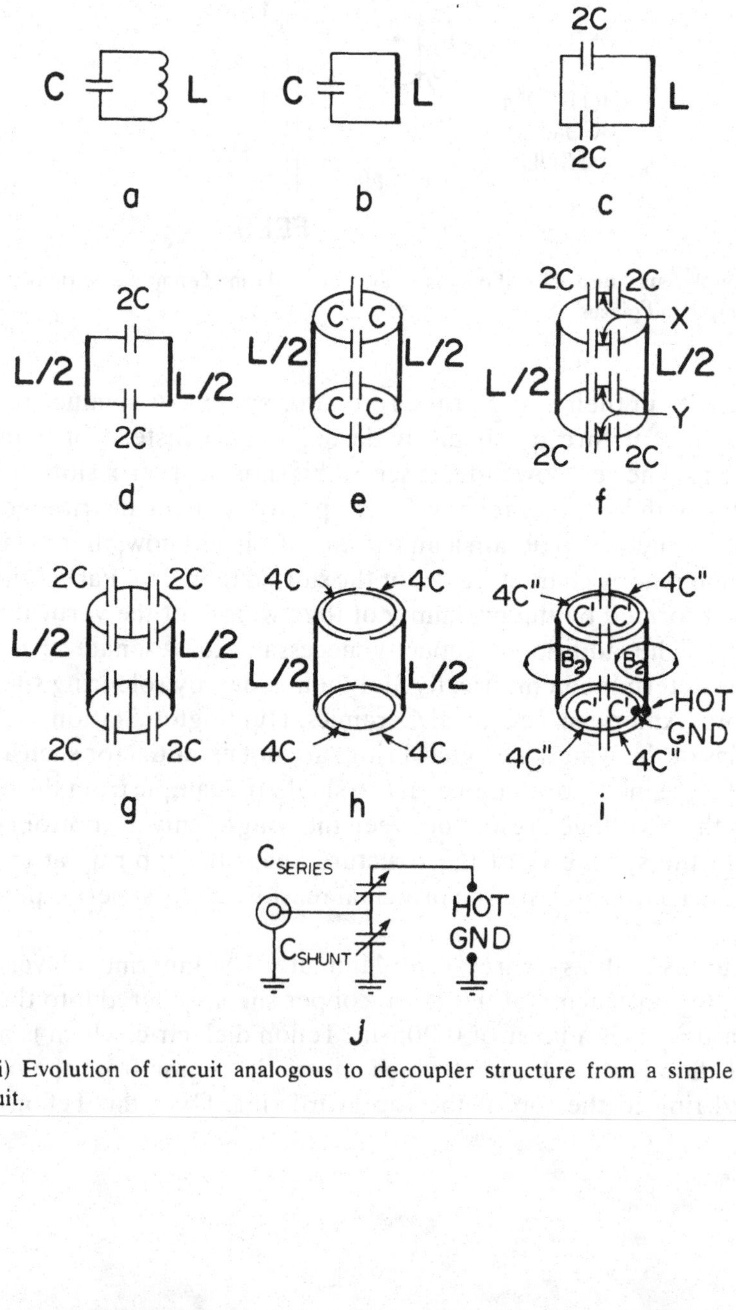

FIG. 2. (a–i) Evolution of circuit analogous to decoupler structure from a simple LC circuit. (j) Matching circuit.

the decoupler structure, evolves from the simple LC circuit of Fig. 2a. Figure 2b is the same simple LC circuit except with the inductor L represented as a broad straight line. Figure 2c splits the capacitor into a pair of equivalent series capacitors. Figure 2d rearranges the inductor into two inductors placed alternately with the capacitors. This is legitimate because the order of circuit elements in a series string is irrelevant. Figure 2e splits apart the two capacitors, each into equivalent parallel combinations. Figure 2f then replaces each of these four with equivalent series combinations. Now because of the symmetry of the circuit the points labeled X and Y in Fig. 2f are always at the same potential so they may be connected together as in Fig. 2g. This interconnection may now be split and merged with the capacitors, and the capacitors themselves combined to produce Fig. 2h. Finally Fig. 2i results from taking a linear combination of Figs. 2e and h. To preserve the resonant frequency the capacities C' and C'' of Fig. 2i must be such that $C' + C'' = C$. The inductance of the circuit is best thought of not as the inductances $L/2$ due to the individual conductors, but rather as the inductance due to the field B_2 linking the circuit in the direction shown in Fig. 2i. When the circuit is resonating a relatively large voltage appears between the points labeled GND and HOT in Fig. 2i. This situation is similar to that of a parallel resonant circuit and thus the structure can be matched to a 50-Ω line in a similar way, that of Fig. 2j. At 300 MHz the values of the various capacitors used are $C' = 10$ pF, $C_{SERIES} = 0.3$ to 3.0 pF, and $C_{SHUNT} = 4.0$ to 40 pF. The value of $4C''$ is not known exactly because the copper is not everywhere tightly against the Teflon dielectric. But if it were, $4C''$ would be equal to 44 pF. Using this approximate value C is 10 pF + 44 pF/4 = 21 pF. At 300 MHz this corresponds to an inductance of 13.4 nH. On the basis of the behavior of the 50-Ω match as a function of frequency, the Q of the structure is quite high, over 300.

Because of the uncertainties in the rate at which heat is removed from the sample by the variable-temperature gas flow it is difficult to give a completely quantitative assessment of the decoupler structure, but the following comparison can be made with the performance of the old coil, which is of the conventional "$\frac{1}{2}$ turn Helmholtz" design. At 1 W of power absorbed by the structure, as measured by a Bird Electronics Model 43 directional wattmeter with the appropriate insert, the $(\gamma/2\pi)B_2$ of the structure is 4.20 ± 0.15 kHz as measured by the method of Pachler (7) using a chloroform sample. This is 2.84 ± 0.14 times higher than the value of 1.48 ± 0.05 kHz measured under identical conditions for the original decoupler coil. In a 10-mm 0.05 *M* NaCl sample the temperature rise at the center of the sample, which was measured by immersing a small thermocouple in it immediately after it was removed from the probe, and which was shown to be proportional to the incident power by measuring it at various power levels, is 4.2 ± 0.2°C/W with the new structure and 4.6 ± 0.2°C/W for the original decoupler coil with the same 30 SCF/hr variable-temperature gas flow. Thus, for the same value of $(\gamma/2\pi)B_2$ the structure dissipates only $(4.2/4.6)/(2.84)^2 = 0.11 \pm 0.02$ times as much energy in the sample as the original decoupler coil.

Figure 3 is a 75-MHz carbon-13 spectrum of 20m*M* lysozyme which exhibits excellent decoupling across the full proton chemical shift range. It was obtained with 2 W of decoupler power in a 10-mm tube with only a 12°C rise in the sample temperature and a 6°C temperature differential within the sample at a variable-temperature N_2 gas flow of 30 SCF/hr.

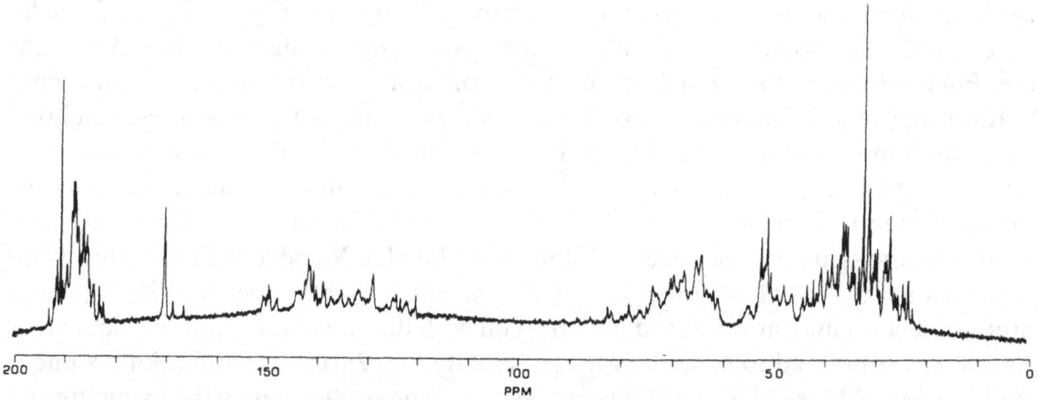

FIG. 3. Carbon-13 spectrum (75 MHz) of 20 m*M* hen egg-white lysozyme, EC No. 3.2.1.17 (Sigma Chemical No. L6876) in 15% D_2O–85% H_2O at pH = 4.0. The spectrum is 15,000 hz wide, contains 16,384 points, and was obtained in 18,000 acquisitions with a 5.36-sec repetition time using an 8.5-μsec, 90°, pulse. An exponential weighting time constant of −0.2 sec was used for sensitivity enhancement. The proton decoupler power was 2 W, corresponding to $(\gamma/2\pi)B_2 = 5940$ Hz, modulated with a swept square wave of bandwidth approximately 3 kHz. The temperatures of the 10-mm sample were 29, 35, and 34°C at the bottom, middle, and top, respectively. The temperature of the variable-temperature control N_2 gas was 23°C and its flow rate was 30 SCF/hr.

A significant advantage of this new decoupler structure is that though it is something like a cavity resonator in its design, it can be easily built in place of conventional decoupler coils and can be matched with the same circuit.

It is possible that a further reduction in the energy dissipated in the sample for a given $(\gamma/2\pi)B_2$ can be achieved by employing a Faraday screen, which in this case would consist of a series of wire loops coaxial with the sample tube and the decoupler structure between the vertical bands and the sample tube.

Structures of this type used as receiver coils should also improve the signal-to-noise ratios of superconducting spectrometers.

ACKNOWLEDGMENTS

The support of D. K. Dalling, who helped prepare and run the lysozyme sample, is gratefully acknowledged. This work was supported by NIH Grant RR 00574-08.

REFERENCES

1. J. J. LED AND S. B. PETERSEN, *J. Magn. Reson.* **32,** 1 (1978).
2. J. B. GRUTZNER AND R. E. SANTINI, *J. Magn. Reson.* **19,** 173 (1975).
3. D. I. HOULT AND P. C. LAUTERBUR, *J. Magn. Reson.* **34,** 425 (1979).
4. D. G. GADIAN AND F. N. H. ROBINSON, *J. Magn. Reson.* **34,** 449 (1979).
5. S. KAN, P. GONORD, C. DURET, J. SALSET, AND C. VIBET, *Rev. Sci. Instrum.* **44,** 1725 (1973).
6. H. J. SCHNEIDER AND P. DULLENKOPF, *Rev. Sci. Instrum.* **48,** 68, 832 (1977).
7. K. G. R. PACHLER, *J. Magn. Reson.* **7,** 442 (1972).

DONALD W. ALDERMAN
DAVID M. GRANT

Department of Chemistry
University of Utah
Salt Lake City, Utah 84112

Received July 23, 1979

A Large-Inductance, High-Frequency, High-Q Series-Tuned Coil for NMR

BRUCE COOK

Department of Biological Sciences, Carnegie–Mellon University, Pittsburgh, Pennsylvania 15213

AND

I. J. LOWE

Department of Physics, University of Pittsburgh, Pittsburgh, Pennsylvania 15260

Received April 20, 1982

Coils with a large inductance normally have a low self-resonance frequency because of resonances between the inductance and distributed stray capacitances. In NMR probes, one normally uses coils whose self-resonant frequency is well above the frequency at which they are to be used. This can be a problem when one wishes to use a large-inductance coil at high frequencies for purposes of impedance matching or containing large sample volumes. We recently wished to build such a large-inductance, large-volume coil to be used at 300 MHz.

Our solution to eliminating the self-resonance problem was to build a coil similar to the one described in Ref. (*1*), but with a completely different intent. We constructed a coil of many sections in series, each section consisting of one or more turns of wire, each section series compensated by a capacitor so that the net reactive impedance of each section was small. The series resistance component for each section is also small. Thus, in normal operation, the voltage drop across each section is small and the effect of distributed capacitance between sections is minimized in comparison to a design where only one capacitor is used to series resonate the total inductive reactance of the coil. For a current I flowing through a coil, an N section coil will have a voltage difference between the ends of the coil that is a factor of N smaller than a similar coil tuned by only one capacitor. Furthermore, since the voltage at any point in the N section coil is much smaller than a similar single-section coil, detuning effects due to the dielectric properties of the NMR sample are much reduced in comparison to the single-section coil. Further, because of the small electric fields inside the coils of this design, this coil should compete favorably with that of Alderman and Grant (*2*), for irradiating conductive samples for the purposes of decoupling, when heating effects need to be kept to a minimum.

The schematic for such a coil is pictured in Fig. 1, along with definitions for various symbols. End effects of the coil are ignored so that all sections of the coil are assumed to be identical. The analysis of the coil is much simplified with this

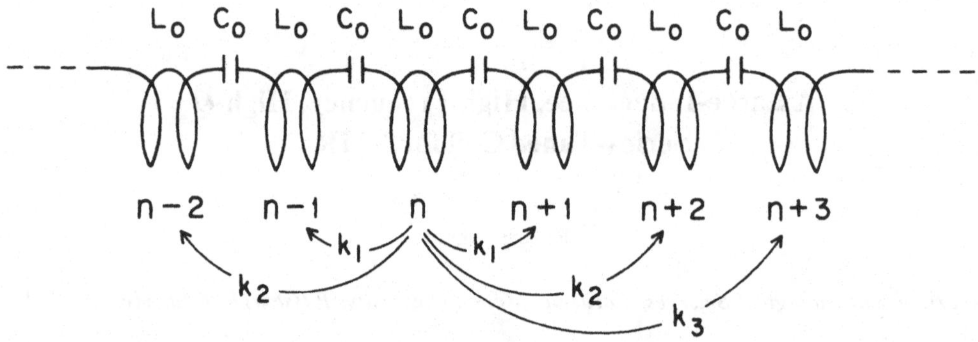

FIG. 1. An N section coil: Current I is assumed to flow through each section; L_0 = inductance of each section; C_0 = coupling capacitor in each section; k_n = magnetic coupling of each section with its nth nearest neighbor = M_n/L_0; R_0 = loss term of coil and capacitor section (not pictured).

assumption. It is also assumed that a current I at angular frequency ω flows through each of the N sections of the coil. There is a propagation time for the current through various sections that leads to a current phase shift between different sections. For this simplified analysis, this will be ignored. The voltage drop across the nth coil plus capacitor combination, ΔV_n, when a current I at angular frequency ω flows through the section is

$$\Delta V_n = IR_0 + \iota\left(-\frac{I}{\omega C_0} + I\omega L_0 + 2I\omega M_1 + 2I\omega M_2 + \cdots\right), \qquad [1]$$

$$\Delta V_n = I\left\{R_0 + \iota\left(-\frac{1}{\omega C_0} + \omega L_0(1 + 2k_1 + 2k_2 + \cdots)\right)\right\}, \qquad [2]$$

$$M_n = k_n L_0 .$$

Series resonance occurs when the reactive term for each element vanishes, which occurs when

$$L_0 C_0(1 + k_T) = \omega^{-2}, \qquad [3]$$

$$k_T = 2(k_1 + k_2 + k_3 \cdots).$$

If a similar analysis were carried out for the N sections without the series capacitors, and interwinding capacitive effects were ignored, the total coil inductance L_T would be

$$L_T = NL_0(1 + k_T). \qquad [4]$$

The coil which we constructed for operation at 300 MHz had 20 sections and is pictured in Fig. 2. Each section was made of one turn of 20-gauge Formvar-coated wire with 1-mm spacing between turns. The coil was 1.5 cm in diameter and 3.0 cm long. Each section was coupled to its neighbor by a pigtail capacitor (Ref. (*1*)) made up of 0.7 cm of the tightly twisted 20-gauge wire. Measurements at 60 MHz yielded values of L_0 = 0.025 μH and C_0 = 3 pF. The coil series resonated at 300 MHz and had an impedance of Z = 50 Ω. Fine tuning of the coil was carried out

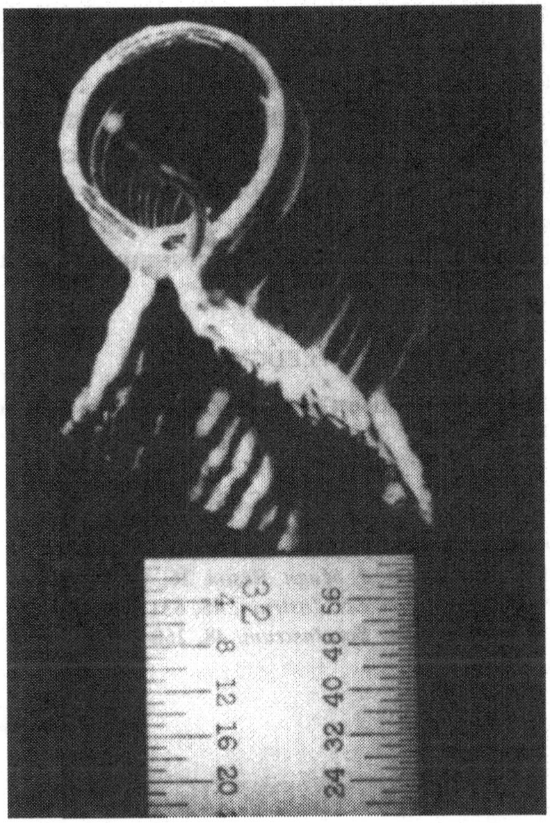

FIG. 2. Photograph of 300-MHz coil. Left part of scale is 32 divisions per inch.

by trimming the length of the pigtail capacitors and the use of a series piston capacitor with a range of 1 to 10 pF.

An identical 20-turn coil without the 20 series capacitors was self-resonant at 50 MHz with a resultant impedance greater than 100 kΩ. Its measured inductance at 1 MHz was 2.4 μH.

From Eq. [3] and the values listed above for L_0, C_0, and ω, the value for k_T is calculated to be 2.75. This implies good magnetic coupling between nearest-neighbor turns only, and poor coupling for turns farther away—which is consistent with our previous experiences. Using this value of k_T in Eq. [4], the total inductance is calculated to be 1.9 μH as opposed to the measured value of 2.4 μH. This seems to be reasonable agreement, considering the accuracy with which L_0 and C_0 were measured—and the ignoring of end effects in analyzing the coils. The Q of our composite coil at 300 MHz is estimated to be $Q = \omega L_T/Z = 90$.

A similar coil resonating at 282 MHz has been used for some time in our laboratory in the study of oriented fluorinated phospholipids in various concentrations of water. Two advantages of the design were made evident. First, sample dielectric detuning effects were lessened. Second, the large volume of the coil allowed for sufficient numbers of nuclei of interest (despite the glass plates used to orient the sample) for a good signal to be demonstrated.

We would like to end this note with two concluding statements. First, because of a finite propagation time, the current in a very long coil operating at high frequencies is not exactly in phase in various parts of the coil; the rf field produced by the coil would then be less homogeneous than if the current was everywhere in phase (*3*, *4*). Second, while our coils have been constructed to operate at high frequencies, we would like to point out that the same idea can be used to construct very large coils to operate at low frequencies. A need for such large diameter coils has arisen in NMR imaging measurements to accommodate patients undergoing diagnostic analysis.

ACKNOWLEDGMENTS

Supported by National Science Foundation Grant No. DMR 78-15441-02 and National Institutes of Health Grant No. NL 24525-03.

REFERENCES

1. R. F. KARLICEK, JR. AND I. J. LOWE, *J. Magn. Reson.* **32,** 199 (1978).
2. D. W. ALDERMAN AND D. M. GRANT, *J. Magn. Reson.* **36,** 447 (1979).
3. I. J. LOWE AND M. ENGELSBERG, *Rev. Sci. Instrum.* **45,** 631 (1974).
4. I. J. LOWE AND D. W. WHITSON, *Rev. Sci. Instrum.* **48,** 268 (1977).

An *in Vivo* NMR Probe Circuit for Improved Sensitivity

JOSEPH MURPHY-BOESCH

Department of Pharmaceutical Chemistry, School of Pharmacy, University of California, San Francisco, California 94143

AND

ALAN P. KORETSKY

Chemical Biodynamics Division, Lawrence Berkeley Laboratory, University of California, Berkeley, California 94720

Received May 18, 1983

Success in using NMR to study metabolic processes and to produce well-resolved images of living tissues has led to a rapid increase in its application to problems in biology and medicine (*1*, *2*). A serious problem with all of these studies is the degrading effect that intact tissues have upon probe sensitivity. The desire to perform these kinds of experiments has therefore necessitated studies of loss mechanisms associated with conductive samples (*3*, *4*). We have recently discovered a tuning scheme for coils which greatly minimizes these losses when used for *in vivo* experiments. The success of this scheme demonstrates the importance of coil-to-ground parasitic losses. We describe here a model for this loss mechanism and show how a simple modification of the normal tuning scheme can minimize its effects. The improvement in sensitivity afforded by this scheme is illustrated using model circuits and ^{31}P spectra obtained *in vivo*, using implanted coils.

Because of their high conductivity, biological tissues can produce large losses in the tuned circuit of an NMR probe, losses reflected in a reduction in the circuit Q. These losses are caused primarily by currents induced in the tissue by the electric and magnetic fields of the coil. While little can be done to reduce currents generated by the magnetic field, shielding of the electric field from the tissue can reduce dielectric losses.

Previous workers have developed circuit models for dielectric losses assuming that the paths of electric fields which penetrate the sample extend in a distributed manner from one side of the coil to the other (*3*, *4*). Their models of these coil-to-coil parasitics can be reduced to a capacitance C_d and a resistance R_d connected in parallel with the coil, as shown in Fig. 1. R_d and C_d have the effect of lowering both the resonance frequency and the Q of the tuned circuit. This description gives good quantitative explanations for the behavior of the losses incurred when biological samples are placed within a standard probe of a high resolution spectrometer (*4*).

For the types of coils used to study animal organs *in situ*, additional paths for the

Reprinted with permission from Journal of Magnetic Resonance **54**, 526–532 (1983); ©Academic Press.

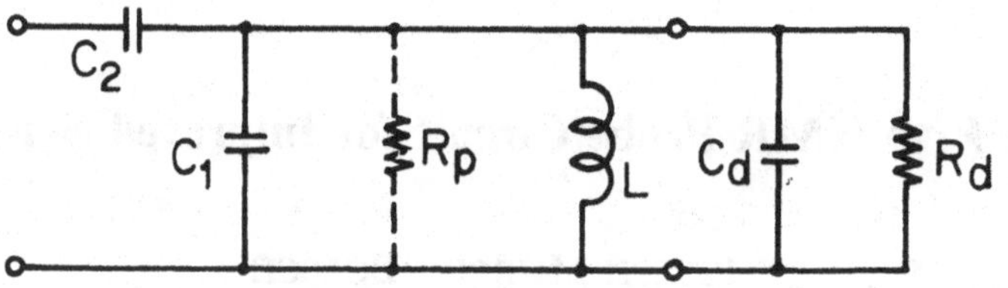

FIG. 1. A probe circuit employing standard tuning and matching which models coil-to-coil influences of the sample. L is the inductance of the sample coil, C_1 and C_2 are the tuning and matching capacitors, and R_p models the coil resistance. C_d and R_d model the coil-to-coil parasitics.

electric fields exist. In our probe, a laboratory rat with an implanted coil is supported by a metal cradle. For proper tuning and shielding, the cradle, the probe casing, and the tuning circuit must all be grounded. Hence, paths for electric fields exist between the coil within the animal and the probe ground. The influence of these coil-to-ground parasitics can be modeled with two lumped element branches between each side of the sample coil and ground, as indicated in Fig. 2a. For simplicity, each branch consists only of a resistance in series with a capacitor, with the capacitance modeling both the insulation about the coil wire and the reactive influence of the sample. In this circuit, the second branch involving R_{d2} and C_{d2} can be neglected, since it is shorted by the ground lead of the tuning circuit. As indicated in Fig. 2b, the remaining branch transforms to the equivalent parallel components R_α and C_α, given by

$$R_\alpha^{-1} = \frac{\omega_0^2 R_{d1} C_{d1}^2}{1 + \omega_0^2 R_{d1}^2 C_{d1}^2} \qquad [1]$$

$$C_\alpha = \frac{C_{d1}}{1 + \omega_0^2 R_{d1}^2 C_{d1}^2} \qquad [2]$$

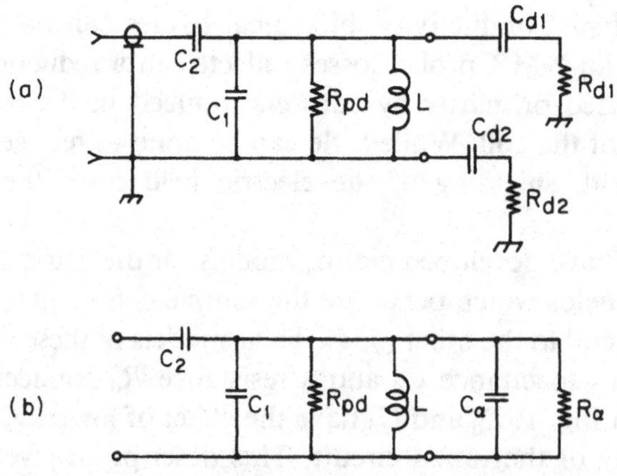

FIG. 2. Probe circuits employing standard tuning and matching which model (a) the additional influences of coil-to-ground parasitics, and (b) an equivalent parallel circuit. The tuned circuit elements are defined as in Fig. 1, except that R_{pd} represents the parallel combination of R_p and R_d, and C_1 incorporates C_d into the tuning. The branches R_{d1} and C_{d1}, and R_{d2} and C_{d2}, model the distributed influences of the coil-to-ground parasitics. R_α and C_α are given by Eqs. [1] and [2].

where $\omega_0/2\pi$ is the resonance frequency of the tuned circuit. The influence of these additional components on the tuned circuit is the same as that of the coil-to-coil parasitics. The total Q of the circuit can then be expressed as

$$\frac{1}{Q_s} = \frac{1}{Q_{pd}} + \frac{1}{Q_\alpha} \qquad [3]$$

where $Q_{pd} = R_{pd}/\omega_0 L$ and $Q_\alpha = R_\alpha/\omega_0 L$.

Once recognized, the influence of coil-to-ground parasitics can be minimized with a balanced tuning circuit containing matching capacitors of equal size on both sides of the coil, as indicated in Fig. 3a. Since the matching capacitors are essentially in series, their values must be approximately twice as large as the single matching capacitor of the standard circuit.

For the purposes of illustration only, two assumptions are made which permit simplification of the circuit. First, we assume that the Q of the sample circuit is high enough to make the impedance of the matching capacitors small compared with the 50 ohm input impedance. Thus, the parasitic elements link in series through ground, reducing the circuit to that of Fig. 3b. Next, we assume that the two branches are identical, that is, that $C_{d1} = C_{d2}$ and $R_{d1} = R_{d2}$. This is approximately correct since the sample coil is generally situated symmetrically within an NMR probe. The coil

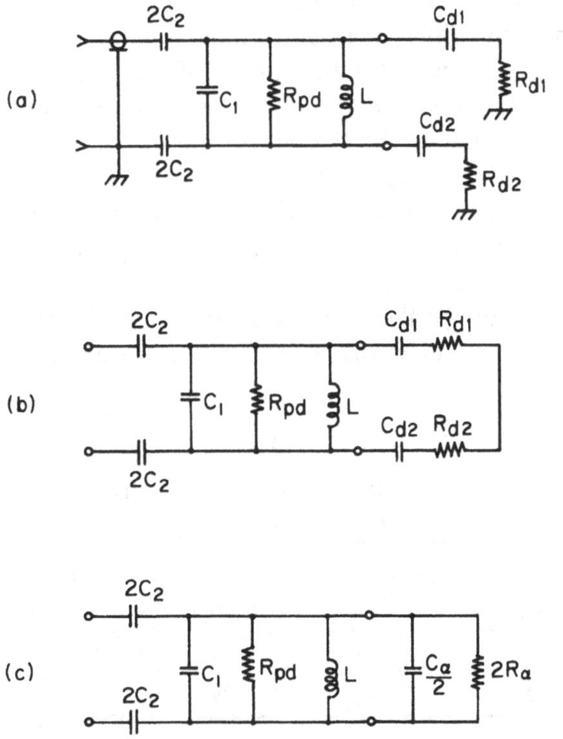

FIG. 3. Circuit models showing (a) the influence of coil-to-ground parasitic elements on the balanced matching and tuning scheme, (b) the reduced circuit assuming a high circuit Q, and (c) the equivalent parallel circuit. The circuit elements are defined as before.

is thereby balanced with respect to ground. Using transformation equations similar to Eqs. [1] and [2], the circuit reduces to that of Fig. 3c. The total Q for the balanced circuit can be expressed as

$$\frac{1}{Q_b} = \frac{1}{Q_{pd}} + \frac{1}{2Q_\alpha}. \qquad [4]$$

Comparing Eqs. [3] and [4], it is apparent that the new circuit configuration has reduced the influence of the coil-to-ground parasitics by approximately one-half. Thus, the balanced matching circuit yields a higher circuit Q. Since sensitivity varies as $Q^{1/2}$ for a fixed frequency (*5*), we can also expect an improved signal-to-noise ratio.

To demonstrate the advantage of using the balanced tuning scheme in the presence of strong coil-to-ground parasitics, we constructed a series of model circuits, all employing a two-turn half-inch diameter coil. Ceramic capacitors, variable from 3 to 20 pF, were used to tune and match the circuits. All parasitic elements consisted of a 22 ohm carbon resistor placed in series with a 5 pF silver mica capacitor. These particular component values were chosen to simulate the variations in tuning and Q typically observed for our *in vivo* experiments. Q measurements were performed near 100 MHz with a Wavetek Model 1062 sweep generator and a 20 dB directional coupler used to measure the reflected power. The Q's were calculated by dividing the resonance frequency of the circuit by the width of the resonance curve at the half-power points. The oscilloscope used for these measurements was calibrated with 1 MHz frequency markers from the sweep generator.

The variation in Q for different loss mechanisms and tuning schemes are shown in Table 1. The first column shows that the Q of the unloaded coil is insensitive to the matching scheme. The second column in Table 1 shows that the balanced matching configuration does not reduce the effects of coil-to-coil parasitics. Only when the loading is placed from each side of the coil to ground does the balanced matching scheme significantly improve the circuit Q and reduce the influence upon tuning. The improvement is greater than predicted by our simple model indicating that some additional factors are at play.

We have been performing ^{31}P spectroscopy on rat organs by implanting the sample coil around the organ of interest (*6*). The circuit we use, shown in Fig. 4, is a mod-

TABLE 1

MEASUREMENTS OF CIRCUIT Q FOR THE STANDARD AND BALANCED CONFIGURATIONS

Tuning scheme	Lossless[a] circuit	Coil-to-coil[b] loading	Coil-to-ground[c] loading
Standard	59	29	26
Balanced	59	29	45

[a] Tuned circuit with no loading elements.
[b] Simulated with a single parasitic element across the coil.
[c] Simulated with two parasitic elements between each side of the coil and ground.

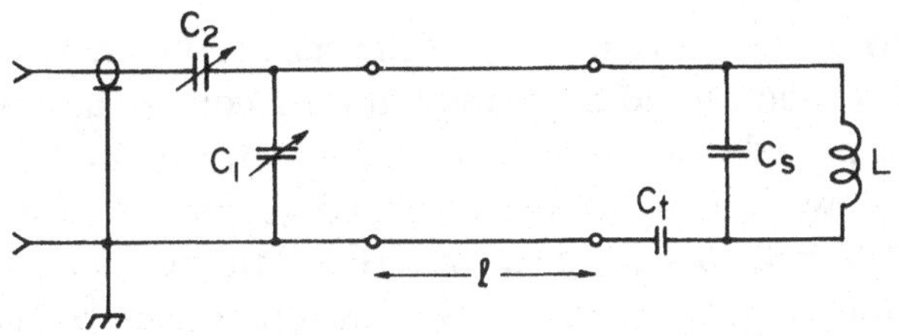

FIG. 4. A modified, balanced tuning circuit for *in vivo* experiments where a length of transmission line is required between the coil and the tuning and matching capacitors C_1 and C_2, respectively. C_s and C_t are mounted near the coil and perform both partial tuning and partial matching of the circuit. C_t is placed in the ground lead to balance the coil.

ification of the balanced matching capacitor circuit just described. The circuit differs from that of Fig. 3 in several respects. First, the tuning capacitor C_1 is no longer placed directly across the coil leads, but instead has one of its leads connected directly to ground. C_s and C_t are small chip capacitors (American Technical Ceramics) placed as close to the coil as possible and insulated from the tissue with silicone sealer (Dow Corning). C_s helps to confine the large circulating current of the tuned circuit to the vicinity of the coil, thus improving the filling factor. C_t performs a partial transformation to a lower impedance, reducing the rf voltage and, therefore, the conductive losses across the transmission line. C_t is chosen somewhat larger than C_2 to offset the imbalancing effect of C_1 and to assure that C_2 will always be within range, regardless of the variations in sample coil inductance. The net result is that (1) the sample coil remains relatively balanced with respect to ground, (2) the transmission line does not have undue influence upon the filling factor or the circuit Q, and (3) the entire circuit can be tuned and matched outside the animal.

To demonstrate the improvement in Q obtainable with this scheme, measurements were made with a coil implanted around the kidney of a laboratory rat. The sample coil consisted of two half-inch diameter loops separated by approximately one-quarter inch. The coil was wound with #22 gauge copper wire insulated with polyethylene tubing (Clay Adams PE-100). The transmission line consisted of approximately 10 cm of this wire in twisted pair. The capacitors C_s and C_t were insulated with silicone sealer; their values were 13 and 27 pF, respectively. The Q of this coil–capacitor arrangement was typically 60 prior to implantation. The arrangement was surgically implanted around a rat kidney following the procedure of Koretsky *et al.* (*6*). Two days after surgery the animal was anesthetized and positioned in the probe, and the leads were connected to the capacitors C_1 and C_2. The values for these capacitors were in the range of 5 to 15 pF. The measured Q of the tuned circuit was 38. The transmission line leads outside the animal were then reversed so that the capacitor configuration resembled that of the standard tuning circuit of Fig. 1. In this configuration the capacitor C_t only performs a partial impedance match. The Q for this arrangement was found to be 13. This dramatic drop in Q is evidence that the coil-to-ground parasitics are the dominant loss mechanism for implanted coils.

We attempted to obtain spectra from a single rat kidney using the two circuit arrangements described above to compare their sensitivities. However, because of (predicted) excess capacitance on the coil, the latter (low Q) configuration could not be tuned to 97.3 MHz, the ^{31}P frequency of the spectrometer. We therefore obtained a spectrum from a second rat kidney using a tuning coil containing only a single, 10 pF tuning capacitor placed close to the coil. The spectrum obtained with this scheme is shown in Fig. 5a, while that for the balanced scheme is shown in Fig. 5b. The signal-to-noise ratio for the β ATP peak was 16 for the former and 41 for the latter. This is a greater improvement in sensitivity than is predicted by the change in circuit Q (13 vs 38) and probably results from the increased filling factor obtained with the partial matching capacitor for the balanced scheme.

While employing a Faraday shield may be the most desirable technique for eliminating dielectric losses (*3*), this is not always a practical solution, as is the case with

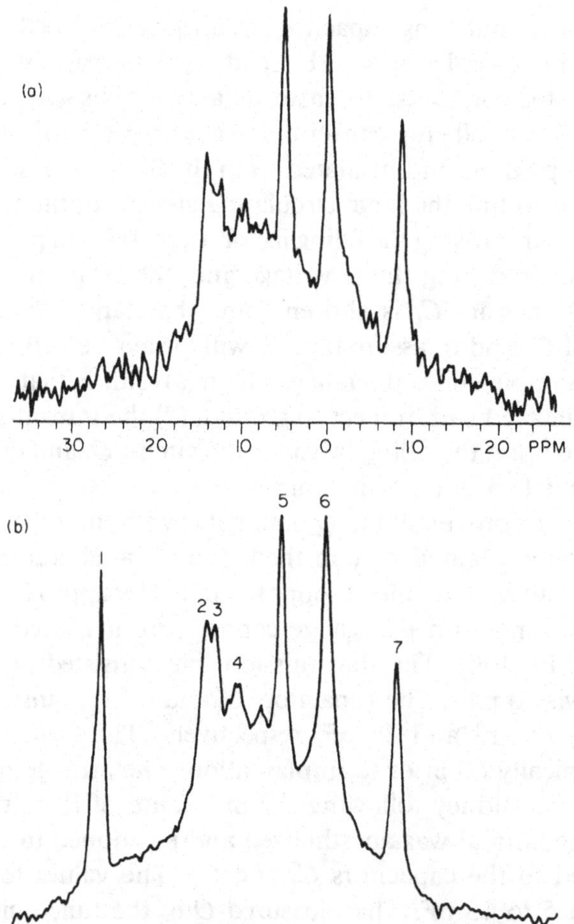

FIG. 5. ^{31}P spectra of two rat kidneys at 97.3 MHz using (a) an implanted coil with only a single tuning capacitor mounted near the coil, and (b) an implanted coil employing the modified, balanced tuning scheme. The experiments were performed on a homebuilt spectrometer using a Nicolet 1180 data system and a Cryomagnet Systems wide bore magnet. The spectra were obtained in 2600 scans using 45° pulses and 130 msec recycle times. A 30 Hz exponential filter was applied to the FID before Fourier transformation. The peaks shown are (1) methylene diphosphonic acid (Sigma Chemical Co.), pH 8.9, in a capillary mounted on the coil, (2) sugar phosphates, (3) inorganic phosphate, (4) urine phosphate and phosphodiesters, (5) γ-ATP, (6) α-ATP,NAD (H), and (7) β-ATP.

implanted coils. Our balanced matching scheme is an easily implemented alternative to reduce the losses associated with living tissue. Since the balanced matching circuit actually performs better than predicted, our model and assumptions only serve as a starting point for a more complete description.

Since the coils used for our performance tests are similar to those used as surface coils (*7*), we expect an improvement in sensitivity similar to that obtained for implanted coils. Furthermore, similar tuning and loss considerations hold for imaging configurations, and therefore the balanced matching circuit should yield improvements for whole body imaging and spectroscopy.

ACKNOWLEDGMENTS

We gratefully acknowledge the support of Dr. Michael W. Weiner, Dr. Thomas L. James, the Cardiovascular Research Institute, and the Medical Services of the Veterans Administration. We also thank Dr. Sam Wang for performing the coil implantations and Dr. Melvin Klein for some stimulating discussions.

REFERENCES

1. David G. Gadian, "NMR and Its Applications to Living Systems," Clarendon, Oxford, 1982.
2. P. Bottomley, *Rev. Sci. Instrum.* **53**(9), 1319 (1982).
3. D. I. Hoult and P. C. Lauterbur, *J. Magn. Reson.* **34**, 425 (1979).
4. D. G. Gadian and F. N. H. Robinson, *J. Magn. Reson.* **34**, 449 (1979).
5. D. I. Hoult and R. E. Richards, *J. Magn. Reson.* **24**, 71 (1976).
6. A. P. Koretsky, W. Strauss, V. Basus, J. Murphy, P. Bendel, T. L. James, and M. W. Weiner, "Acute Renal Failure" (H. E. Eliahu, Ed.), pp. 42–46, John Libbey, London, 1982.
7. J. J. H. Ackerman, T. H. Grove, G. C. Wong, D. G. Gadian, and G. K. Radda, *Nature (London)* **283**, 167 (1980).

Printed in the United States
By Bookmasters